EE 844

THIN LAYER
CHROMATOGRAPHY

THIN LAYER CHROMATOGRAPHY:

QUANTITATIVE ENVIRONMENTAL AND CLINICAL APPLICATIONS

Edited by

JOSEPH C. TOUCHSTONE

Department of Obstetrics and Gynecology
School of Medicine
University of Pennsylvania, Philadelphia

DEXTER ROGERS

Kontes
Vineland, New Jersey

A Wiley-Interscience Publication

JOHN WILEY & SONS

New York • Chichester • Brisbane • Toronto

Library of Congress Cataloging in Publication Data

Main entry under title:
  Thin layer chromatography.

  "A Wiley-Interscience publication."
  Includes index.
  1. Thin layer chromatography—Congresses.
2. Biological chemistry—Technique—Congresses.
3. Chemistry, Clinical—Technique—Congresses.
I. Touchstone, Joseph C.   II. Dexter Rogers.
[DNLM: 1. Chromatography, Thin layer—
Congresses.   QD79.C8 T443 1979]

QP519.9.T55T45   543°.08956   80-36871
ISBN 0-471-07958-8

Printed in the United States of America

10 9 8 7 6 5 4 3 2 1

Contributors

R. E. Allen
Celanese Corporation
Charlotte, North Carolina

Thomas E. Beesley
Whatman, Inc.
Clifton, New Jersey

D. T. Braymen
National Animal Disease Center
Ames, Iowa 50010

James J. Chan
Department of Health Services
Berkeley, California 94704

Milos Chvapil
Department of Surgery
Arizona Health Sciences Center
Tucson, Arizona 85724

David Y. Cooper
Department of Surgery
School of Medicine
University of Pennsylvania
Philadelphia, Pennsylvania
19104

Charles K. Cross
Canada Packers Ltd.
Toronto, Canada

Gregory Cuca
Center for the Biology of
Natural Systems
Washington University
St. Louis, Missouri

C. M. Davis
Texas Research Institute of
Mental Sciences
Houston, Texas 77030

L. J. Deutsch
Celanese Corporation
Charlotte, North Carolina

M. F. Dobbins
School of Medicine
University of Pennsylvania
Philadelphia, Pennsylvania
19104

Donald T. Downing
Department of Dermatology
College of Medicine
University of Iowa
Iowa City, Iowa 52240

David J. Edwards
Department of Psychiatry
School of Medicine
University of Pittsburgh
Pittsburgh, Pennsylvania 15261

David C. Fenimore
Texas Research Institute of
    Mental Sciences
Houston, Texas   77030

Joseph H. Fleisher
Department of Surgery
Arizona Health Sciences Center
Tucson, Arizona   85724

M. Y. Fukayama
Department of Environmental
    Toxicology
University of California
Davis, California   95616

Melvin E. Getz
Science and Education
    Administration
U. S. Department of Agriculture
Beltsville, Maryland   20705

Louis J. Glunz
Regis Chemical Company
Morton Grove, Illinois   60053

M. E. Grace
The Wellcome Research
    Laboratories
Burroughs Wellcome Company
Research Triangle Park,
    North Carolina   27709

James G. Hamilton
Roche Research Center
Hoffman-LaRoche, Inc.
Nutley, New Jersey   07110

Gerald J. Hansen
School of Medicine
University of Pennsylvania
Philadelphia, Pennsylvania
    19104

Joyce J. Henry
Center for the Biology of
    Natural Systems
Washington University
St. Louis, Missouri

Charlene Z. Hirsch
Wistar Institute
Philadelphia, Pennsylvania
    19104

Dietrich Hoffman
Naylor Dana Institute for
    Disease Prevention
American Health Foundation
Valhalla, New York   10595

Dennis P. H. Hsieh
Department of Environmental
    Toxicology
University of California
Davis, California   95616

Carol C. Irwin
Eunice Kennedy Shriver Center
Waltham, Massachusetts   02154

L. N. Irwin
Eunice Kennedy Shriver Center
Waltham, Massachusetts   02154

Toshihiro Itoh
Department of Neurology
Yale University
New Haven, Connecticut   06510

Hiroshi Kaneko
Division of Chemistry
School of General Studies
Kitasato University
Sagamihara Kanagawa 228
Japan

Justus G. Kirchner
Coca Cola Company
Atlanta, Georgia

David Kritchevsky
Wistar Institute
Philadelphia, Pennsylvania
    19104

Sidney S. Levin
Harrison Department of
    Surgical Research
School of Medicine
University of Pennsylvania
Philadelphia, Pennsylvania
    19104

Robert E. Levitt
School of Medicine
University of Pennsylvania
Philadelphia, Pennsylvania
    19104

George A. Maylin
Diagnostic Laboratory
College of Veterinary Medicine
Cornell University
Ithaca, New York   14853

P. Madyastha
Center for the Biology of
    Natural Systems
Washington University
St. Louis, Missouri

T. Allen Merritt
Department of Obstetrics and
    Gynecology
School of Medicine
University of Buffalo
Buffalo, New York

R. K. Miller
Department of Obstetrics and
    Gynecology
School of Medicine
University of Buffalo
Buffalo, New York

Suba Nair
Center for the Biology of
    Natural Systems
Washington University
St. Louis, Missouri

Stanley Nesheim
Division of Chemistry and
    Physics
Food and Drug Administration
Washington, D. C.   20204

Danute E. Nitecki
Department of Microbiology
    and Immunology
University of California
San Francisco, California
    94143

James E. Oliver
Pesticide Laboratory
U. S. Department of Agriculture
Beltsville, Maryland   20705

D. W. Rice
Department of Environmental
    Toxicology
University of California
Davis, California   95616

Victor W. Rodwell
Department of Biochemistry
Purdue University
Lafayette, Indiana   47907

Irwin Schmeltz
Naylor Dana Institute for
    Disease Prevention
American Health Foundation
Valhalla, New York   10595

Peter M. Scott
Food Directorate
Health and Welfare Canada
Ottawa, Canada

N. P. Sen
Food Research Division
Health Protection Branch
Ottawa, Canada

Joseph Sherma
Department of Chemistry
Lafayette College
Easton, Pennsylvania   18042

Carl W. Sigel
The Wellcome Research
    Laboratories
Burroughs Wellcome Company
Research Triangle Park,
    North Carolina 27709

R. H. Smith
Department of Chemistry
Western Maryland College
Westminster, Maryland

J. R. Songer
National Animal Disease Center
Ames, Iowa 50010

H. Michael Stahr
Veterinary Diagnostic
    Laboratories
Iowa State University
Ames, Iowa 50011

R. D. Stephens
Department of Health Services
Berkeley, California   94704

R. D. Stephens
Department of Health Services
Berkeley, California 94704

Jane Strauss
School of Medicine
University of Pennsylvania
Philadelphia, Pennsylvania
    19104

Masamichi Tanaka
Division of Chemistry
School of General Studies
Kitasato University
Sagamihara Kanagawa 228
Japan

Lawrence D. Tobias
Roche Research Center
Hoffman-LaRoche, Inc.
Nutley, New Jersey   07110

Joseph C. Touchstone
School of Medicine
University of Pennsylvania
Philadelphia, Pennsylvania
    19104

E. Tuley
Center for the Biology of
    Natural Systems
Washington University
St. Louis, Missouri

Alvin Wenger
Naylor Dana Institute for
    Disease Prevention
American Health Foundation
Valhalla, New York   10595

M. Whitlow
Center for the Biology of
    Natural Systems
Washington University
St. Louis, Missouri

Henry Weiner
Department of Biochemistry
Purdue University
West Lafayette, Indiana   47907

J. L. Woolley
The Wellcome Research
   Laboratories
Burroughs Wellcome Company
Research Triangle Park,
   North Carolina   27709

J. J. Wong
Department of Environmental
   Toxicology
University of California
Davis, California   95616

N. L. Young
School of Medicine
Cornell University
New York, New York   10021

Robert K. Yu
Department of Neurology
School of Medicine
Yale University
New Haven, Connecticut   06510

E. Zomora
Department of Obstetrics and
   Gynecology
School of Medicine
University of Buffalo
Buffalo, New York

Preface and Overview

A symposium dealing with "Clinical and Environmental Applications of Quantitative Thin Layer Chromatography" was held in Philadelphia on January 15 to 17, 1979. The purpose of this symposium was to bring together the practitioners of thin layer chromatography to discuss methodology and applications. The need for this symposium was perceived by several persons who foresaw the benefits to be gained from uninterrupted discourse among practitioners. In retrospect, the purpose of this symposium was fully justified. The meeting went smoothly. Great satisfaction was expressed by many for both personal and professional reasons. Hopefully, this symposium will become an established scientific event.

The timing of this symposium was, in effect, determined by the current level of sophistication of quantitative thin layer chromatography, when theory can be communicated, practice guided by sound principles, and applications attempted with reasonable expectations of success. The instrumentation of thin layer chromatography has also reached a high level of development, sometimes by design, sometimes by chance. As James B. Conant concludes in his thoughtful essay, "Science and Common Sense" (Yale University Press, New Haven, 1952), both theory and technology are required to unfold our knowledge of Nature by reducing the "degree of empiricism." This is clearly true of thin layer chromatography because the applications described in this symposium demonstrate the versatility and sensitivity of thin layer chromatography and reveal the ingenuity with which the practitioners have devised methods for toxins and clinically significant substances.

Thin layer chromatography represents one of several means for resolving mixtures of trace organic substances. Thin layer chromatography is generally applicable, except for volatile or reactive substances. The only requirements are suitable mobile phases and sorbents with which competing equilibrium systems can be

established for solutes. Improvements in the technology of
sorbents have led to the development of high performance,
reversed phase, and biphasic plates. These developments have
increased the efficiency and application of thin layer chromatography.

     The low cost of thin layer chromatography has always been
emphasized, perhaps obscuring our appreciation for its analytical
effectiveness. Low capital costs, low material costs, and
reduced personnel requirements have been listed. Perhaps our
value system has misled us by associating low cost with low performance. ("We get what we pay for.") We assert, however, that
low cost and analytical performance are not mutually exclusive.
"Cost-effective" may be a better epithet to describe thin layer
chromatography.

     In practice, the potential for thin layer chromatography may
not be realized because the total analytical process is not taken
into consideration, particularly sample preparation and cleanup,
and because chromatography is not performed adequately. Sample
preparation and cleanup are special problems relating to the compound of interest and the milieu in which it is found. Problems
encountered in thin layer chromatography usually arise from
imperfect understanding of the methodology, such as the selection
of the mobile phase and sample application. Many times, it seems,
a recipe for a mobile phase has been handed down from one generation of chromatographers to the next or concocted from available
solvents on the shelf. In order to separate a mixture of compounds, the mobile phase should be selected according to principles of chromatography. The mobility of the compound of interest
should be related to solvent strength and adjusted between an R_f
of 0.3 to 0.7. In situ quantitation is effective when material
has undergone displacement equivalent to this R_f range. Selectivity should then be established to resolve the compound from any
overlapping compounds, while maintaining a constant solvent
strength. Selectivity can be introduced by considering the structural differences among the compounds to be resolved.

     Sample application must be accurate, uniform, and reproduci-
ble. It requires considerable care and patience. A simple, convenient device which meets these requirements is the channeled
plate with preadsorbent zone. All too frequently, chromatographic
plates are overloaded with sample. THINK SMALL! There is no need
to apply much material because of improvements in efficiency of
resolution and sensitivity of quantitation. Unfortunately, short-
comings of this type are compounded by carelessness.

     The reason for performing thin layer chromatography has also
been changing over the years. Thin layer chromatography was used
originally to analyze natural mixtures with the emphasis placed

on identification and characterization of resolved compounds.
From this need developed the many fluorimetric and colorimetric
reactions which can be applied directly to chromatograms. More
recently, the trend has been toward quantitative analysis of
trace organic substances. Quantitative elution, still the more
common procedure for quantitation, is both time-consuming and
prone to systematic errors. In situ quantitation has since been
developed with scanning densitometers becoming available for this
purpose. A well resolved chromatogram can be scanned rapidly and
reliably, provided the compound of interest is detectable by
fluorescence or absorbence. Fluorescence can be used to detect
picogram quantities of material. Absorbence, in both visible and
ultraviolet ranges, can detect low nanogram quantities. More
importantly, the scanning densitometer can provide objective data.
Since there are applications of thin layer chromatography to
life-or-death situations, "eyeballing" a chromatogram is no longer
justified!
    Unfortunately, there seems to be neglect surrounding quanti-
tative thin layer chromatography. For example, a reference text
dealing with analytical methods for steroids, which was published
as recently as 1978, does not refer to quantitative thin layer
chromatography. This oversight is surprising because in situ
quantitation of steroids already contributes to inexpensive clinical analyses. Another example is the prevailing notion that the
capabilities of high performance liquid chromatography exceed
those of thin layer chromatography. The advantages of high performance liquid chromatography are considered to be (1) more
efficient resolution, (2) modest preparative capacity, and (3)
automated operation. However, this perception of high performance liquid chromatography may be breaking down because of the
considerable expense of instrumentation and its limited analytical capacity. Despite its promise, not everyone can gain access
to high performance liquid chromatography. In sociological terms,
thin layer chromatography may prove to be a more popular form of
instrumentation, which can be afforded by a greater proportion of
the scientific world.
    The efficiency of high performance thin layer chromatography
and high performance liquid chromatography are essentially equivalent, when they are determined on a common basis. For preparative
purposes, thin layer chromatography has always been recognized as
a practical and effective means for purifying material for
further instrumental analysis. The amount of material that can
be purified by thin layer chromatography is enough for mass spectrometry, the most commonly used of instrumental methods for
identification and structural determination. Automated thin layer
chromatographic systems have never been practical, but development of effective automated systems may be coming.

chromatographic systems have never been practical, but development of effective automated systems may be coming.

The future development of thin layer chromatography, we predict, will see these scenarios: (1) Sample application will be
reduced in size to 0.1 mm from the present 1-2 mm, and the amount
of material applied will be reduced correspondingly. A preadsorbent zone now provides a fine line of application measuring about
0.1 mm, which broadens substantially as the material passes the
junction with the sorbent. (2) Chromatography will be completed
with 1-5 cm, instead of the present 10-20 cm. Developments 1 and
2 will maximize the efficiency of thin layer chromatography. (3)
The resolution of scanning densitometers may become improved and
their output coupled to high speed calculators. (4) Thin layer
chromatography will become automated, and hundreds of quantitative
analyses can be performed each day, as well as numerous analytical
procedures. In this dynamic era thin layer chromatography will
finally realize its full potential.

We thank the participants of this symposium who prepared
their presentations for publication. In most instances, the editors have allowed the authors to tell their story in their own
style. We wish to express our thanks to our colleague, Dr. H. M.
Stahr, who shared in the responsibility for organizing this symposium, and to Kontes for providing much help. The support of the
Department of Obstetrics and Gynecology of the School of Medicine,
University of Pennsylvania, also contributed to the success of the
meeting.

Joseph C. Touchstone

Dexter Rogers

Philadelphia, Pennsylvania
Vineland, New Jersey

Contents

Contents

    T. Itoh, M. Tanaka, and H. Kaneko

THIN LAYER
CHROMATOGRAPHY

CHAPTER 1

History of Thin Layer Chromatography

Justus G. Kirchner

Beyerinck, the Dutch biologist, recorded the first use of thin
layer chromatography (TLC) in 1889 (1). In fact, this work
predates all other chromatographic work except that of Runge
(2,3), Schoenbein (4), and some of Goppelsroeder's work (5,6).
By allowing a drop of a mixture of hydrochloric and sulfuric
acids to diffuse through a thin layer of gelatin, he found that
the hydrochloric acid traveled faster than the sulfuric acid
and thus formed a ring around the latter. He was also the first
to apply a visualizing agent in thin layer chromatography. The
hydrochloric acid zone was detected by brushing on a solution of
silver nitrate, and the sulfuric acid was made visible with a
barium chloride solution. Then in 1898 Wijsman (7) showed that
there were two enzymes present in malt diastase by using the
same technique. He also showed that only one of these enzymes
split off maltose from soluble starch. Wijsman was the first
person to use a fluorescent phenomenon for detecting a zone on a
thin layer. He prepared a gelatin layer containing starch and a
fluorescent bacteria obtained from sea water. By then allowing
the amylase mixture to diffuse in the layer, he obtained a
fluorescent band only where the β-amylase reacted with the
starch. This is a very sensitive test and he was able to detect
1/28,000,000 of a milligram of maltose; this is about 40 pg.
   Izmailov and Schraiber (8) came along 49 years later and
made use of a thin layer of aluminum oxide spread on a glass
plate. The layer did not contain a binder and was used for circular chromatography of galenical preparations by placing a drop
of the solution on the adsorbent and developing into concentric
rings with drops of solvent, the drop method of thin-layer
chromatography. They pointed out the usefulness of this

1

technique for testing adsorbents and solvents for use in column chromatography.

Two years later in 1940 Lapp and Erali (9) used a loose layer of aluminum oxide 8 cm long on a glass slide that was supported on an inclined aluminum sheet. This sheet was cooled at its upper end and heated at the lower end. The sample was placed at the top of the adsorbent layer and gradually developed with drops of solvent. The use of heat at the lower end of the layer increased the evaporation rate of the solvent so that increased development could take place.

By placing adsorbent in cups of spot plates Crowe, in 1941 (10), selected the adsorbent and the solvent using a technique similar to that of Izmailov and Schraiber. The adsorbent selected was then formed in a wedge-shaped layer in a petri dish that was tilted. Here again the drop method was used.

Békésy in 1942 (11) prepared a chromatographic column by pouring a slurry between two glass plates held apart by a cork gasket. He prepared a micro column in the same manner by placing a glass plate over another plate containing a shallow groove; the two plates were held together with agar. In this case again the micro column was filled with an adsorbent slurry.

Returning just a little to 1939, Brown (12) demonstrated the use of a technique for circular paper chromatography by placing filter paper between two glass plates. The sample and the developing solvent were applied through a small hole in the upper plate. To obtain a stronger adsorbent, he proposed the use of a thin layer of alumina between two sheets of paper. Williams in 1947 (13) used the same idea but eliminated the paper entirely.

In separating inorganic ions, Meinhard and Hall (14) used a mixture of aluminum oxide, filter aid, and starch on microscope slides. This was the first instance of the use of a binder to hold the adsorbent to the glass support. The filter aid was used to prevent checking of the surface of the finished layer. Circular chromatography was employed, using a special developing paper to slowly apply the solvent.

In isolating and identifying the flavoring components in citrus juices, Kirchner and his associates needed a micro-chromatographic method for the purification and identification of the terpenes and other compounds. Paper chromatography was employed at first, but it was soon evident that the paper was too weak an adsorbent to be of any great use. Gas chromatography was just beginning to be developed during this period, but had not advanced sufficiently to be useful. Kirchner and Keller in 1950 (15) developed the use of silica-impregnated paper. Although this was an improvement over paper

chromatography the preparation of the impregnated paper is
rather tedious, and the product still has a limited capacity.
The work of Meinhard and Hall seemed a technique that could be
modified so as to combine the advantages of column and paper
chromatography. A suitable starch binder was needed in order
to obtain a nonchecking layer without the use of the filter aid.
At the same time this yielded a stronger adsorbent, because it
was not diluted with the filter aid. Sixteen different adsorbents were evaluated and silica gel was selected as the most
satisfactory and universal adsorbent. It was necessay to sieve
the available material so that only the finer particles were
used (finer than 149 microns). New visualizing agents were
developed not only to locate very unreactive compounds but also
to indicate the types of compounds that were present. Because
some of the visualizing agents reacted with the starch binder,
gypsum was introduced as a binder whenever necessary. Larger
plates ($5\frac{1}{4}$ × $5\frac{1}{4}$ in.) were used to allow for greater development
and consequently more effective separation. Two-dimensional
chromatography was also introduced at this time to achieve
greater versatility. The most significant contribution to thin
layer chromatography was the development of the layers in a
closed tank analogous to the ascending paper chromatographic
technique. The first paper on this important work was published
by Kirchner et al. in 1951 (16). In the following years, 1952-
1957, additional work on modifications and applications of the
method appeared in a series of papers (17-24). Although many
workers from 1951 to 1958 used the method, it attracted little
attention, probably because of its use in a specific field.
Similarly Tswett's work in a specific field was relatively
unnoticed for a period of time.
      Stahl et al. published a paper on thin layer chromatography
in 1956 (25), and asserted that they had developed a "new
method" by eliminating the difficulties of "complicated paste
production with binders and the troublesome preparation of
plaster strips." This was accomplished by using an extremely
fine-grained silica gel (0.5-5 microns) which, they claimed,
eliminated the need for binders. After claiming credit for
creating the new method of thin layer chromatography, Stahl
abandoned it and returned in 1958 (26) to the uniform gypsum-
bound layers of Kirchner et al., although claiming that it was
his method. It is indeed interesting to note that he has
credited Gaenshirt with the development of the gypsum binder.
It was only after the wide publicity given the thin layer
method by Desaga and Merck in advertising the availability of
equipment and adsorbents that the method became popular. It

was also during this period that Chemical Abstracts began to
index thin layer chromatography, thus making it easier to
locate information on the method.

The larger plates used by Kirchner (16) were adopted by
Reitsema (27). In using these he originated the term
"chromatoplate."

Because assay methods for compounds in solution were
already available, and the instrumentation and techniques for
direct thin layer plate analysis had not been developed, the
first quantitative thin layer work evolved around elution
techniques. Quantitative thin layer chromatography was introduced by Kirchner et al. in 1954 (22) with an elution method
for the determination of biphenyl in citrus fruit and products.
The average error in this analysis was found to be +2.8%, thus
showing the reliability of the chromatographic method at that
time when conditions and procedures were carefully standardized.
Equipment for automatic elution from the thin layer plate has
been designed (28) and recoveries of better than 99% have been
reported (29).

With time, equipment and techniques have been developed so
that direct analysis on the thin layer plate is practical (30).
Analyses can now be accomplished more quickly and the possibility of loss in transferring eluates is eliminated. These techniques are described in other chapters.

During thin layer's history many other special techniques
and applications have been introduced. These include horizontal, continuous, multiple, stepwise, and polyzonal development.
Each of these methods has its place. There is also electrophoresis and isoelectrofocusing on thin layers, both with a number
of advantages over thicker slabs. There are multiple layers
using two or more adsorbents on the same layer. In addition,
a number of automatic techniques have been introduced, ranging
from those for starting or stopping development through automatic scrapers and spot extractors to more sophisticated apparatus for carrying out the entire procedure. These and many
others have added to the history of TLC.

The most recent addition to the thin layer story has been
the development of high-performance thin layer chromatography
(HPTLC) (31). A number of workers presaged this development,
which is based on the use of a uniform, very fine grained (60 Å)
particle. For example: Thoma (32) in 1968 recommended the use
of very finely divided adsorbent to obtain better resolution.
The sorbent for this method was ultimately developed by Halpaap
and Ripphahn (33) and, although the rate of development is
decreased so that smaller plates must be used to keep development times reasonable, this is more than offset by the increased
resolution.

There is now scarcely any field where thin layer chromatography has not been applied to advantage, and although HPTLC is pushing hard on the heels of TLC, it cannot displace the latter completely, for who has the finances to buy and operate 100 HPTLC instruments in order to perform 100 separations at one time? All in all these two techniques supplement one another and should continue to do so.

REFERENCES

1. M. W. Beyerinck, Z. Phys. Chem. 3, 110 (1889).
2. F. F. Runge, Annalen der Physik und Chemie XVII, 31, 65 (1834).
3. F. F. Runge, "Farbenchemie," III, 1850.
4. C. F. Schoenbein, Verhl. Naturforsch. Ges. Basel 3, 249 (1861).
5. F. Goppelsroeder, Verhl. Naturforsch. Ges. Basel 3, 268 (1861).
6. F. Goppelsroeder, Mitt. k. k. Tech. Gewerbemuseums Wien N.S. 2, 86 (1888).
7. H. P. Wijsman, De Diastase, beschouwd als mengsel van Mattase en Dextrinase (Amsterdam, 1898).
8. N. A. Izmailov and M. S. Schraiber, Farmatsiya (Sofia), 1938, 1.
9. C. Lapp and Erali, Bull. Sci. Pharmacol. 47, 49 (1940).
10. M. O'L. Crowe, Anal. Chem. 13, 845 (1941).
11. N. V. Békésy, Biochem. Z. 312, 100 (1947).
12. W. G. Brown, Nature 143, 377 (1939).
13. T. L. Williams, Introduction to Chromatography (Blackie, Glasgow, 1947), p. 3
14. J. E. Meinhard and N. F. Hall, Anal. Chem. 21, 185 (1949).
15. J. G. Kirchner and G. J. Keller, J. Am. Chem. Soc. 72, 1867 (1950).
16. J. G. Kirchner, J. M. Miller, and G. J. Keller, Anal. Chem. 23, 420 (1951).
17. J. M. Miller and J. G. Kirchner, Anal. Chem. 23, 428 (1951).
18. J. G. Kirchner and J. M. Miller, Ind. Eng. Chem. 44, 318 (1952).
19. J. M. Miller and J. G. Kirchner, Anal. Chem. 24, 1480 (1952).
20. J. G. Kirchner and J. M. Miller, J. Agric. Food Chem. 1, 512 (1953).
21. J. M. Miller and J. G. Kirchner, Anal. Chem. 25, 1107 (1953).
22. J. G. Kirchner, J. M. Miller, and R. G. Rice, J. Agric. Food Chem. 2, 1031 (1951).

23. J. M. Miller and J. G. Kirchner, Anal. Chem. 26, 2002 (1954).
24. J. G. Kirchner and J. M. Miller, J. Agric. Food Chem. 5. 283 (1957).
25. E. Stahl, G. Schroeter, G. Kraft, and R. Renz, Pharmazie 11, 633 (1956).
26. E. Stahl, Chem. Ztg. 82, 323 (1958).
27. R. H. Reitsema, Anal. Chem. 26, 960 (1954).
28. H. Faulk and K. Krummen, J. Chromatogr. 103, 279 (1957).
29. R. K. Vitek, C. J. Seul, M. Baier, and E. Lau, Am. Lab. 6, 109 (1974).
30. J. C. Touchstone and J. Sherma, Eds., Quantitative Thin Layer Chromatography (Wiley-Interscience, New York, 1979).
31. A. Zlatkis and R. E. Kaiser, HPTLC High Performance Thin-Layer Chromatography (Elsevier, Amsterdam, 1977).
32. J. A. Thoma, "Polar Solvents, Supports, and Separation," in Advances in Chromatography, edited by J. C. Giddings and R. A. Keller (Marcel Dekker, New York, 1968), Vol. 6, p. 61.
33. H. Halpaap and J. Ripphahn, "High Performance Thin-Layer Chromatography: Development, Data, and Results," in Ref. 31, p. 95.

CHAPTER 2

Technology of TLC Sorbents

Thomas E. Beesley

Previously, the predominant problems of thin layer chromatography (TLC) involved the use of stationary phases as adsorbents. In the general theory of chromatography today, the adsorption process has taken on new meaning and therefore the word "sorbent" has come into use to cover not only the wider range of substrates for the great variety of applications but also the changes in thinking on the mechanism of separation.

Undoubtedly, the early workers in TLC were fascinated by the spectrum of materials that could be coated onto a piece of glass and rapidly tested for its suitability for their particular problem. In many early works on TLC, lists of adsorbents ranging from charcoal to starch were presented in a given order to act as a guide in their effective use. These lists were compiled on the basis of the ability of these materials to bind organic molecules by electrostatic forces, these being the same forces that are also responsible for holding together the crystal lattice. It is therefore not surprising that the order of these lists follows Moh's hardness scale. The binding forces for organic molecules on these surfaces is generally due to ion-dipole or dipole-dipole forces. These electrical surface forces can induce dipole moments in nonpolar compounds and increase the dipole moment in polar compounds and therefore demonstrate catalytic activity. Cassidy (1) recognized this potential problem and listed sorbents by their catalytic activity. Other mechanisms as the ion exchange properties of these materials and hydrogen bonding effects led to an infinite variety of potential separations on TLC.

Several factors have reduced drastically the number of sorbents currently in use. It would appear that the dominant factor in this reduction was the impetus to reproduce the system.

Consequently, every effort was made to choose a model that was
.readily available and inexpensive. In addition, the model should
have minimum catalytic activity and widest flexibility as a
medium for separation. The ability to produce this material in
pure and predictable form surely was a major consideration.

The development of these sorbents has been a curious blend-
ing of theory and practice. For all of the analytical tests
currently performed on a material suitable for TLC, the ultimate
test of that suitability was then and is now its chromatographic
behavior. As an example of our limited knowledge, it is curious
to note that more than one chromatographic test must be run to
insure its identity to previous samples. There are, however, a
number of facts we do understand, and a number of meaningful
analytical tests we do perform, all demonstrating the intricate
nature of chromatographic separations. This understanding to
date has led to improved chromatographic performance in terms of
selectivity, resolution, and reproducibility. This then is how
one manufacturer views the technology of sorbents.

The following sorbents are discussed in order of decreasing
use and not in order of potential because in considering the
large variety of available silica gels and the magnitude of
silica gel applications, the usefulness of these other materials
is lessened by their own obscurity.

SILICA GEL

Silica gel is synonymous with chromatography. It is a porous,
amorphous structure available in a wide range of forms with
unsurpassed capacity for both linear and nonlinear isothermal
separations. The linear capacity of an activated silica exceeds
that of an activated alumina by one hundredfold. Optimally,
deactivated silicas exceed 5 to 15 times the capacity of deacti-
vated alumina. The structure of silica gel places it very near
the middle of Moh's hardness scale, with very little potential
for catalytic activity. Being somewhat acidic in nature, the
result of which is the potential to behave as a cation exchange
material (2), the hydroxylated surface utilizes hydrogen bonding
forces to effect separations.

The process for making the most common chromatographic
silica gels is simple. Dilute hydrochloric acid is added to a
solution of sodium silicate; the solution solidifies and the sol
undergoes changes dictated by the rate of addition, and the final
pH of the solution before solidification occurs. The pH at which
gelation occurs strongly affects the surface area. The dehydra-
tion, i.e., condensation, steps that subsequently occur continue

and the final pore structure is determined by the length of this aging process and the final washing procedures used, as well as the drying procedures.

Because the surface supplies the necessary hydroxyls for the separation mechanism and because increasing the porosity increases the surface, the goal is to create more porous gels without losing structural integrity. At the same time, however, it is important to control the pore diameter, which affects the transfer rates of solutes through a particle and thus affects the efficiency. This compromise generally approaches that of gels with surface areas 200 to 500 m^2/g containing 4 to 6 hydroxyls/nm^2.

It is helpful to visualize silica gel as an aggregate of elementary particles $\sim$ 100 Å in diameter. Grinding larger particles down to smaller particles does not affect the surface area. The combined nature of the pore structure and the surface area determines the capacity factors (k') of the adsorbent, that is, the measure of the degree of retention of a solute compared to the solvent front. For TLC it is calculated as

$$\frac{(1 - R_f)}{R_f}$$

SURFACE AREA

Much work has been conducted in the areas of pore diameter, volume, and distribution, regarding both the control and measurement of these parameters (3). The relationship of diameter, volume, and surface area is expressed as:

$$d_p = 4 \frac{V_p}{SA} ,$$

where d_p = average pore diameter (Å)

V_p = pore volume (ml/gm)

SA = surface area (m^2/g)

Considering the need for good mass transfer in and out of the pore, the diameter must be held within certain limits. Although controversy exists on the exact nature of the chromatographic response resulting from hydroxyls of wide pore or narrow pore silica, the divergent response of various organic compounds measured as resolution, to 4, 6, or 8 nm average pore diameters

has been noted in our laboratory. Because the average pore
diameter results from a fairly broad range, and the chromatographic response is different with narrow and wide pore structures, it is imperative to measure the distribution and to keep
the identically shaped plot of pore diameter versus percent pore
volume, assuring no shift in concentration of high or low pore
diameter configurations.

The typical method of analyzing these characteristics is
the nitrogen adsorption isotherm, which can be extended to
produce surface data by the BET method at low partial pressures
and pore volumes at higher partial pressures and finally some
indication of pore shape by the plotting of the corresponding
desorption curve. This adsorption-desorption plot is referred
to as a hysteresis loop. This type of data is generally the
first to be obtained on a typical sorbent for chromatography.
The second analysis required is the bulk density determination,
which indicates any changes in internal formation of the particle. This requirement dictates the packing density on a layer
or in a column. The third analysis is for heavy metals and iron
because their presence can contribute to or detract from a separation. The fourth phase of the investigation is to determine
the pore distribution by passing soluble polystyrenes ranging in
diameter from 8 to 3000 Å (M.W. 78 to 110,000) through a packed
column of material that indicates the percent of chromatographically available pores. Data indicate that on the average, only
65% of the pores determined by the nitrogen adsorption method
are available for chromatography, the smallest molecule that was
tested being benzene.

Photomicrographs have been taken from time to time and have
shown that the particles of silica generally have some crystal-
linity and that certain particles that were later identified as
wall scale in the reaction vessel were quite amorphous and had a
higher water solubility and weaker wall structure. This then
would contribute to column failures under pressure or silica
solubility in recovery experiments. This problem is approached
by taking representative samples of silica in water and assaying
for free soluble silica.

HYDROXYL CONTENT AND ACTIVITY

As the separation process on silica is somehow a function of the
surface hydroxyl content, which is responsible for selective
adsorption onto the particle, the hydroxyl content must also be
controlled. Whether, as Snyder (4) indicates, there are two
different reactive hydroxyls, or whether, as Scott and Kucera (5)

indicate, the interactions are solute-solvent not solute-surface, the activity of the adsorbent can be affected by the number of silanol groups present and the amount of adsorbed water on the surface. It has been noted on silicas that water adsorption is greater with narrow pore silicas and, in fact, one measure of the reactivity of an adsorbent is to add water dropwise, while shaking in a centrifuge tube the powder that had been dried at 150° C for several hours, noting at what point the powder begins to aggregate. The assumption is that the pores fill before reacting on the surface to increase hydrogen bonding between particles. The higher the water uptake, the more active the adsorbent. Values of 2×10^{20} OH/g of silica have been determined.

The convincing arguments of Scott and Kucera state that, at temperatures up to 200° C, two molecular layers of water still exist on the surface and the interactions are primarily displacement of the second layer by either more polar solutes or solvents. They indicate that only compounds having very high k' of 20 or more actually interact with the surface silanol group.

Heating silicas up to 400° C can cause hydroxyls in close proximity to condense and form the more hydrophobic siloxane bond, which will not contribute to the separation mechanism. Exposure to water or alcohols can cause the cleavage of these bonds, some of which may be under greater stress than others, leading to erratic behavior of the material.

Heating beyond 400° C begins to affect the surface area and adsorption isotherms indicate a continued decreased capacity for solvents as ethyl acetate as silica is heated to 1000° C.

Conditioning of an activated TLC plate to 40% RH (6.67% H_2O by weight) gives a more reproducible chromatographic system and seems to suggest the deactivation of very active sites and a more uniform hydrogen-bonding mechanism (6). Scott and Kucera calculate this to three layers of water on the surface and the mechanism is an interaction of the solute with the first or second layer of water. For the purposes of characterizing an adsorbent the material must be looked at chromatographically at stages from full activation (no surface water) to maximum deactivation.

The other parameters of the layer can be evaluated by keeping relative humidity constant. It has been found that at 40% RH the separation mechanism on a given TLC plate is reproducible with very narrow limits for certain types of plates.

The standard deviation from R_f values was determined for a variety of TLC plates under these conditions and fell into three categories:

Type of TLC Plate	R_f-- Standard Deviation (%)
Silica Gel G Plates	8.4
Polymer Bound Silica Plates	2.5
Preadsorbent Silica Plates	0.9

Grinding Processes

Analysis of particle size and size distribution. The method of grinding has been found to affect layer performance, from the standpoint of both particle shape and contamination. Current processing equipment is lined with ceramic and combines grinding and air classification so that particles are not reground but are centrifugally removed as they are broken down into the proper size range. The resulting particles appear to have faceted surfaces that result in good flow characteristics.

This area of product development has undergone more drastic changes than all the others. The battles of efficiency, permeability, and resolution are mainly fought here and the differences between column chromatography and TLC are brought home From relatively broad to today's narrower distribution of particles, great strides in plate performance and plate-to-plate consistency has been noted. Only as the particle decreases in mean diameter can the particle distribution be made narrower since the interstitial volumes of the larger particles cause a decrease in efficiency.

The following results were obtained from plates spotted with 20 ng of quinidine:

Plate Type	Range (Microns)	Mean	H (Micrometers)
HP-TLC	3- 8	5	15
K5	5-20	10	19
K5A	14-20	17	44

Data from the Coulter particle size analyzer have been shown to be misleading as an absolute test of particle distribution insofar as the grinding method changes particle shapes, which is not taken into consideration in the Coulter analysis. Consequently, particle distribution curves showing virtually identical Coulter curves can show chromatographic differences based on particle shape, which undoubtedly results from a different orientation in the layer, leading to a greater or lesser particle packing density. Sedigraphs provide more accurate, chromatographically related curves and, to an experienced eye, one good microscope provides enough data to judge the material.

   REVERSE PHASE TLC

The ability to change the selectivity of a media was never more possible than at this stage of development in chromatography. Silica gel and alumina retain solutes almost exclusively on a basis of polarity. Separations of close polarity even with different molecular weights are therefore difficult. Selectivity can only be effected by the mobile phase in these cases. Changes in the surface are then required to dramatically change selectivity.
   The reaction of surface hydroxyls with organic silanes, especially when employing octadecyl chains, provides selectivity not only between solutes of very similar polarities but different molecular weights.
   The problem of mass transfers and solvent velocity on TLC with reverse phase (RP) materials was resolved by this manufacturer by spacing the octadecylsilyl chains yielding approximately 100 Å pores and reacting the remaining hydroxyl groups with a smaller silane.
   There are a number of possible arrangements including the utilization of the unreacted hydroxyls so that both adsorption and partition can function. The problem is that the analyst may or may not be aware of what is being worked with.
   The pertinent analytical facts that are required include:

      1. Chain length of the organic silane that is used.
      2. Percent carbon, hydrogen, and oxygen.
      3. Retention data for nitrobenzene.

Since the chromatographic mechanism for reverse phase separation is the retention of nonpolar compounds by the organic stationary phase an increase in the polarity of the mobile phase will result in increased retention of the more nonpolar species. An increase in carbon loads therefore results in lower polarity solvents being required for retention, resulting in faster developing systems of higher efficiency. It can be seen now that simple adjustments in mobile phase conditions can easily reproduce a separation even if percent carbon varies.
   The retention of nitrobenzene is used to monitor the number of available hydroxyl groups, which will indicate reproducibility in mobile phases whose composition would allow adsorption to take place.

Using similar reverse phase mechanisms on silicas employed in HPTLC results in the direct correlation of a KC_{18} plate with a TLC column. A plot of the capacity factors for both demonstrates the similarity in selectivity and a method for calculating the capacity factors for the column from TLC, R_f data.

ALUMINUM OXIDE

As stated previously, a change in the surface causes more dramatic changes in selectivity. The mechanism of aluminum oxide separations can involve both hydroxyls and the aluminum ion and therefore is to be considered when a change in selectivity is required. Aluminum oxide is similar to silica in production methods: the particles are formed by the condensation of hydrated aluminum cations. The resulting structure is slightly different from silica, having pores rigidly fixed by a crystalline lattice structure, virtually cylindrical at 27 Å in diameter.

The work of Borello et al. (7) is useful for an understanding of the hydration and chemisorption of water onto the surface. Density determinations indicate a high degree of reproducibility in its manufacture and layers may be repeatedly activated and deactivated to give separations of amazing consistency. The possibility of catalytic activity is high, which indicates the binding forces may be more electrostatic in nature. However, at high water contents separations similar to silica may be obtained. The simple two-dimensional development experiment showing stability of the organic sample to the system should be conducted.

Analytical tests are reserved to bulk density determinations and pH of aqueous suspensions. The activity is determined chromatographically with dyes at various stages of heating from 110° to 200° C.

MAGNESIUM SILICATE (FLORISIL)

The selectivity of this sorbent has been determined and is claimed to preferentially adsorb bases. Its activity lies between alumina and silica and is less prone to sample reactions than the former. Many samples tend to chemisorb on Florisil (8) which limits its usefulness. Nevertheless it has been extensively used in recent years for certain difficult lipid and pesticide samples. It was then tested chromatographically for its suitability in this area.

CELLULOSE

It would be unfair to ignore this material because the availability of microcrystalline cellulose was a major step from paper chromatography and it allowed partition chromatography to flourish. Even today, with the availability of reverse phase materials, certain biologically sensitive substances can only tolerate a media as gentle as cellulose. The difference in the structure of helical natural fibers and the more crystalline chemically treated fibers can cause separation differences dependent on the molecular retentive properties of the former.

Water regain values are the primary analytical measure of reproducibility and control of fiber dimensions. A chromatographic test used for evaluation is the preparation of food dyes in partition systems.

CONCLUSION

As can be seen from the complexity of the system--pore structure, size, and shape of the particle, conditions of activity influenced by water adsorption--it is not surprising that an understanding of the solute and solvent interactions is infinitely more difficult.

Changes in selectivity are possible from a varied choice of sorbents or by modification of the surface of silica gel by reaction with organic silanes of varying chain length, controlling the residual hydroxyl content, as well as reacting with various groups including amino, cyano, or other ion exchange functions. Other available materials offer selectivity not exactly predictable but reasonable for investigation.

Current analytical determinations aid in stabilizing performances of current sorbents but carefully controlled chromatographic tests are essential to establish similarity to a previously prepared product.

SUMMARY OF A REPRODUCIBLE QUANTITATIVE TLC SYSTEM

In order to measure the performance of an analytical TLC plate for quantitation the following operating parameters are offered as a guide:

1. Adjust layer to 40% RH condition equivalent to 6.7% H_2O W/W on surface of silica (K6 $\sim$ 400 M^2/g).

2. Spotting: 2, 3, 4, 5, or 10 µl applications on preadsorbent TLC with Drummond Microcaps. Perform serial
   dilution of standards and spot all samples at constant
   volume from a solvent of low-to-moderate polarity. Keep
   concentrations at 0.1 to 5 µg. Spot at same distance
   from plate edge, usually 1.5 cm, with 0.5 cm solvent
   depth in tank.
3. Solvents: fresh glass distilled solvents free of water
   and peroxides (pass through dried Al_2O_3) accurately
   measured.
4. Tank environment: unsaturated, maintaining 2 cm distance between silica surface and glass or porous surface
   for partition systems.
5. Temperature: controlled 68° ± 2° C. Colder temperature
   reduces diffusion, slows migration.
6. Length of development: effective region of TLC plate
   lies in R_f region 0.2 and 0.8. For quantitation,
   regions below 0.2 are too compact and those above 0.8
   are too diffuse.
7. Post development: the plate must be allowed to evaporate before placing in an oven for drying.
8. Visualization: accurate and successful application of
   reagent is accomplished by dipping. Concentration of
   reagent and maximum instrument response must be determined. Wipe back of chromatogram.
9. Standard curve/internal standards required for accuracy.
10. Scan 90° to direction of development for maximum
    response.

REFERENCES

1. H. G. Cassidy, Fundamentals of Chromatography, Technique of
   Organic Chemistry (Interscience, New York, 1957), Vol. 10.
2. I. Jane, J. Chromatog. 111, 227 (1975).
3. J. Ripphahn and H. Halpaap, J. Chromatog. 78, 63 (1973).
4. L. Snyder, Principles of Adsorption Chromatography (Marcel
   Dekker, New York, 1968).
5. R. P. W. Scott and P. Kucera, J. Chromatog. 149, 93 (1978).
6. L. Snyder, ibid., p. 145.
7. E. Borello et al., J. Catal. 35, 1 (1974).
8. L. Snyder, J. Chromatog. 12, 488 (1963).

CHAPTER 3

Preparation of Samples

Joseph Sherma

In quantitative densitometry for quantitation in thin layer
chromatography (TLC), one or more samples are applied as spots
or streaks onto the origin of a layer along with three or four
standard zones to which the samples will be compared after
scanning. The compound of interest (the analyte) in the sample
should separate as a discrete zone with the same regular shape
and R_f value as the standards, and the amount of analyte should
be bracketed by the standards and fall within the linear portion
of the calibration curve. These are general requirements and
techniques of densitometry that are followed by most workers,
although exceptions are recognized. For example, it may be
adequate to apply only one standard spot against which to compare multiple samples (1), and the fluorescence mode of scanning
can be quite spot-shape independent (2). However, sample
preparation methods are normally designed to provide a final
spotting solution of adequate concentration and purity to meet
these requirements.

The types of sample preparation procedures carried out for
TLC are very similar to those required for gas chromatography
and high-performance column liquid chromatography, namely,
extraction, solvent partitioning, column chromatography on various adsorbents and resins, centrifugation, evaporation, and
dissolving of residues.

J. C. Touchstone has introduced the idea that the
degree of required cleanup is governed by the analyte/junk
(impurity) ratio:

$$\text{success of chromatography} \approx \frac{\text{amount of analyte}}{\text{amount of junk}} \cdot$$

For very pure and/or high concentration samples, the ratio
is favorable and a low degree of sample purification is
required. If the analyte is present in trace amounts and/
or there are many interfering compounds present, more
elaborate cleanup procedures will be needed. These techniques are discussed in this chapter.

DIRECT APPLICATION OF SAMPLES

If the analyte is present in sufficient concentration and purity, it is sometimes possible to directly apply the sample with
no required extraction or cleanup. The concentration must be
such that the normal volume (e.g., 5-1000 nl for HPTLC and 1-5
μl for TLC) will give a detectable zone. The content of impurity must not retain the analyte at the origin or distort the
shape (i.e., tailing) or change the R_f (higher or lower) of the
analyte spot.
      As a specific example of this approach, we have determined
1 ppm levels of p-aminobenzoic acid (PABA) with recovery of 100
+ 8% by directly applying 25 μl of urine containing 25 μg of
PABA. The minimum detectable level of PABA as a red spot after
spraying with N-(1-naphthyl) ethylene diamine reagent was 10 ng,
and 50 μl of urine could be applied before extraneous materials
at the origin interfered with the migration of PABA. The minimum detectable level was, therefore, 10 ng/50 μl or 0.2 ppm.
The PABA from the sample and the standard both migrated with R_f
0.34 on silica gel developed with benzeneglacial acetic acid
(5:1 v/v). Because of its amphoteric nature, PABA could not be
successfully extracted from urine (or water) at any pH with any
organic solvent tested (3).
      Caffeine was determined in cola, tea, and coffee by direct
application of the different beverages. One to six μg of caffeine was spotted on a silica gel F layer from a 2 μl microcap
micropipet along with 10 μl of cola sample or 5 μl of instant
coffee or tea. Development with ethyl acetate-methanol (85:15
v/v) separated caffeine as a circular zone at R_f 0.6 from the
remainder of the samples remaining as residue at the origin.
The amount of caffeine was determined by scanning by densitometry the unknown and reference caffeine in the fluorescence
quench mode (4).
      Lipid profiles were determined by direct application of
0.5 μl of plasma from capillary blood or 0.5 μl of liver homogenate to a 10 × 20 cm HP silica gel layer. The samples were
applied to the origin over a 15 μl spot of absolute methanol,
and the sample spot was covered immediately with a few

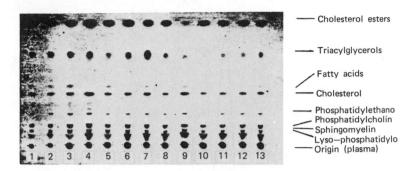

Figure 1. Separation of lipids in capillary blood without extraction.

microliters more of methanol. After drying under a stream of cold air, the plate was developed with chloroform-methanol-water (65:30:5 v/v) twice for 3.7 cm each and then with hexane-diethyl ether-acetic acid (80:20:1.5 v/v) to within 1 cm of the top of the plate. After drying for 1 hour in warm air, fluorescent spots were produced by treatment with $(NH_4)HCO_3$ and heating for 10 hours at 150° C. The resulting fluorescent spots were scanned across the axis of development with 366 and 430 nm excitation and emission wavelengths respectively, allowing quantitation by densitometry in the nanomole range. Separations were sharp and reproducible without lipid extraction prior to chromatography, as shown in Figure 1 (5).

APPLICATION OF SAMPLE SOLUTIONS OR EXTRACTS

When the analyte is a macro constituent of the sample, simply dissolving it in an appropriate amount of solvent, followed by direct spotting of an aliquot of sample along with bracketing standards, is adequate for densitometric quantitation. This approach was reported (6) for the determination of caffeine in APC tablets on C_{18} reversed phase layers containing fluorescent indicator. The active ingredients of the tablet were dissolved in 25 ml of methanol-chloroform (1:1 v/v), and 1 µl portions applied with 0.5-2.5 µg of standard caffeine. After development with methanol-0.5 M NaCl (1:1 v/v) to separate the aspirin, phenacetin, and caffeine, the quenched fluorescence of the caffeine zones ($R_f \simeq 0.5$) was scanned. The filler in the tablet did not dissolve.

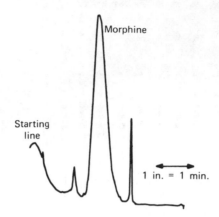

Figure 2.   Morphine separated from urine.

Amiben containing 92% 3-amino-2,5-dichlorobenzoic acid was
analyzed in a similar manner after dissolving and dilution; the
derivative formed with N-(1-naphthyl) ethylene diamine dihydro-
chloride reagent was scanned for quantitation (7) after separation by TLC as mobile phase on silica gel G.

     Extracts of trace constituents from some types of samples
can also be directly spotted, if any coextracted impurities are
resolved from the analyte spot by the thin layer chromatographic
development or not detected by the visualizing method. To minimize the amount of interfering coextractives, the least polar
solvent that will extract the analyte should be used. Many
polar impurities, which are the most troublesome in biological
samples, will then be unextracted.

     As an example, residues of the sulfoxide of the organophos-
phorus insecticide phorate were determined in tap water at the
10 ppb level by extraction with ethyl acetate, drying of the
combined extracts by passage through Whatman PS phase separating
paper, evaporation to dryness, and dissolving of the residue in
ethyl acetate. Determination was carried out on silica gel
microslides developed with hexane-acetone-chloroform (65:30:5
v/v); detection of phorate was as an orange spot after reaction
with TCQ (N,2,6-trichlorobenzoquinoneimine) reagent (8). An
alternate common method for drying extracts is by adding anhydrous sodium sulfate; traces of water should be removed prior to
spotting for best TLC results.

     The drugs morphine and amphetamine were extracted from
urine buffered to pH 9.5 by the classic Davidow method, using
chloroform-isopropanol (96:4 v/v) as the extraction solvent (9).

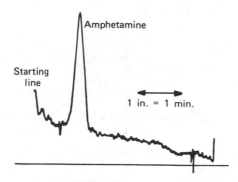

Figure 3.   Amphetamine separated from urine.

chloroform-isopropanol (96:4 v/v) as the extraction solvent (9).
After filtration and concentration, the extract was spotted and
separated, and the drugs were detected as fluorescent spots
after reaction with the specific reagents. Figure 2 shows the
densitometer scan for morphine originally present at 0.50 ppm in
the urine; two impurities also were present but did not interfere with quantitation. Figure 3 shows the scan for amphetamine
extracted from urine at 0.25 ppm, with no interfering peaks
being found.

CLEANUP BY PARTITION OF EXTRACTS

Extracts are often too impure for direct application to TLC and
may require partitioning with an inmiscible solvent as a cleanup
step. The idea of such differential partitioning steps is to
leave impurities behind in one solvent, while extracting the
analyte into another. For example, nonpolar organochlorine pesticides extracted from fatty foods into hexane may be partitioned into acetonitrile, the acetonitrile diluted with water or
salt solution, and the residues re-extracted back into hexane.
     A general differential partition scheme for separation and
recovery utilizing pH control is illustrated in Figure 4. Such
a scheme becomes easier to apply as more information about the
nature of the sample is known. The principle involved is that
acids are converted into salts and are soluble in aqueous solution at high pH, whereas at low pH the acids are un-ionized and
extractable into organic solvents. Basic compounds are
extracted into organic solvents at high pH and into water in
their salt forms at low pH.

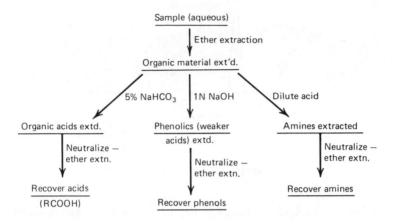

Figure 4. Schema for extraction with control of pH. After three extractions, neutral compounds left in ether. Next hydrolyze with enzyme, acid, or base to break conjugates (glucoride, sulfate). Then repeat extractions.

Application of differential partitioning has been made for cleanup of estriol from pregnancy urine and some chlorophenoxy acid herbicide residues from tap water (free acid and ester forms). The procedure for estriol (10) is shown in Figure 5 and that for the herbicides (11) in Figure 6. The methods obviously involve many experimental manipulations that require care if losses at each step are to be minimized. Very pure solvents and reagents should be used so that interferences are not introduced during the sample workup. Radioactive reference material when available should be added to assess recovery.

Residues of the herbicide Amiben (3-amino-2,5-dichlorobenzoic acid) in crops at 0.1 ppm level can be determined by densitometry of TLC after a similar partitioning cleanup scheme. The initial step is a basic hydrolysis (reflux) to hydrolyze the glycoside conjugate, after which the residue is purified by transfer between an ether phase as the free acid (pH 1) and an aqueous phase as a salt (pH 5.8) (7).

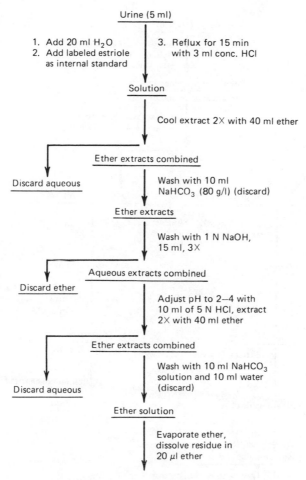

Figure 5.   Procedure for extraction of estriol from urine.

CLEANUP BY COLUMN CHROMATOGRAPHY

Column chromatography is often valuable for purification of solvent-partitioned extracts or for direct cleanup of extracts without partitioning. The former approach has been used for the isolation of steroids from nonsaponifiable lipids during the investigation of algal cells. The first step was cell disruption in ethanol or water solution by ultrasonication. The solution

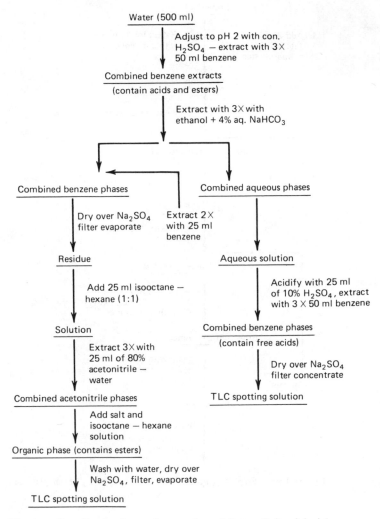

Figure 6.   Procedure for extraction of herbicides.

was made strongly alkaline with 10% KOH and allowed to stand
overnight. In this step, the ester linkages (triglycerides)
were hydrolyzed, and the free acids were neutralized by the
strong base. Extraction with n-heptane then removed the
nonsaponifiable lipids (sterols), leaving behind the bulk of

cellular material. The cells were removed by centrifugation, and then the fatty acids were recovered by acidification of the aqueous phase, followed by a second heptane extraction. To separate the sterols from the neutral lipids and associated impurities such as pigments and alkanes, the mixture was applied to a silica gel column. Elution with n-heptane eluted the alkanes, followed by benzene to elute the sterols and a few pigments. This effluent, now stripped of very nonpolar and very polar material, was concentrated and separated on a silica gel layer with diethyl ether-n-heptane to separate the sterols from interfering pigments (12).

The partitioning between hexane and acetonitrile described earlier may not be adequate for cleanup of residues of organo-chlorine pesticides extracted from some sample matrices. In these cases, the final hexane solution is concentrated and applied to a Florisil column, and the pesticides are further purified by elution with ethyl ether-petroleum ether (6:94 v/v) (13). The column effluent is concentrated and subjected to TLC.

A column (1 cm i.d.) containing 1 g of silica gel deacti-vated with 20% water has proved valuable for the cleanup and fractionation of multiresidues of different classes of insecti-cides. Samples are extracted with acetonitrile, and residues are partitioned into methylene chloride. The methylene chloride is concentrated and applied to the silica gel column, which is eluted with a stepwise gradient composed of the solvents shown in Figure 7. Chlorinated insecticides elute in fractions 1 and 2, organophosphorus insecticides in fractions 2 and 3, and carba-mate insecticides in fraction 3(14). The fractions containing the compound(s) of interest are concentrated and separated by TLC. The position of elution is important evidence for confirma-tion of residue identity.

Macroporous silica gel SI-200 (15) packed as a 5-g micro-column in a glass tube has been used for the direct cleanup of tap water extracts containing 8 ppm residues of phorate oxygen analog or carbofuran without intervening solvent partitioning. The column was eluted with a stepwise gradient of 2, 5, and 10% acetone in hexane, and was reusable after regeneration of the column by washing in turn with methanol, acetone, and hexane. The third fraction, containing the residues, was concentrated and spotted on laned, preadsorbent-containing silica gel TLC plates prior to development, detection with chromogenic reagents, and densitometry.

The degree of cleanup that can be achieved by column chroma-tography is clearly illustrated by the determination of thiourea in Figure 8. The first part shows the scan of an acetone extract of orange peel after thin layer chromatography without column

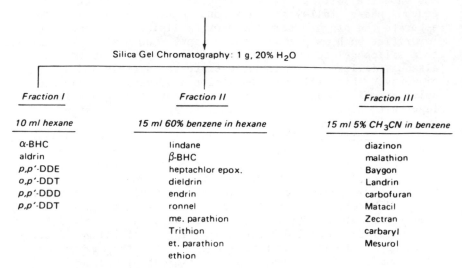

Figure 7. Silica gel column separation of insecticides.

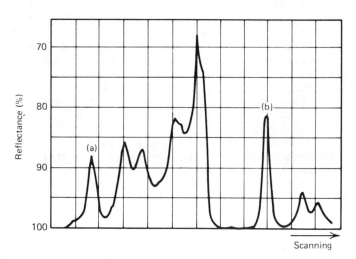

Figure 8. Thiourea (b) in orange peel after cleanup by column chromatography.

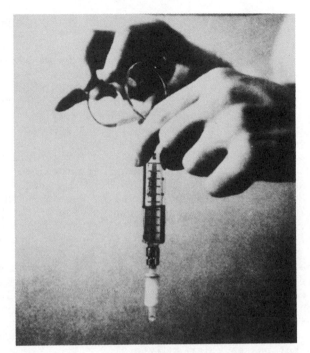

Figure 9. Use of the Waters Sep-Pak cartridge.

cleanup. Part (b) is the scan of a chromatogram after column
cleanup of the extract (17).

MODERN CLEANUP SYSTEMS

Examples of contemporary, commercial sample preparation systems
composed of prepacked, disposable tubes or cartridges are the
Sep-Pak cartridges (Waters Associates). These are designed for
cleanup prior to HPTLC but serve equally well for TLC. The
cartridge is fitted to a syringe with the barrel removed, sample
is poured into the syringe barrel (Figure 9), and the plunger is
returned to force the solution into the cartridge. The process
is repeated with an appropriate solvent or solvents, to purify
and remove the analyte.
      The rationale behind the choice of solvents for the silica
gel and C_{18} reversed phase (RP) cartridges that are commercially
available is as follows. The silica gel cartridge eliminates
low polarity compounds by washing with a nonpolar solvent, while
the analyte(s) is retained on the cartridge. The cartridge is
next washed with a more polar solvent to remove the analyte(s);

the least polar solvent should be used, so that very polar
impurities still remain on the cartridge. The C_{18} cartridge
eliminates high-polarity compounds by washing with a polar
solvent, and then the analyte(s) is removed by washing with a
less polar solvent (but as polar as possible). The effluent
containing the analyte(s) is brought to a known volume for final
TLC analysis. Preliminary studies are required to assure quantitative results, that is, complete recovery of the analyte
with the elution solvent and no elution but adequate cleanup of
the analyte with the other solvent.

A specific use of the C_{18} Sep-Pak is for isolation of anticonvulsant drugs from serum. The drugs are retained on the
cartridge when serum acidified with 0.01 M $NH_4H_2PO_4$ is passed
through. Polar impurities are removed by washing with 1-30 ml
of salt solution, depending upon the retention of the polar
components. Drugs are eluted with 5 ml of acetonitrile; any
proteins or lipids that are present will be precipitated by this
solvent.

Another way to use these cartridges is to pump the sample
through continuously. Passage of a sample through a silica gel
cartridge causes retention of polar impurities and allows passage of nonpolar components. For quantitative recovery of the
analyte, final elution with a nonpolar solvent is required.
Passage of a sample through an RP cartridge retains nonpolar
impurities but passes polar analyte(s).

The Sep-Pak cartridges evolved in the Waters laboratories
from small columns of sorbents dry packed into glass disposable
pipets plugged with glass wool. Sample and solvents were forced
through by pressure from a rubber bulb on the top. These
"cleaner columns" can easily be prepared by individual workers
in the laboratory with silica gel or C_{18} packings or any other
available sorbent material that might provide the required sample cleanup.

The Extube (formerly "Jet" technique) method was designed
for isolation of drugs by the principle of liquid-liquid extraction. The sample to be analyzed (urine, whole blood, stomach
lavage, bile, etc.) is introduced into a disposable Extube, which
is packed with an inert fibrous matrix of large surface area.
When the selected solvent is poured through, it interphases film-
on-film with the sample. Analyte is extracted into the organic
solvent, which passes freely through the matrix, while impurities
such as water, pigment, particulate matter, and other polar components are retained by the matrix. A pure and relatively dry
extract for TLC is produced in about 30 minutes. A plunger can
be introduced to increase pressure and flow rate, if necessary.
Broad-spectrum extractions with a single solvent at one pH,

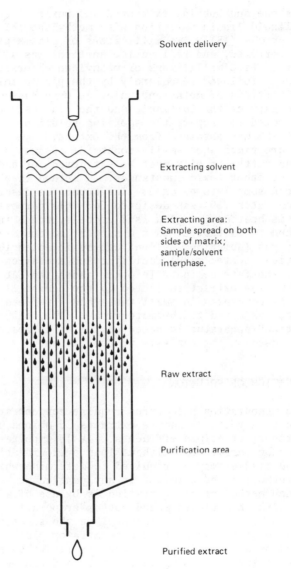

Solvent delivery

Extracting solvent

Extracting area:
Sample spread on both
sides of matrix;
sample/solvent
interphase.

Raw extract

Purification area

Purified extract

Figure 10. Illustration of cleanups by use of the Extube.

29

selective extraction at an optimum pH, and multiple extractions
of the same sample at different values can be carried out (Figure 10).

Touchstone and Dobbins extracted and purified cortisol from
plasma by liquid-liquid partition with methylene chloride in an
unbuffered Extube (18). Two milliliters of plasma plus 8 ml of
water were vortexed, and radioactive cortisol was added to monitor recovery. Five milliliters of methylene chloride was poured
into the tube, followed immediately by the plasma solution.
Four 10-ml portions of methylene chloride were passed through
the tube to extract the cortisol, and the eluates were combined
and concentrated to prepare the spotting solution. An average
recovery of 79% was obtained from the Extube.

There are times when small amounts of tissues or other samples such as butter or milk must be extracted for analysis of
pesticides or other toxic substances. A method involving codistillation can sometimes be used. The apparatus described by
Storherr and Watts (19) was designed for this purpose. While
the sample is heated, gas glow is directed through in the specially designed tubes that contain a support to contain the sample.
The flow of gas (sometimes solvent can be injected in the bottom
of the vertical heated tube) facilitates evaporation of the sample leaving the fats behind. In other cases the fat may be
extracted and the extract subjected to the codistillation. The
distillate is condensed in small tubes and after evaporation of
any solvent, subjected to chromatography. The Kontes "Sweep
Codistillation" apparatus is based on the Watts design and has
been widely used in the analysis of pesticides.

EVAPORATION OF SOLUTION

Most sample preparation procedures require concentration or
evaporation to dryness of sample extracts, combined partitioning
solvent batches, or column effluents. It is important that
evaporations be carried out without loss or degradation of the
analyte, and studies may be required to determine which of the
available methods is best to use in any particular situation.

A common method of concentration is by use of a rotary
evaporator with an attached round bottom evaporative flask (Figure 11). A helpful variation is to place the solution in a
Kuderna-Danish (K-D) evaporative concentrator flask and lower
tube instead of a round bottom evaporator flask, so that the
concentrated solution ends up in the tube and can be applied to
chromatogram without transfer (Figure 12). The incline of the
assembly should be ca. 20° from the vertical so that no

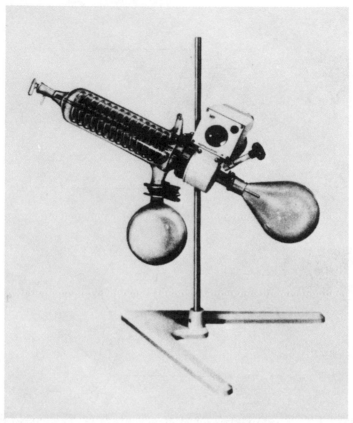

Figure 11.   Rotary evaporator used for concentrating extracts.

condensate collects in the lower part of the K-D flask, with the
concentrator tube about half immersed in a water bath adjusted
to a temperature appropriate to the solvent and lability of the
analyte. When the solvent is reduced to a small volume, the
flask and joint are rinsed with a little solvent, and final concentration is carried out in the tube under a gentle stream of
nitrogen.
      Nitrogen blowdown is another common method for concentration
of rather small volumes of solutions. A manifold facilitating
delivery of the gas flow is shown in Figure 13. The base of the
apparatus is a tube heater that holds nine 12-ml extended-tip
K-D concentrator tubes. When used without gas flow with an

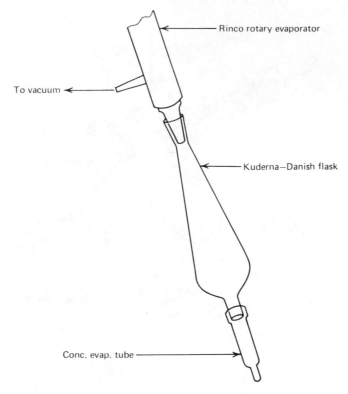

Rinco rotary evaporator

To vacuum

Kuderna—Danish flask

Conc. evap. tube

Figure 12.   The Kuderma-Danish evaporative concentrator.

ebullator in each tube and a reflux column attached to each,
simultaneous evaporation of solutions to less than 1 ml can be
quickly carried out in the tube heater without attention (Figure
14). Depending upon the opening set in the elevating table,
more or less solution is left unevaporated in the tubes.

DISSOLVING OF EVAPORATED RESIDUES

It is common practice to evaporate solutions just to dryness and
then to dissolve the residue in a known volume of the same or a
different solvent, from which a definite portion or the total can
be spotted on the thin layer plate. The least polar solvent in

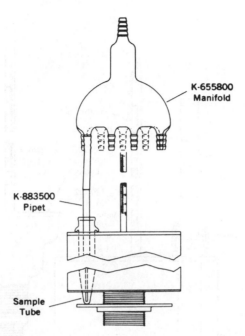

K-655800
Manifold

K-883500
Pipet

Sample
Tube

Figure 13.   The Kontes concentrator for multitube operation.

which the analyte is soluble should be used. In this way, purification can be achieved if some highly polar impurities are
left undissolved in the residue (selective solvation). Solvents
with too high a boiling point or high polarity are difficult to
remove from the sorbent during application. If a small amount
of solvent is retained on the sorbent after application, it can
adversely affect the separation by causing zone spreading or
deformation, or a different R_f value. Care must be taken, however, since hot air used to dry the solvent at the origin can
decompose labile substances on the surface of an active sorbent.

REFERENCES

1. H. Bethke and R. W. Frei, Anal. Chem. 48, 50 (1976).
2. U. Hezel, Am. Lab. 1978, 91 (May).
3. J. Sherma and L. Dorflinger, Am. Lab. 1976, 63 (January).

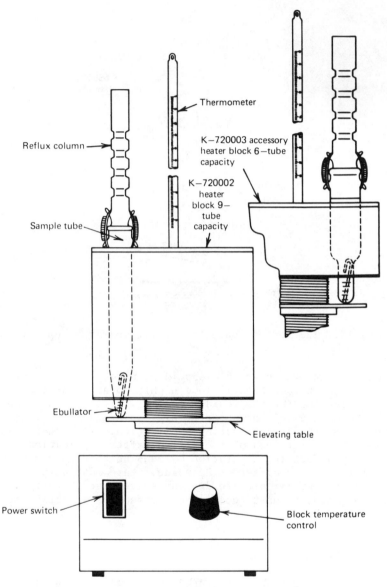

Figure 14. The Kontes tube heater for concentration of extracts.

34

4. J. Sherma and D. A. Grey, unpublished results.
5. R. Kupke and S. Zeugner, J. Chromatog. 146, 261 (1978).
6. J. Sherma and M. Beim, J. High Resolut. Chromatog. Chromatog. Commun., in press.
7. J. Sherma and J. C. Touchstone, Chromatographia 8, 261 (1975).
8. J. Sherma and K. Klopping, Am. Lab. 1977, 66 (December).
9. J. Sherma, M. F. Dobbins, and J. C. Touchstone, J. Chromatog. Sci. 12, 300 (1974).
10. W. Wortmann, B. Wortmann, C. Schnabel, and J. C. Touchstone, J. Chromatog. Sci. 12, 377 (1974).
11. J. Sherma and J. Koropchack, Anal. Chim. Acta 91, 259 (1977).
12. D. J. Peters, J. M. Hayes, and G. M. Hieftje, A Brief Introduction to Modern Chemical Analysis (W. B. Saunders Co., 1976), pp. 316, 355.
13. P. A. Mills, J. Assoc. Off. Agr. Chem. 42, 734 (1959).
14. J. Sherma and T. M. Shafik, Arch. Environ. Contam. 3, 55 (1975).
15. M. E. Getz, Talanta 22, 935 (1975).
16. J. Sherma, Am. Lab. 1978, 105 (October).
17. B. Mandrou, S. Brun, and A. Kingkate, J. Assoc. Off. Anal. Chem. 60, (1977).
18. J. C. Touchstone and M. Dobbins, in Densitometry in Thin Layer Chromatography, edited by J. C. Touchstone and J. Sherma (John Wiley and Sons, New York, 1979), p. 654.
19. R. W. Storherr and R. R. Watts, J. Assoc. Off. Agr. Chem. 48, 1154 (1965).

CHAPTER 4

Application of the Sample in Thin Layer Chromatography

Joseph C. Touchstone, Robert E. Levitt, and Gerald J. Hansen

The methods described for application of samples to thin layer chromatographic plates are numerous. Care must be taken to select the proper technique in order to ensure reproducibility, because this is essential to accuracy in quantitative determinations. In this chapter the various manual and automatic methods available at present, the relative ease and rapidity of each, and the importance of correct sample application will be discussed.

It is quite important in high-performance thin layer chromatography (HPTLC) that proper sample application be followed. A prerequisite is that the zone of application be small, no larger than 1-2 mm in diameter or, if streaking is used, no more than 0.1 - 2 mm in width. Resolution in HPTLC is largely dependent on the ability to reproduce these small zones of application, as mobile phase travel is short.

Samples should be applied as a 0.01 - 1.00% solution in the least polar solvent in which they are soluble. Solvents with too high a boiling point or high polarity are difficult to remove from the sorbent during application. If solvent is retained in the sorbent after the sample has been applied, it will adversely affect the separation, due to spreading of the sample in the matrix.

Usually applications of 0.1 - 5 $\mu g/\mu l$ solution concentrations are practical. The environment for applying the prepared samples to the chromatographic layer should be clean and free of dust, because dust particles may interfere with fluorometric evaluation of the plates. Further, the area should be well ventilated and free of chemical fumes that may alter the sample or be absorbed onto the plate while the sample is being applied. In order to ensure adequate air circulation for evaporation of the solvent, a stream of filtered air from a hair dryer may be

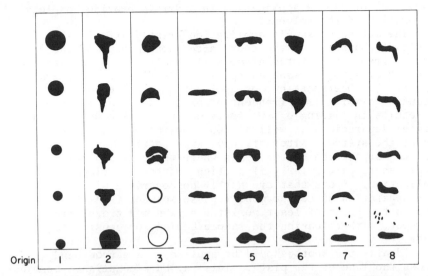

Figure 1.   Separation by TLC illustrating results of differences
in sample application.
    Lane 1:   Application of a single spot producing 4 similar
              spots in the developed chromatogram.
    Lane 2:   Application of too highly concentrated sample
              showing "tailing" effect in the final chromato-
              gram.
    Lane 3:   Application of sample without adequate drying
              between successive drops results in open "rings"
              in the final chromatogram.
    Lane 4:   Properly streaked sample, results in thin bands
              across the lanes in the developed chromatogram.
    Lane 5:   Concentration of the sample near the edges of the
              lane results in tailing.
    Lane 6:   Concentration of the sample near the center of the
              lane results in the developed chromatogram.
    Lanes 7 and 8: Distortion of spots due to excessive dis-
              turbance of the surface of the adsorbent during
              spotting.

directed across the area of sample application. Care must be
taken, however, to prevent evaporation of the solvent before the
sample is applied onto the plate, especially when some of the
automatic spotting techniques are employed. The stream of air
may be directed at the plate for several minutes after

completion of sample application in order to provide complete
evaporation of the solvent.

The method of sample application greatly affects the
resultant character of the spots after development of the
chromatogram. Faulty technique will result in the formation of
"rings," arcs, or "V"-shaped spots in the final chromatogram
(Figure 1). Overspotting, or application of an excessive
amount of sample, is a frequent error in sample application;
it results in tailing of the spots during development and sub-
optimal separations as well as poor separation due to overload-
ing of the system. Alternatively, overspotting may result in
spurious separation of two zones where there should only be a
single spot, particularly if a "ring" pattern develops during
application. Spots that are highly concentrated will also give
poor separation since the mobile phase solvent tends to flow
through the point of least resistance, and may travel around
the spot, either leaving the sample behind, or causing tailed
zones.

Ideally the spots should be applied in a narrow band
across the lane of the plate. This may be accomplished by manually streaking the sample across the lane or by moving the plate
across the path of the automatic continuous spotter. A streak-
type application will produce sharper separation, allowing
improved reproducibility and more reliable quantitation. Moreover, the mobile phase is less likely to flow around a streak
than a spot during development. Some spreading of the sample
through the gel during application is avoidable, but with careful technique this will be minimal.

Samples should be applied near the "bottom" of the layer,
usually about 1.5 - 2 cm from the edge, so that the layer makes
contact with the mobile phase but the sample zone itself is not
immersed in the liquid. This avoids dissolution of the sample
in the developing phase. In practice a sheet of white cardboard can be used as a background while the sample is applied to
the chromatoplate. A dark line can be drawn on this background
so that when the plate is positioned on it the line will be
seen through the layer 2 cm (or the predetermined distance) from,
and parallel to, the edge. This serves to provide a uniform
marker across the layer to guide the sample application.

The bandwidth or spot diameter must be as small as possible,
because, as the chromatogram develops, the areas occupied by the
separated zones tend to increase and the band or spot widens due
to diffusion. Nyborn (1) has observed that there is no direct
relationship between the amount per spot and the spot size,
variations being due to characteristics of the separated substances and to layer thickness within limits. The load applied

is critical and affects the shape and position of the zone after development. Overloaded sample applications produce cometlike vertical streaks. If the adsorption isotherm is convex, the R_f measured from the midpoint of the streak will be greater than normal. If the absorption isotherm is concave, the R_f value will be less than normal. The ends of the streak will overlap other spots on the chromatogram, rendering isolation and resolution poor. This is particularly true when one compound is present in larger quantity than another with a similar R_f value. Herein lies one of the most common faults of the practice of thin layer chromatography. It should be remembered that this is a sensitive technique, and the smallest sample size that can show visualization of all desired components is the one to be applied. This is one of the prerequisites for reproducible R_f values. This general rule also holds true for sample application in quantitative work. Too concentrated a zone will tend to give nonlinear calibration curves.

The technique of applying microliter amounts of samples requires practice. It is necessary to prevent contact between the tip of the sample applicator and the gel so that the layer does not become marked or scratched. However, it is equally important to avoid applying the sample in large drops as this will lead to the ring effect described earlier. Marks, streaks, or holes in the layer will cause distorted spots in the final chromatogram. Illustrated in Figure 1 are some of the types of spots obtained because of faulty technique. These may be avoided by slowly, carefully streaking the sample across the lane; larger volumes may require multiple streaks with adequate drying before restreaking the same area.

MANUAL SAMPLE APPLICATORS

For quantitation, it is essential that a reproducible volume be delivered onto the layer each time. There are a variety of calibrated disposable and nondisposable syringe type devices available. These are well suited to manual as well as automatic spotting techniques. However, for most purposes calibrated disposable microcapillaries are preferred because they allow rapid sample application and avoid contamination of samples. A number of companies produce capillary pipets for general use. Disposable uncalibrated glass capillaries are widely used for applying samples to layers. The "Microcaps" produced by Drummond Scientific Co. can be obtained both calibrated and uncalibrated (Figure 2).

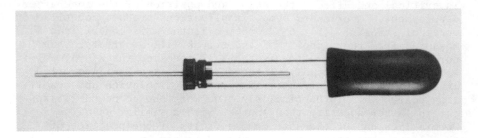

Figure 2.   Glass capillary micropipet in holder.
Courtesy A. H. Thomas Co.

These pipets are simple to use. They fill readily and,
once touched to the layer at a right angle, the capillary action
of the layer draws out the liquid completely. Because they are
disposable, the time-saving factor is considerable and contamination between samples is eliminated.

It is necessary to gain experience using these capillaries
before reproducibility can be assured; however, the time
required to become proficient is not excessive. The quality of
the spots will be influenced by the size of the pipet employed.
For example, it is much easier to obtain thin linear streaks
when 2 μl capillary pipets are used, rather than 5 μl pipets
because the "run off" of the sample into the plate is more
easily controlled by the operator when the pipet volume is
smaller.

We as well as others have determined that glass microcapillary pipets give reproducible results for a given worker and for
different workers.

Emanuel (2) more recently reported results of a survey of
reproducibility in delivery by a number of micropipets. It
appears from this report that simple glass capillary pipets may
be the most accurate microdelivery system. As they are disposable and relatively inexpensive, they are useful in applying
samples to thin layer plates. Table I shows the results of this
study. These experiments evaluated reproducibility of delivery
by the pipets, not the ease of delivering the sample to the
layer.

The microcapillary pipets are supplied with a glass dropper
with a rubber bulb at one end and a rubber stopper at the other,
through which the pipet is inserted. The rubber bulb has a
small hole in the end so that suction is not inadvertently

Figure 4. Microsyringe with capacity of 1 μl.
Courtesy A. H. Thomas Co.

which takes a stainless steel plunger. The plunger tip travels
the length of the bore and is used as the reference point for
readings. The boxes (capillaries) and plunger are inexpensive
enough to justify discarding after a single use to avoid contamination of samples and the time-consuming cleanup. The finger
loop handle of the plungers allows one-hand operation (Figure 3).
Microsyringes such as those manufactured by Hamilton Co., Uni-
metrics, Labcrest, and others, have been widely accepted and
can be used for delivery of calibrated volumes of solution.
They usually are made with glass barrels and a stainless steel
or tungsten plunger. Some have a replaceable needle; others
have permanently attached needles. They are available in volumes of 1 μl to 100 μl and are calibrated to deliver as little
as 0.05 μl. They were originally designed for use in gas chromatography and are equally adaptable to use in TLC.   These
syringes require some practice in use for reproducibility of
sample delivery in quantitative work. Figure 4 shows an example
of this type of syringe. With care and proper preparation of
standard solutions it is possible to obtain reproducibility in
the separations with a precision better than + 1% with the same
syringe. These pipets must be meticulously cleaned between the
application of repetitive samples.
     Various devices are available to aid in proper spacing and
alignment of samples on the plates. Their use is a matter of
individual preference and, as mentioned previously, a simple
straight line drawn across a piece of cardboard may suffice,
especially with prescored chromatographic plates. Many plates
are distributed individually wrapped in thin cardboard sheets
into which a depression is fashioned for the plate. These are
calibrated along the edges, providing a reproducible location
for the desired origin for sample application.
     Also available are several other template devices, with
notched edges evenly spaced, that create a space for sample
application. One inexpensive model is available from Camag Inc.
and is illustrated in Figure 5.

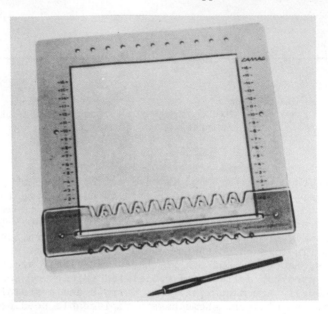

Figure 5. Plastic spotting template for evenly spacing sample
application. Courtesy Camag, Inc.

    More recently introduced by Sindco Corporation is a unique
apparatus known as a "spot-dry sampler" (Figure 6). This
instrument is a template with provision for gas flow (cold or
warm) to the point of application. The apparatus is designed as
a hand rest for chromatographic spotting and features a drying
system that allows the gas flow to be easily stopped while
applying the sample. When the hand is raised between sample
applications, the gas flows over the zone to quickly evaporate
the solvent. In this way the application zone (starting spot)
is kept as dry as possible, a prerequisite for good resolution
as well as uniform final zones. This instrument is particularly
valuable when application using high boiling-point solvents is
required. It also assures that the application zone is dry
before the plate is placed in the development chamber. This
apparatus is inexpensive and simple to use.
    With plates that are prescored the cardboard device
described above is probably the simplest and most inexpensive
method. This method is used almost exclusively in our laboratory and the results are found to be reproducible and accurate.
    Kaiser (3) has discussed the importance of optimal sample
application in HPTLC. Volumes of 10-100 nl solution are

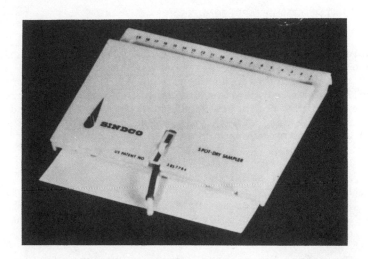

Figure 6.   Spotting template with drying system.
            Courtesy Sindco Corp.

recommended if manual application is used. The use of a solvent
in which the solutes have an R_f of less than 0.1 also facilitates sample application because it will assure that the zones
do not have spurious travel in the layer. These low volumes
dictate concentrated sample solutions. The Hamilton 7101 (1 mμ)
syringe has been used for this purpose. This syringe in conjunction with a micrometer can conveniently be used to apply 5 μl to
HPTLC layer.
     Camag has available capillary applicators designed to
deliver nanoliter volumes. These are available from Camag or
V-Tech. They are self-loading platinum-iridium sample capillaries designed for use in HPTLC. The maximum loading in these
capillaries is 33 nl/mm of capillary. The capillary is cleaned
by flaming. One disadvantage is that there are no disposable
parts and care must be taken to clean them between samples.
     The Merck applicator type 10226 as used in conventional TLC
has a 758 nl volume and uses disposable capillaries. With
proper technique (warming the layer and use of an air blast) it
could conceivably be used for application to HPTLC. It is possible to apply 50-75 nl using this applicator.

AUTOMATIC SAMPLE APPLICATORS

Automatic devices have been developed to relieve the technician
of the burden of multiple repetitive sample applicators. In
general these are useful; however, they do have limitations.
For small volumes and numbers of samples, these devices are not
necessary, and may not be as accurate. However, when used with
volumes of 10 μl or greater, they may be very helpful. These
devices are of four general types:

1. Holder for x number of disposable capillaries.
2. Syringe-type sample streak applicators.
3. Large capacity tube fitted with a capillary delivery
   system to deliver a single sample.
4. Multiple syringe holders with provision for automatic
   depression of the plunger.

1. The Thomas-Morgan sample spotter consists of a bar supported in springs. It holds 19 individual capillaries and
applies these at the same time to the plate on individual lanes.
2. The Camag Chromato charger is a carriage-mounted syringe
pipet on a track designed to provide linear delivery of the sample. The syringe pipet delivers up to 100 μl through a 65-mm-
long needle. This device is not practical when many samples are
to be applied, as reloading and repositioning are needed with
each sample.
3. The sample applicator recently introduced by Kontes
Instruments Division, shown in Figure 7, essentially consists of
large-bore glass tubes fitted with syringe needles. The needle
is positioned just above the layer in order to gradually deliver
the solute to the layer by capillary action. The essential
feature is that the needle is surrounded by a template with ori-
fices that permit compressed gas flow over the layer to evaporate the solvent as the solution percolates into the layer. The
gas (air, nitrogen, etc.) can be cooled or heated, or the plate
can be heated. Because of this, regardless of the volume of
solvent, a small spot will result. This type of sample applicator can be very useful when dilute samples must be applied as
a small zone to the layer. Several applications of up to six
samples can be performed simultaneously. Up to 5 ml of solution
can be applied by this method. It is widely used in pesticide
work where cleanup results in dilute samples.
4. The automatic spotters made by Analytical Instrument
Specialties (Figure 8) allow application of up to 10 or 19
samples simultaneously depending on the model chosen. The glass

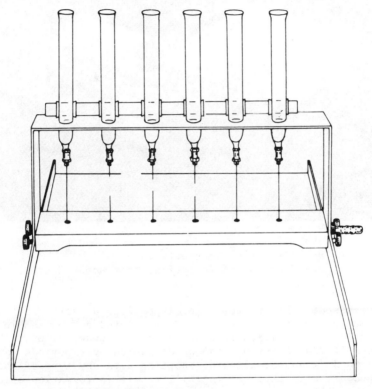

Figure 7.   Kontes Chromaflex spotter with manifold for drying sampler under air or nitrogen.   Courtesy Kontes Glass Co.

syringes provided accommodate either 1, 5, 10, 50, or 100 µl and consist of a blunt typed needle coated with Teflon to prevent sample residue creep-back and deposition on the tip of the needle during delivery.   Samples are drawn up into the syringes, which are placed in the instrument with the tips of the syringes aligned over the origins of the lanes of the TLC plates.   At the opposite end of the syringe a motor-driven bar slowly depresses the plungers of all syringes simultaneously, and the rapidity of sample delivery is regulated by the speed with which the plungers

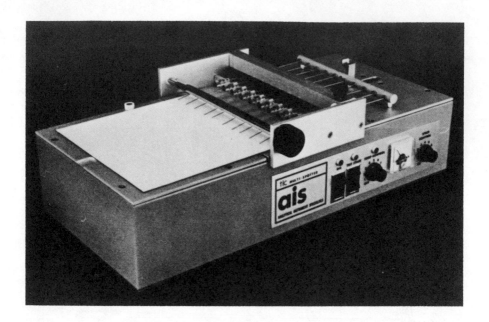

Figure 8.   AIS automatic spotter.
            Courtesy of Analytical Instrument Specialties.

are depressed.   The spotter is equipped with a heater so that
the temperature of the layer can be regulated during sample
application; in addition this helps solvent evaporation.   The
contents of 100 µl syringe can be delivered over periods varying
from 2 to 30 minutes, allowing spotting with any type of solvent.
Unattended use of this machine will result in formation of circular spots on the layer.   However, streaking may be achieved by
slightly shifting the plate several times during sample application so that the tips of the syringe deliver the sample to.
several spots across the base of each lane.   A hair dryer positioned above the plate facilitates evaporation of the solvent
during application.   Provision for heating is also provided.
     The reproducibility and accuracy of this automatic has been
tested.   Cholesterol in a constant concentration of 10 ng/µl was
spotted in varying volumes by manual methods using Drummond
microcapillaries, and by automatic (AIS) spotter.   Development
was carried out in Hexane-Ether-Acetic acid (86:13:1).   Spots
after charring were scanned with a Schoeffel model 3000 spectrodensitometer at 400 nm equipped with a computer to convert the

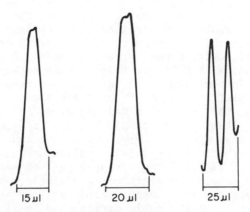

Figure 9. Peak area obtained from scan vs. amount cholesterol; manual application of sampler.

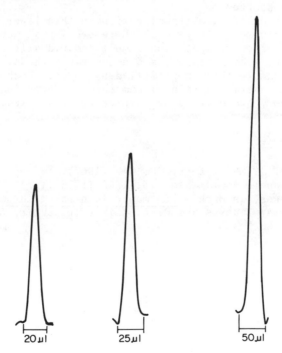

Figure 10. Peak area obtained from scan vs. amount cholesterol, applied with the AIS multi-spotter.

photomultiplier response to optical density units. The response
was channeled into a Schoeffel SDC 300 density computer recorder
in order to provide a continuous recording of the photomultiplier response. Peaks were obtained for each sample and quantitation was performed. The results of this study are shown in
Figures 9 and 10. With manual streaking up to 15 µl of sample
could be applied, with a linear relationship between the amount
of cholesterol in the sample and the recorder response. With
application of maetran, 15 µl of the scans showed "double peak-
ing," indicating that the spots had migrated, as shown in lane
3 in Figure 1. That is, the sample had diffused to the periph-
ery of the circular spots and migrated as two bands. In contrast, up to 250 µl could be safely applied using the AIS automatic spotter, still retaining a single peak on scanning, indicating that the sample had migrated as a single band. It is not
generally appreciated that larger sample volumes are best
applied with use of an automatic device, and small volumes best
applied manually. This study indicates the importance of selecting the proper method of sample application depending upon the
volume to be spotted.

The methods of sample application in thin layer chromatography are discussed. In general there are manual methods, which
require the worker to spot each sample individually using microcapillary pipets or syringes, and automatic methods, which
deliver the sample through prefilled syringes. Each of these is
indicated for use in certain circumstances, depending upon the
volume, number, and quality of samples to be applied.

REFERENCES

1. N. Nyborn, J. Chromatogr. 28, 447 (1967).
2. C. F. Emanuel, Anal. Chem. 45, 1568 (1973).
3. R. E. Kaiser, in High Performance Thin Layer Chromatography,
   edited by A. Zlatkis and R. E. Kaiser (Elsevier, Amsterdam,
   1977), p. 85.

CHAPTER 5

Derivative Formation for TLC

David J. Edwards

The term chromatography originates from the work of Michael Tswett, the Russian botanist who in 1906 separated plant pigments on a column of calcium carbonate. However, the original term has become somewhat of a misnomer since the application of chromatographic techniques extends far beyond the analysis of colored compounds. Although compounds separated by chromatographic methods may be detected by their native absorption or fluorescent properties, compounds that are not ideally suited for such methods of detection may be conveniently detected after their conversion to derivatives that are readily detectable. Derivative formation, therefore, extends the usefulness of chromatographic techniques to the analysis of the vast majority of compounds that are not themselves sufficiently absorbing or fluorescent.

Derivatization refers to any process by which a compound is converted to a different chemical compound. This includes such widely used procedures in TLC as charring and exposure to iodine vapor. However, in this chapter the term derivatization will be used to describe reactions that result in an increase in molecular weight of the original compound, such as by acetylation or esterification.

The formation of derivatives may occur either prior to or following the chromatographic separation of the original compound. The choice of these two alternatives depends on the particular chromatographic technique used, the nature of the compounds and their derivatives, and the specific advantages sought by derivatization. In gas chromatography, derivatization is generally used to volatilize compounds to aid in the chromatography, and derivatives are therefore almost always formed prior to injection of the samples onto the column. In other column chromatographic techniques, such as high-performance liquid chromatography (HLPC)

and ion-exchange chromatography, derivatization may occur either
before or after the chromatographic step. Amino acid analyzers
use post-column derivatization techniques in which the amino
acids are first separated by ion-exchange chromatography and are
subsequently reacted with ninhydrin or fluorescamine so that they
may be detected. On the other hand, amino acid analyses have
also been performed by converting the amino acids to the corresponding phenylthiohydantoin, dansyl, or dinitrophenyl derivatives prior to their separation. Since this chapter will deal
primarily with those derivatization techniques used in TLC, the
reader is referred to other reviews for their use in gas chromatography (1, 2) and HPLC (3).

In TLC, derivatization may also be carried out either before
or after chromatography. Derivatization is usually carried out
subsequent to TLC by the use of spray reagents for enhancing the
detectability of the separated compounds. Alternatively, the
compounds may be scraped from the TLC plate, eluted, and then
reacted in the test tube with the appropriate reagent.

The formation of the derivatives prior to TLC may be useful
for substances when the derivatizing agent results in high background staining of the TLC plates or for more accurately controlling the reaction conditions. Perhaps more importantly, derivatization prior to TLC offers several advantages besides improving
the detectability of the compound, which may be important to a
particular analysis. For example, derivatization may be used to
alter the chromatographic properties of compounds that are difficult to separate. Also it can offer the advantage of increasing
the stability of unstable compounds. Volatile compounds that
undergo large losses on TLC may be derivatized to less volatile
derivatives in order to increase their recovery. Derivatization
may also be used to reduce the reactivity of compounds that react
with the stationary phase or to reduce the polarity of compounds
that cannot be displaced from the origin of the chromatogram.
Furthermore, by the selection of specific derivatizing agents and
by the use of cleanup steps to isolate the desired derivatives
before they are spotted on TLC plates, the specificity of the
procedure may be improved. Finally, the use of one or more
derivatizing agents may increase the certainty by which an
unknown compound is identified.

This chapter will focus on the various derivatizing reagents
that are useful or may potentially be useful in TLC, their specificity, advantages and disadvantages, and applications. The
emphasis will be given to techniques in which the derivatives are
formed prior to TLC, since extensive lists of spray reagents for
visualization of compounds separated by TLC are available else-
where (4,5). The structures of some of the commonly used

reagents and typical reactions that they carry out are shown in
Tables I and II. Obviously in a limited space it is impossible
to include all of the reagents that have been used. The inten-
tion is simply to describe those that are most useful and react
with a broad range of functional groups.

REAGENTS FOR INTRODUCING CHROMOPHORES

Table I includes a list of some of the more commonly used derivatizing reagents for introducing chromophores. Although the examples shown illustrate some of the reactions that occur, most of
these reagents are capable of reacting with more than one type of
functional group.

1-Fluoro-2,4-dinitrobenzene (Sanger's Reagent) (FDNB)

FDNB is frequently used for preparing 2,4-dinitrophenyl (DNP)
derivatives of amines and alcohols. The reagent is highly reactive, and the reaction is typically carried out at room temperature in a slightly alkaline solution for 1 or 2 hours. Since
many of the DNP-derivatives, especially the DNP amino acids, are
photosensitive, they should be kept in the dark as much as possible. Most of the DNP derivatives are bright yellow in color.
The molar extinction coefficients of DNP-amines, for example, are
approximately 1.6×10^4 (moles/1)$^{-1}$ cm^{-1} at 360 nm (6). However,
certain derivatives, such as 0-DNP-tyrosine and im-DNP-histidine,
are colorless.
     FDNB reacts with a relatively broad range of functional
groups, including primary and secondary amines, alcohols, phenols, sulfhydryls, and imidazoles (7,8). Compounds with two or
more of these functional groups will incorporate two or more DNP-
groups. Thus, di-DNP derivatives will be formed from histidine,
lysine, cysteine, and tyrosine, although a longer reaction time
may be necessary to completely react histidine (9). If desired,
the DNP groups can be displaced from the imidazole, sulfhydryl,
and phenolic groups, but not from the amino group, by treatment
with β-mercaptoethanol (10).
     Various reaction conditions are detailed in several sources
(6,9,11-13). Typically, amino acids in a protein hydrolysate are
dinitrophenylated in either aqueous solution or in ethanol. In
the first case, the amino acids are dissolved in water, adjusted
to pH 8.9 with NaOH, and FDNB is added. After the reaction is
complete, the excess reagent is removed by extracting with
peroxide-free ether. In this procedure, considerable amounts of
dinitrophenol may be formed, which may interfere with the TLC

Table I Reations of Reagents Forming Chromophoric Derivations

1. 1-Fluoro-2,4-dinitrobenzene (Sanger's reagent) (FDNB)

$$FDNB + RNH_2 \longrightarrow DNP\text{-amine} + HF$$

2. 2,4-Dinitrobenzene sulfonic acid (DNBS)

$$DNBS + RNH_2 \longrightarrow DNP\text{-amine} + H_2SO_3$$

3. 2,4-Dinitrophenylhydrazine (DNPH)

$$DNPH + R\overset{O}{\underset{\|}{C}}R' \longrightarrow DNP\text{-hydrazone} + H_2O$$

4. 2,4,6-Trinitrobenzene sulfonic acid (TNBS)

$$TNBS + RNH_2 \longrightarrow + H_2SO_3$$

5. 3,5-Dinitrobenzoyl chloride

$$+ ROH \longrightarrow + HCl$$

6. p-Nitrophenacyl bromide

$$+ RCOOH \longrightarrow$$

54

Table I (continued)

7. p-Nitrobenzyl bromide

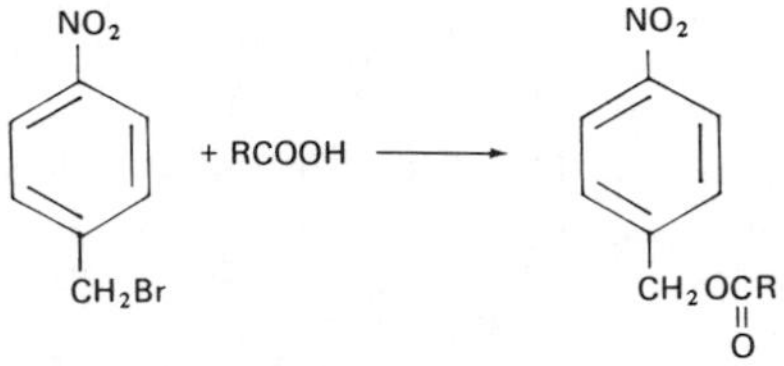

8. Phenyl isothiocyanate

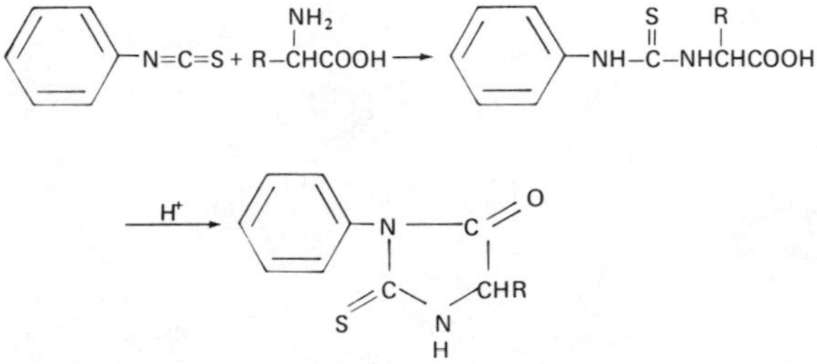

9. 4-N,N-Dimethylaminonaphthylazobenzene-4'-isothiocyanate (DANABITC)

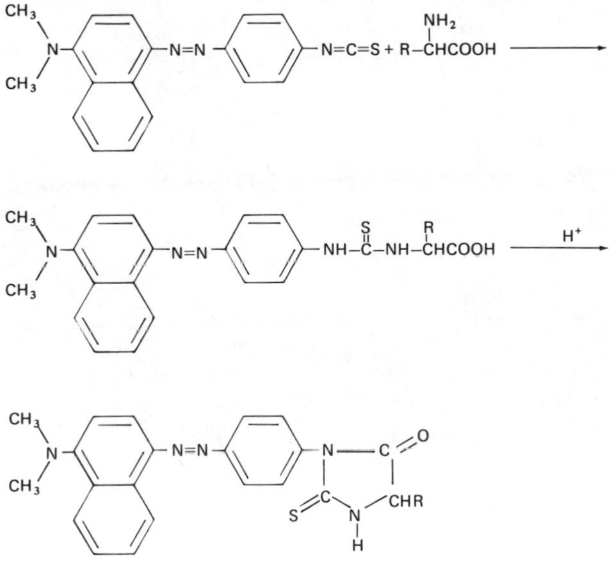

10. 4-Dimethylaminoazobenzene-4'-sulfonyl chloride (Dabsyl Cl)

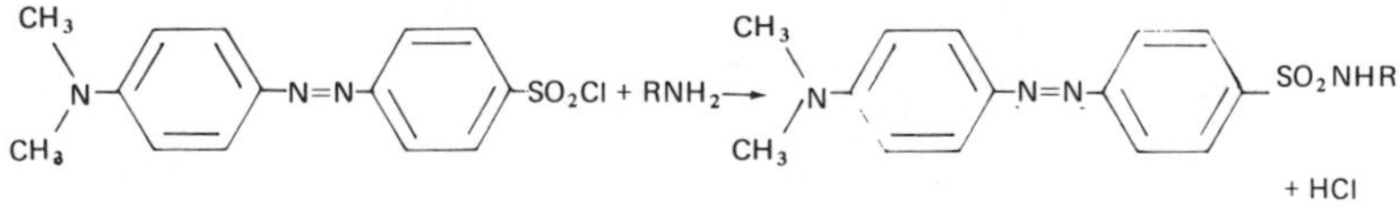

55

Table II Reactions of Reagents Forming Fluorescent Derivations

1. 1-Dimethylamino-naphthalene-1-sulfonyl chloride (Dansyl-Cl)

CH$_3$–N–CH$_3$ + RNH$_2$ ⟶   CH$_3$–N–CH$_3$ + HCl

SO$_2$Cl   SO$_2$NHR

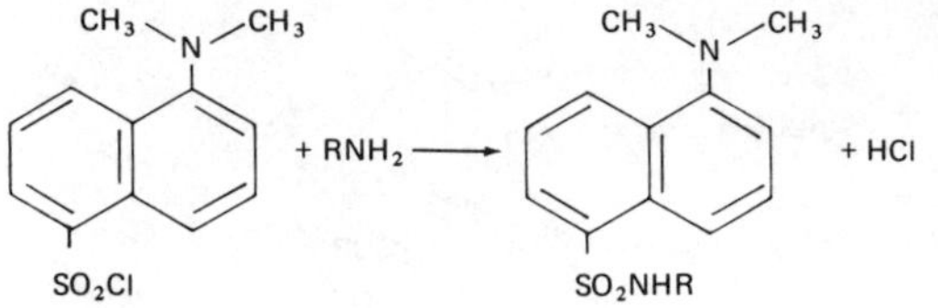

2. 4-Chloro-7-nitrobenzo[c]-1,2,5-oxadiazole (NBD-Cl)

NO$_2$ + RNH$_2$ ⟶   NO$_2$ + HCl

Cl   NHR

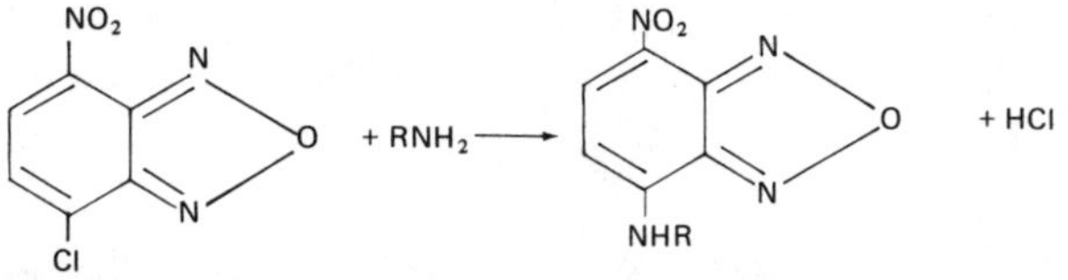

3. Fluorescein isothiocyanate

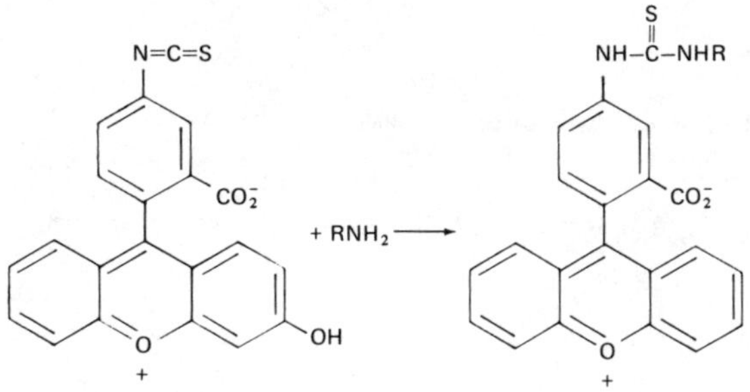

N=C=S   + RNH$_2$ ⟶   NH–C–NHR (S)

CO$_2^-$ CO$_2^-$

OH

4. Fluorescamine

+ RNH$_2$ ⟶

RNN ... O
OH
COOH

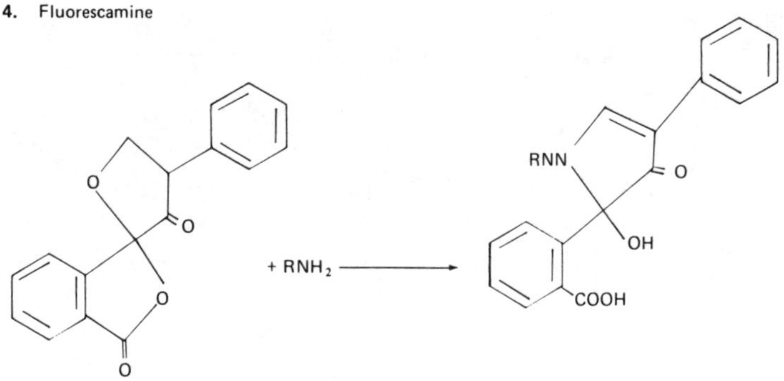

5. 4-Bromomethyl-1-7-methoxycoumarin

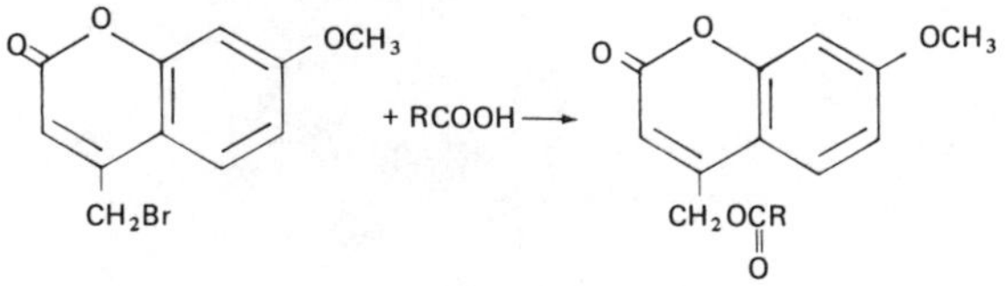

O   OCH$_3$ + RCOOH ⟶   O   OCH$_3$

CH$_2$Br   CH$_2$OCR (O)

56

separation, but if necessary may be removed by sublimation or solvent extraction (11).

When dinitrophenylation is carried out in ethanol, the amino acids are dissolved in 1 ml of 2% FDNB (11). After the excess reagent is removed by ether extraction, the aqueous solution is acidified with 6N HCl. The aqueous layer is then extracted with ether to remove all the DNP-amino acids except DNP-arginine, di-DNP-histidine, and DNP-cysteic acid. The latter water-soluble amino acids may be chromatographed directly or extracted with sec-butanol/ethyl acetate (1:1) (9).

Usually, the ether-soluble and water-soluble DNP-amino acids are chromatographed separately. Comprehensive reviews are available for TLC chromatography of DNP-amino acids or silica gel (9,13), cellulose (14), and polyamide (15).

The water soluble DNP-amino acids may be separated on silica gel G layers using n-propanol-34% ammonia (7:3) as solvent (9). The complete separation of all the ether-soluble DNP-amino acids usually requires two-dimensional TLC. Good separation on silica gel G of all the ether-soluble DNP-amino acids can be achieved, for example, by using two-dimensional TLC with toluene-pyridine-2-chloroethanol-0.8 M ammonia (10:3:6:6) in the first direction and benzene-pyridine-acetic acid (40:10:1) in the second direction (16) (Figure 1.)

In addition to amino acids, FDNB has also been used to dinitrophenylate various other compounds, including peptides, aliphatic amines, phenylethylamines, and phenols. Peptides may be dinitrophenylated by procedures that are essentially the same as those used for free amino acids (13,17). Short peptides, such as dipeptides may be directly separated by TLC (18). For end-group analysis of peptides and proteins, the DNP product is hydrolyzed to the individual amino acid residues and the N-terminal amino acid residue is identified by the presence of the amino acid residue with a DNP moiety on the α-amino group. Other DNP derivatives will be formed from residues with reaction side chains, such as ε-DNP-lysine, O-DNP-tyrosine, and im-DNP-histidine. Of course, hydrolysis of peptides subsequent to their dinitrophenylation can result in the formation of DNP derivatives of amino acids occurring at sites other than the N-terminal. In this case, ε-DNP-lysine, ODNP-tyrosine, and im-DNP-histidine will be found in the hydrolysate. Because these derivatives still have an underivatized α-amino group, they all will be extracted with the water-soluble DNP-amino acids.

Day et al. (19) used FDNB to form the DNP derivatives of eleven primary and secondary amines containing one to four carbon atoms. All of the derivatives could be separated by TLC using silica gel G plates and hexane/ether (70:30) as the solvent.

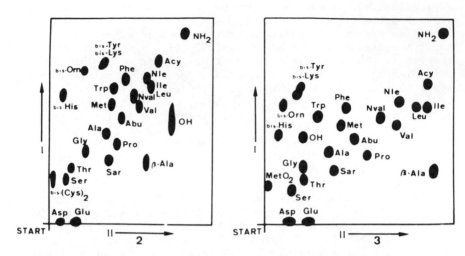

Figure 1.   Two-dimensional TLC on DNP-amino acids on silica gel G.
Solvent systems are (1) toluene-pyridine-2-chloroethanol-0.8 M
ammonia (10:3:6:6); (2) chloroform-benzyl alcohol-acetic acid
(70:30:3); and (3) benzene-pyridine-acetic acid (40:10:1).
Abbreviations include: bis-, bis-DNP-derivatives; NH_2, 2,4-
dinitroanaline; and OH, 2,4-dinitrophenol.   From Chromatography:
A Laboratory Handbook of Chromatographic and Electrophoretic
Methods, 3rd ed., edited by Erich Heftmann.   (Copyright 1975 by
Litton Educational Publishing, Inc. Reprinted by permission of
Van Nostrand Reinhold Co. and Dr. A. Niederwieser.)

Movement on the TLC plate appeared to be dependent only on the
molecular weight and not on the extent of branching. Aliphatic
DNP-amines have also been separated by TLC using ether-ethyl
acetate (4:1) as the solvent (20).
     FDNB has also been used as a reagent for phenols. Cohen et
al. (21) compared various reaction conditions for preparing the
DNP derivatives of several phenols. They found that the phenols
could be spotted on TLC plates and then sprayed first with a
saturated solution of sodium methoxide in methanol, followed by
4% FDNB in acetone, to produce a good yield of the corresponding
derivatives.
     FDNB has been applied to a broad range of analytical prob-
lems, from N-terminal analysis to pesticide analysis. This is a
result of its rapid reaction with several functional groups and
the advantages the DNP moiety offers not only because of its high

molar extinction coefficient (an advantage for both TLC and HPLC) but also because of its electron capturing properties. The latter property enables these derivatives to be analyzed in subnanogram quantities by using gas chromatography with electron capture detection. Only a few examples can be given here for applications of FDNB as a derivatizing agent. Of course, as already noted, the most popular use of FDNB is for the analysis of amino acids. Shank and Aprison (22) used FDNB to measure the concentrations of amino acids in as little as 10 mg of nervous tissue. Quantitation was carried out by eluting the DNP derivatives from the silica gel with 0.01 N sodium bicarbonate and measuring the absorbance at 360 nm. A similar procedure but one using paper chromatography was applied to the pyrrolidine in rat brain (23). Gregerman and Kowatch (24) developed a double isotopic assay for renin in human plasma by using [^{3}H] FDNB, combined with a two-dimensional TLC purification step. Cox (25) analyzed three N-nitrosamines in food by electromechanical reduction to the corresponding amines followed by their conversion to the DNP derivatives and separation by HPLC. Cohen and Wheels (26) used FDNB for the analysis of substituted urea and carbamate herbicides. The herbicides were first separated by TLC, then hydrolyzed to the corresponding aromatic amines and converted to the DNP derivatives by spraying with 5% HCl followed by FDNB (4% w/v in acetone). The DNP-amines were eluted and analyzed by gas chromatography with electron-capture detection. Similar procedures for the analysis of carbamate insecticides involved their hydrolysis to the corresponding phenol and their subsequent conversion to DNP-ethers (27).

2,4-Dinitrobenzene Sulfonic Acid (DNBS)

DNBS is another reagent for preparing DNP derivatives. Although it is far less commonly used than FDNB, it offers two important advantages that make it a useful reagent in certain applications. First, DNBS is relatively specific for primary and to some extent secondary amino groups; unlike FDNB, DNBS does not react with hydroxyl groups, sulfhydryls, or imidazoles. Second, DNBS is water soluble. DNBS may be the reagent of choice for derivatizing amines, because the products can be readily separated from the excess reagent by solvent extraction. Thus, DNP-amines are extracted by benzene, leaving the excess DNBS in the aqueous phase. On the other hand, FDNB is the preferable reagent for derivatizing amino acids, because the DNP-amino acids can be extracted from the excess reagent by aqueous solutions. Generally, DNBS reacts with amino groups more slowly than does FDNB. Longer reaction times and elevated temperatures (usually 100° C)

are required for complete reaction to occur. Crawhall and
Elliott (28) first described the use of DNBS to form the N-DNP
derivatives of various amino alcohols (e.g., ethanolamine,
alaninol, leucinol, etc.). The separation on silicic acid layers
of the N-DNP derivatives of several amino alcohols in urine and
tissues has been reported (29). DNBS has also been used to
derivatize phenylethylamines in tissues in order to analyze them
by gas chromatography with electron-capture detection (30) and by
gas chromatography/mass spectrometry (31).

2,4-Dinitrophenylhydrazine (DNPH)

DNPH is used to introduce a DNP group on carbonyl compounds. The
reaction usually proceeds rapidly to give hydrazones in excellent
yields. The reaction is usually carried out in aqueous ethanolic
sulfuric acid (32), although in some cases it is necessary to
carry out the reaction in the absence of acid (33). The hydrazones usually crystallize out of solution very readily. For this
reason, DNP-hydrazones were often used even before the days of
chromatography, when compounds were commonly identified on the
basis of the melting points of derivatives.
    Stahl and Jork (34) have compiled extensive lists of the
chromatographic behavior of various DNP-hydrazones on TLC. Published results for TLC of hydrazones of a large number of carbonyl
compounds are also available in several other sources (33,35-37).
Byrne (33), for example, examined the DNP-hydrazones of 41 different carbonyl compounds by TLC on silica gel. Derivatives of 70
carbonyl compounds using TLC on alumina have been described (37).
    The DNP-hydrazones can be detected by their visible color.
The DNP-hydrazones of saturated aliphatic aldehydes and ketones
are yellow or orange, whereas the corresponding derivatives of
α,β-unsaturated carbonyl compounds are red and those of aromatic
compounds are yellow, orange, or red (38).
    To enhance the detectability of these derivatives, advan-
tage can be taken of the fact that they undergo a color change
in alkaline conditions. Exposure of the derivative to either
ethanolamine (33) or to propylamine, isopropylamine, or diethylamine (37) has been used for this purpose. Usually the color
change is reversible and fades as the amine evaporates, although
aliphatic aldehydes may give rise to permanent color changes.
In addition to increasing the sensitivity, exposure to bases
allows different classes of derivatives to be distinguished by
the colors obtained (33). Beyer and Kargl (39) took advantage
of the changes in chromatographic properties of the DNP-hydrazones that occur in the presence of base and certain inorganic

compounds to separate complex mixtures of these derivatives by
TLC on zinc carbonate, using pyridine as the developing solvent.

One application for DNP-hydrazones has been the determination of steroids in physiological fluids. Treiber and Oertel (40)
described a procedure for estimating 11-deoxy-17-ketosteroids in
human urine by separating them by TLC on silica gel as DNP-hydrazones. In this procedure, the derivatives were eluted with chloroform and quantitation was achieved by spectrophotometry.
Treiber et al. (41) and Knapstein et al. (42) similarly measured
ketosteroids in human urine and plasma, respectively, by using a
double beam scanning spectrodensitometer instead of elution techniques for quantitation (Figure 2). By this procedure, as little
as 0.02 µg of a ketosteroid could be detected (41). Separation
of the DNP-hydrazone derivatives of steroids using reversed-phase
TLC on polyamide has been reported by Penzes et al. (43).

DPNH has also been applied to the analysis of ketoacids by
TLC (44). Separation of the DNP-hydrazones of ketoacids from
those of other carbonyl compounds may be obtained by taking advantage of the acidic nature of the ketoacids. This method has been
used for the determination of pyruvate, α-ketoglutarate, and
acetoacetate in urine.

Thompson and Hedin (45) used DPNH to derivatize 15 fatty
acids. The fatty acids were first refluxed with thionyl chloride
to form the acid chloride. After the reagent was removed by
gentle heating under vacuum, DPNH was added to form the corresponding DNP-hydrazides. TLC separation of these derivatives was
achieved on alumina, on silica gel G, and on polyamide.

Alkylquinones have been separated by TLC of their DNP-hydra-
zones(46).

2,4,6-Trinitrobenzene Sulfonic Acid (TNBS)

TNBS has a specificity similar to that of DNBS in that it reacts
with primary amino groups but not with hydroxyl or imidazole
groups (47). However, TNBS is a more highly reactive reagent
than DNBS and derivatization may be carried out under relatively
mild conditions (usually pH 8 and room temperature). One disadvantage of TNBS is that trinitrophenyl (TNP) derivatives are generally far less stable to light and to acid than the corresponding
DNP-derivatives (47). Compared to DNP-derivatives, the corresponding TNP-derivatives have a slightly lower molar extinction
coefficient and an absorption maximum of about 15 nm lower. The
properties of TNP-amino acids on paper chromatography have been
studied (47). However, TNBS is more commonly used as a spray
reagent to visualize compounds already separated by chromatography.

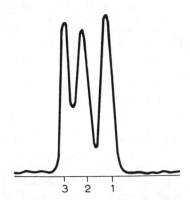

Figure 2.   Separation of about 2.5 µg each (1) DHEA, (2) etiocho-
lanolone and (3) androsterone zones.   From Trieber, Knapstein, and
Touchstone, J. Chromatogr. 37, 83 (1968).   (Reprinted by permission from authors and Elsevier Scientific Publishing Co.)

3,5-Dinitrobenzoyl Chloride

3,5-Dinitrobenzoyl chloride is sometimes used for making colored
derivatives of amines and alcohols for TLC.   The reagent is typically mixed with benzene or benzene plus pyridine and the reaction is carried out by adding the reagent to a solution of amines
or alcohols dissolved in benzene or aqueous ethanol (48).
Neiderhiser et al. (49) applied this reagent to detection of
diethylamine, a metabolite of disulfiram in urine (49).   Their
procedure was specific for secondary amines, because primary
amines were removed by extraction into an aqueous layer.   The
derivative was isolated on silica gel TLC developed in chloroform-
methanol (1:1) (DADB).
        An analogous reagent, 4-dimethylamino-3,5-dinitrobenzoyl
chloride, has been reacted with various amines to give yellow to
red products that can be separated by TLC (50).   These DADB-
derivatives have the advantage of being more intensely colored
and more light stable than the corresponding DNP-amines.

Esterification Reagents

Esterification of carboxylic acids to derivatives absorbing in
the UV may be achieved with several reagents, such as p-nitrobenzyl bromide, p-nitrophenacyl bromide, p-bromophenacyl bromide,

and p-phenylphenacyl bromide (51). These reagents have been
primarily used for analysis of carboxylic acids by HPLC or gas
chromatography. p-Nitrobenzyl bromide has been used to analyze
fatty acids by negative ion mass spectrometry (52) and prostaglandins by HPLC (53).

Phenyl Isothiocyanate (PITC)

PITC is primarily used in peptide sequencing. For review, see
Rosmus and Deyl (12). The reagent reacts with the free amino
group on the N-terminal amino acid to form a phenylthiocarbamoyl
peptide (PTC-peptide). A second step involves cleavage of the
N-terminal residue with acid to yield a 2-anilino-5-thiazolinone
derivative and peptide chain minus the N-terminal residue.
Finally, the thiazolinone is converted to a phenylthiohydantoin
(PTH) in aqueous acid solution. The PTH-amino acids may then be
identified by TLC. The above steps can be repeated with the
remaining peptide in order to further determine the sequence of
the original peptide.

There is an extensive literature on the separation of PTH-
amino acids by TLC. See review, Rosmus and Deyl (12). Generally,
if complex mixtures of amino acids are present, two-dimensional
separations are necessary to resolve them all. In order to
detect the PTH-amino acids on TLC plates, advantage may be taken
of their high molar extinction coefficient and thus using a fluorescent indicator in the layer. Alternatively, they may be
detected by spraying with ninhydrin dissolved with collidine in
ethanol and heating to 110° C. Quantitation can be carried out
in situ or after elution. For quantitation in situ, quenching of
the fluorescent indicator can be measured. For this purpose,
polyamide layers give better results than silica gel layers.

Phenylisothiocyanates for Making Colored Derivatives

Chang and Creaser prepared azo dye isothiocyanates, 4-N, N-dimeth-
ylaminoazobenzene-4'-isothiocyanate (DABITC) (54) and 4-N-N-
dimethylaminonaphthylazobenzene-4'-thiocarbamoyl (DANABITC) (55),
which give colored thiohydantoin derivatives to replace phenyl
isothiocyanate as an N-terminal reagent. Because derivatives
formed from these two reagents were different in color (i.e., red
and purple, respectively), derivatives formed with one reagent
were found to be useful as markers for identifying the unknown
derivatives obtained from the other reagent (55). DABITC was
also used for derivatizing amino alcohols (56).

Similarly, p-phenylazophenyl isothiocyanate (PAPITC) has
also been used to make colored thiohydantoins of N-terminal amino

acids (57). The 3-(p-phenylazophenyl) 2-thiohydantoins of the
amino acids were separated by silica gel TLC.

4-Dimethylaminoazobenzene-4'-Sulfonyl Chloride (DABS-Cl)

DABS-Cl was first synthesized by Lin and Chang (58) as a sensitive chromophoric reagent for labeling amino acids and peptides.
The sensitivity of DABS-Cl is claimed to be about one hundredfold
greater than FDNB and almost as sensitive as the fluorogenic
reagent, dansyl-Cl (see below). As little as 10^{-10} to 10^{-11}
moles of the dabsyl amino acids could be detected on thin-layer
plates.
    Dabsylation was generally carried out by adding DABS-Cl in
acetone to a solution of the amino acids dissolved in carbonate-
bicarbonate buffer (pH 8.9). This mixture was allowed to react
at 70° C for 6 minutes. The dabsyl derivatives were separated by
two-dimensional TLC on silica gel using benzene-pyridine-acetic
acid (80:20:5) in the first direction and toluene-2-chloroethanol-
25% ammonia (100:80:6.7) in the second direction (58). An
improved separation was achieved on polyamide TLC using water-2-
chloroethanol-formic acid (100:60:3.5) in the first direction and
benzene-acetic acid (6:1) for the second direction (54).
    The dabsyl amino acids are very photostable, an advantage
over the corresponding DNP-amino acids. By varying the pH, reaction of either amino groups or hydroxyl groups could be controlled. For example, serine reacted with DABS-Cl to form predominately the N-dabsyl derivative below pH 10 and predominately
the O-dabsyl above pH 12 (Figure 3). N,O-bis dabsyl serine was
formed only in the pH range of 10-12 (58).

REAGENTS FOR MAKING FLUORESCENT DERIVATIVES

Several reagents are now available for preparing fluorescent
derivatives to increase the sensitivity of quantitative analysis,
particularly for TLC and HPLC. Recent advances in measuring
fluorescence with TLC scanners greatly expand the utility for
fluorescent reagents for use in quantitative analysis by TLC. In
addition to increasing the sensitivity of the analysis, fluorescent analysis has the potential to increase specificity. Selectivity can be gained not only by the choice of the reagent but
also by selecting the proper wavelengths for both excitation and
emission of the fluorescent derivative.
    The use of fluorescent derivatives has been recently
reviewed (59). Application of fluorescent derivatization to pesticide residue analysis has also been reviewed (60).

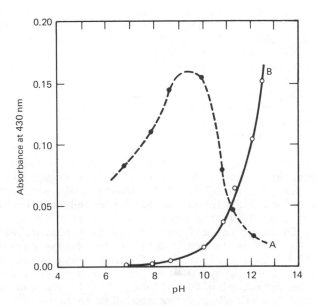

Figure 3. Reaction of DABS-Cl with serine at various pH values. Dabsylation of serine was carried out in 0.1 M phosphate buffer (pH 6, 7, and 8), in carbonate-bicarbonate buffer (pH 9, 10, and 11), and in carbonate-NaOH (pH 12 and 13). Curves A and B indicated the formation of N-dabsyl serine and O-dabsyl serine, respectively. From Lin and Chang, Anal. Chem. 47, 1634 (1975). (Reprinted with permission from authors and American Chemical Society.)

Dansyl Chloride (dansyl-Cl)

Dansyl chloride (1-dimethylamino-naphthalene-5-sulphonyl chloride) is one of the most commonly used reagents for preparing fluorescent derivatives. A major advantage of dansyl derivatives is that, in addition to being intensely fluorescent, they also possess excellent chromatographic properties on TLC. As pointed out by Seiler (61), dansyl amines normally give regularly shaped spots on silica gel G layers and do not tail, even when the plates are heavily overloaded.

Dansyl-Cl reactions are usually carried out in the dark in acetone-water (3:1) saturated with sodium carbonate. Primary and secondary amines, phenols, imidazoles, sulfhydryls, and some alcohols are quantitatively converted to the corresponding dansyl

derivatives at room temperature. Some selectivity may be
obtained by the choice of pH, because amines react at slightly
alkaline pH, whereas a higher pH is required for phenols and
imidazoles (59). Amino acids usually react first with the amino
group but the carboxyl group of the dansyl amino acids formed
tends to react with excess dansyl chloride to yield mixed anhy-
drides, which partially break down to aldehydes, dansyl-OH and
dansyl-NH_2. γ-Amino acids form γ-lactams when dansylated.

Most dansyl derivatives are easily removed from the reaction
mixture by solvent extraction. However, dansyl amino acid derivatives primarily remain in the alkaline phase along with excess
reagent and dansyl-OH, the hydrolysis product formed from dansyl-
Cl. If necessary, excess dansyl-Cl may be destroyed by the addition of excess proline (61). The formation of γ-lactams when
γ-amino acids are dansylated can be exploited for the assay of
γ-aminobutyric acid, as the dansyl derivative of γ-aminobutyric
acid (GABA) is extractable from the reaction mixture with organic
solvents. To more selectively isolate the dansyl derivative of
GABA from the dansyl derivatives of other amino acids and amines,
the lactam ring may be hydrolyzed in 2 M KOH so that dansyl
amines may be extracted with toluene. The lactam can be reformed
in acetic hydride and then extracted with toluene (62).

TLC separation of dansyl derivatives has been widely carried
out on silica gel layers. Mobile phases used for two-dimensional
separation of the dansyl derivatives of various aliphatic amines,
polyamines, catecholamines, and related compounds have been
reviewed by Seiler and Wiechmann (63). These authors have compiled a list of R_f values of 112 dansyl derivatives on silica gel
G in 21 different mobile phases. Dansyl derivatives have been
separated less commonly on layers of alumina (64), polyamide (65,
66), or cellulose (67).

The dansyl amines separated by TLC can be quantitated by
fluorescence scanning or by elution. One caution is that humidity and low pH have been observed to diminish the quantum yield
of fluorescence (61,68). This problem may be circumvented by
drying the TLC plates and spraying with triethylamine (61) or
simply heating the TLC plates in an oven at 130° C (68). Under
ideal conditions, as little as 10^{-12} moles/spot of dansyl derivative can be detected.

Dansylation techniques have been applied to the analysis of
small quantities of amino acids and other amines, steroid hormones, pesticides, and drugs in various types of samples.
Seiler (62), for example, separated by TLC the dansyl derivatives of the amines in rat brain extracts, including GABA,
putrescine, spermidine, and spermine. Quantitation was achieved
by direct scanning of the fluorescent intensity (Figure 4).

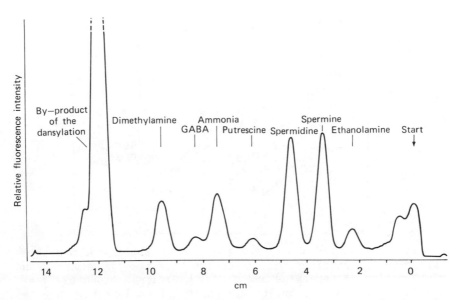

Figure 4. Record of the fluorescence intensity along the length
of the chromatographic path of a thin-layer chromatogram analogous to that shown in Figure 3. (Dansylation products of the
brain extract of a 1-day-old rat.) Approximate amount of GABA
derivative: 50 pmole. From Seiler, Research Methods in Neurochemistry, edited by M. Marks and Rodnight (Plenum Press, New
York, 1975), vol. 3, p. 431. (Reproduced with permission from
the authors and Plenum Press.)

Similarly, Heby and Anderson (69) analyzed putrescine, spermidine,
and spermine in urine by converting them to the corresponding
dansyl derivatives and quantitating them by in situ scanning of
the thin layer chromatograms. At least 50-100 ng of the dansyl
derivative must be present in each spot in order to obtain reproducible results (70). Moreover, the fluorescent intensity tends
to fade with time, even if the chromatograms are kept in the dark.
In order to increase the sensitivity of the method, either mass
spectrometry (71) or autoradiography of ^{14}C-dansyl derivatives
have been used for detection of the products (70). The latter
authors applied their methods to the analysis of free amino acids
and serotonin in individual giant neurons of the snail.

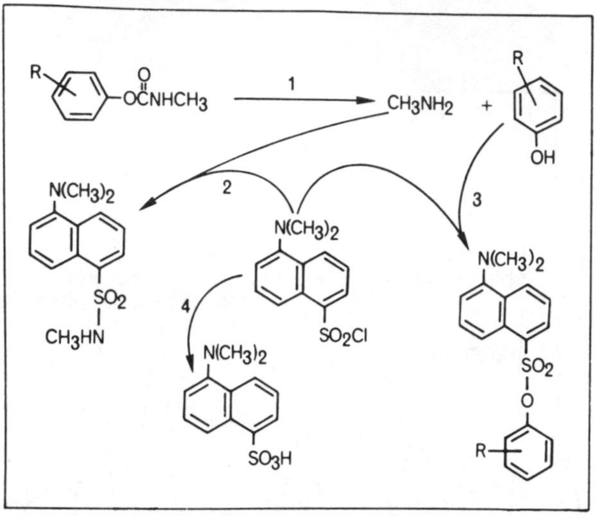

Figure 5. Reaction scheme for the dansyl labeling of N-methylcarbamate insecticides. 1 = hydrolysis of the carbamate; 2 and
3 = labeling of the amine and phenol hydrolysis products; 4 =
hydrolysis of the reagent by carbonate. From Lawrence and Frei,
J. Chromatog. 98, 253 (1974). (Reproduced with permission from
the author and Elsevier Scientific Publishing Co.)

Pesticides have also been analyzed by dansyl labeling techniques
combined with TLC (see reviews by Lawrence and Frei [60] and by
McNeil and Frei [72]). For example, dansylation has been successfully applied to the analysis of low levels of N-methylcarbamate
insecticides in water and soil samples with little or no cleanup
needed. The procedure involves hydrolysis of the carbamate with
0.1 M Na_2CO_3 to form a phenol and methylamine (Figure 5). These
hydrolysis products are subsequently dansylated, separated by TLC,
and quantitated by in situ scanning spectrofluorometry (Lawrence
and Frei [60]). Limits of detection are about 5-10 ng. Similarly, N-phenylcarbamate and urea herbicides may be hydrolyzed to
yield anilines, which may be determined by TLC of their dansyl
derivatives (60).
        A more extensive list of some of the many applications of
dansyl chloride is provided in chapters by Seiler and Demisch (59)
and Seiler (73).

Dansyl Hydrazine

A reagent that is related to dansyl chloride is the one formed by
substitution of the chlorine atom by a hydrazine group. This
reagent, dansyl hydrazine, has been used for the introduction of
the dansyl moiety onto carbonyl compounds, such as 17-ketoster-
oids (74).

5-di-n-Butylaminonaphthalene-1-sulfonyl Chloride (BNS-Cl)

BNS-Cl is an analogue of dansyl-Cl in which the two methyl
groups on the nitrogen have been replaced by butyl groups. This
reagent produces derivatives that are somewhat more fluorescent
than the corresponding dansyl derivatives, although the differences are rather small (75). However, since the BNS-derivatives
are more lipid soluble, they may be extracted in less polar
reagents, resulting in a better cleanup of the samples and, in
addition, to an increased fluorescence. All the BNS-amino acid
derivatives can be extracted from the reaction mixture with ethyl
acetate (75). The separation of the BNS-derivatives in picomole
quantities was achieved by two-dimensional polyamide TLC. Most
of the derivatives appear as yellow-green fluorescent spots but
the derivatives of histidine and tyrosine appear as deep orange
fluorescent spots (75).

4-Chloro-7-nitrobenzo[c]-1,2,5-oxidiazole (NBD-Cl)

NBD-Cl is another widely used reagent for preparing fluorescent
derivatives. This reagent is structurally analogous to FDNB in
that it is an aryl halide with an activated halogen (59). Its
specificity is also similar to FDNB. Primary and secondary
amines are readily derivatized in either aqueous solutions or in
organic solvents. Phenols and sulfhydryl compounds are more
slowly reacted under alkaline conditions. Under certain conditions (pH 7 sodium citrate buffer containing 1 mM EDTA), NBD-Cl
reacts with sulfhydryl groups without reacting with amino groups
(76).
   Amines may be derivatized with NBD-Cl by reacting the amine
solution with 4 volumes of a 0.5 mg/ml solution of NBD-Cl in
methanol. After the addition of 0.1 M NaHCO$_3$, the reaction is
allowed to proceed by heating to 55° C for 1 hour (59).
   NBD-Cl has been used in place of dansyl-Cl for many applications. Although both reagents are similar in many respects,
NBD-Cl offers several advantages over dansyl-Cl (77). First,
since NBD-Cl, unlike dansyl-Cl, is nonfluorescent, interference
from the presence of excess reagent can be avoided. Second,

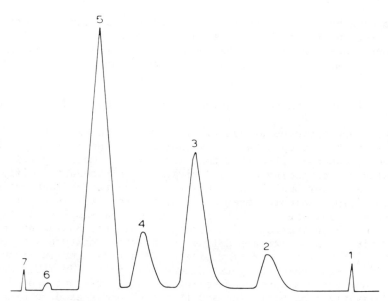

Figure 6.   Recording of an in situ analytical determination of a
derivatized amine mixture separated on Polyamide-11 layer.   1,
Origin; 2, NBD-dimethylamine; 3, NBD-diethylamine; 4, NBD-piperi-
dine; 5, NBD-ethylpropylamine; 6, solvent front.   From Klimisch
and Stadler, J. Chromatogr. 90, 141 (1974).   (Reproduced with permission of the author and Elsevier Scientific Publishing Co.)

NBD-Cl is more stable and more water soluble than dansyl chloride.
Finally, the excitation wavelength of NBD derivatives occurs in
the visible spectrum (450-480 nm) and therefore can be measured
in a fluorometer without the need of a quartz cuvette.   In comparison, dansyl derivatives have excitation maxima of about 360 nm.
NBD derivatives can be detected at a similar level of sensitivity as the corresponding dansyl derivatives.
     One application of NBD-Cl has been for the detection of free
amino acids and of the N-terminal amino acid of peptides (78).
These authors have presented data for the fluorescence maxima and
the R_f on silica gel G TLC of the NBD derivatives of all the protein amino acids.   TLC was carried out with various proportions
of chloroform:methanol, and the best resolution was obtained with
a mixture of 70:30.

Klimisch and Stadler (79) used NBD-Cl for the microdetermination of aliphatic amines, including dimethylamine, diethylamine, and piperidine. Quantitation was achieved by separation on polyamide sheets and subsequent in situ measurements with a Zeiss chromatograph spectrophotometer (Figure 6). A linear calibration curve was obtained over the range of 15 to 150 ng per spot.

The properties of the NBD derivatives of a large number of basic drugs have also been studied (80). Derivatives of such drugs as amphetamine, chlorpheniramine, mescaline, ephedrine, and fenfluramine were separated on silica gel G using ethyl acetate-cyclohexane (2:3) and diethyl ether-benzene (1:1). The sensitivities by which these drugs could be detected ranged from 0.01 to 0.5 µg.

NBD-Cl has also been used in the same way as dansyl-Cl for the analysis of pesticides that hydrolyze to alkalamines (60). These procedures were used to detect carbamate insecticides, including Zectran and Matacil, and triazine herbicides.

Fluorescein Isothiocyanate

Fluorescein isothiocyanate is a fluorogenic reagent analogous to phenylisothiocyanate. This reagent has been used primarily for the analysis of amino acids (59). As in the case of phenyliso-thiocyanate, the derivatives produced from amino acids are usually converted to the corresponding thiohydantoin by treatment with HCl. The fluorescent thiohydantoins may be separated on silica gel G (81).

Fluorescamine

Fluorescamine is a relatively new reagent but has rapidly gained widespread use. It is nonfluorescent and reacts with primary amines to produce highly fluorescent derivatives. Although other compounds such as secondary amines and alcohols are capable of reacting with fluorescamine, the products of these reactions are not fluorescent (82).

Because the reagent is rapidly hydrolyzed by water, it must be dissolved in organic solvents, such as acetone and dioxane. The amines may be reacted in aqueous solution, provided the reaction mixture is rapidly mixed so that the reaction occurs before the reagent is hydrolyzed. The reaction generally proceeds at room temperature with a half-time of a fraction of a second and an optimal pH of 7.5 to 9.

The chief application of fluorescamine has been for the detection of amines, amino acids, and peptides after their separation by ion-exchange chromatography. The sensitivity of this

reagent has been a major advance over the use of ninhydrin for
the detection of micro quantities in amino acid analyzers.
Fluorescamine also has promise in the determination of pesticides
by labeling the amine hydrolysis products of carbamates, ureas,
and triazines (60).

Although fluorescamine has been used as a spray reagent to
detect amines separated by TLC, it has rarely been used to form
derivatives prior to TLC separation. One exception is the
recent report by Nakamura and Pisano (83). These authors developed two methods for the detection of various phenylethylamines,
including the catecholamines and their 3-0-methyl derivatives
(Table III). In one method, the amines were separated by silica
gel TLC and were then detected by dipping the plates in a fluorescamine solution and spraying with 70% perchloric acid. In an
alternate method, the amines were spotted on the plates and
derivatized before TLC by dipping the plates in the fluorescamine solution. After the derivatives were separated by TLC,
they were then visualized by spraying with 70% perchloric acid
to give intense blue fluorescent spots. With either of these
methods, the limit of detection was between 5 and 800 pmole,
depending on the particular compound being analyzed.

4-Bromomethyl-7-methoxycoumarin (Br-Mmc)

Br-Mmc is a new reagent for the fluorescence labeling of carboxylic acids. Dunges (84) used this reagent to analyze trace quantities of fatty acids. The fatty acids and the reagent were
dissolved in acetone (final volume 25 µl) and refluxed for 1 hour
in the presence of K_2CO_3 using a microrefluxer. The Mmc derivatives were separated on silica gel layers. As little as 50
pmole of the derivative of each fatty acid could be detected
(Figure 7). Quantitation was carried out by measuring the fluorescence in situ.

The Mmc derivatives are stable and can be stored at room
temperature for at least 3 months. However, the reagent is
unstable to light and should be protected.

Although this is a new reagent, the authors suggest that it
has promise as a "universal" fluorescence label for monocarbox-
ylic acids. Dicarboxylic acids, however, do not react with this
reagent.

CONCLUSION

This chapter has pointed to the utility of derivatization techniques to chemical analysis of compounds by TLC. As indicated

TABLE III. R_f VALUES AND DETECTION LIMITS OF CATECHOLAMINES AND RELATED PHENYL-ETHYLAMINES AND OF THE CORRESPONDING FLUORESCAMINE DERIVATIVES ON TLC PREPARED BY METHODS I AND II

Compound	Method I*		Method II**	
	$R_f \times 100$	Detection Limit (pmole)	$R_f \times 100$	Detection Limit (pmole)
L-DOPA	50	80	26	75
DL-α-MethylDOPA	52	500	19	250
L-3-0-MethylDOPA	55	8	27	20
Dopamine.HCl	58	90	70	75
3-0-Methyldopamine.HCl	62	5	69	10
4-0-Methyldopamine			69	5
5-Hydroxydopamine.HCl	51	300	1	75
6-Hydroxydopamine.HBr	57	800	3	200
L-Norepinephrine.HCl	54	70	59	200
DL-α-Methylnorepinephrine.HCl			66	350
DL-Normetanephrine.HCl	60	20	67	20

* Solvent system: n-Butanol-acetic acid-water (5:2:3).
** Solvent system: ethyl acetate-n-hexane-methanol-water (60:20:25:10).

From: Nakamura and Pisano, J. Chromatogr. 154, 51 (1978). With permission from Elsevier Scientific Publishing Co.

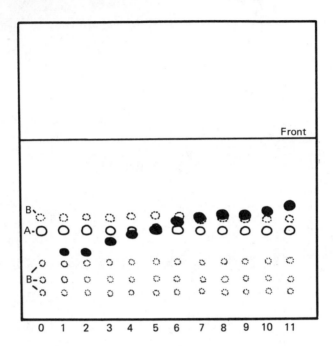

Figure 7. Thin layer chromatogram of directly applied Br-Mmc
reaction mixtures of 11 fatty acids from C_1 to C_{24}:   (1)   formic
acid, (2) acetic acid, (3) propionic acid, (4) butyric acid,
(5) valeric acid, (6) caprylic acid, (7) lauric acid, (8) palmitic acid, (9) stearic acid, (10) behenic acid, (11) lignoceric
acid.   (A) Br-Mmc, (B) byproducts, 5 µl applied of each reaction
mixture containing 20 nmole and 60 nmole Br-Mmc per 100 µl.
Reprinted with permission from Dunges, Anal. Chem. 49, 442 (1977).
(Copyright by the American Chemical Society.)

earlier, no attempt has been made to provide a comprehensive
review of all the derivatizing reagents which might be used.
Rather, the intent has been simply to illustrate the usefulness
of derivatizing reagents in TLC and to indicate the reagents that
may be the most useful.
      Derivatization techniques have application to a wide range
of analytical problems, including the analysis of steroid, amino
acids, drugs, pesticides, and other environmental pollutants.
Whether the reader will find derivatization techniques to be
necessary or desirable will depend on many factors, including the

nature of the compounds being analyzed, the composition of the
sample (what interfering compounds are present), and the sensitivity that is needed.

　　If the samples are to be derivatized, one must then decide
whether the derivatives are to be formed before or after TLC and
what particular reagent is to be used. Many factors must enter
into the choice of the derivatizing reagent. First, one must
consider the functional groups that are present in the compounds
to be analyzed and then select a suitable reagent that will react
with them under conditions in which they are stable and do not
undergo side reactions. The specificity of the reagents must be
considered. In some cases one may not want a reagent that is
able to react with several different functional groups. Finally,
one must consider the properties of the derivatives formed,
whether they are stable to light or air, whether they have good
chromatographic properties, and whether they can be detected with
sufficient sensitivity. If high sensitivity is needed, the formation of fluorescent derivatives will probably be necessary; otherwise, chromatophoric derivatives might suffice. As to whether
the derivatives should be formed before or after TLC, this again
will depend upon the particular analytical problem being considered. Although derivatization reagents are by far more commonly
carried out after TLC by spraying or elution techniques, there
are many cases in which it would be advantageous to carry out the
derivatization step before TLC.

REFERENCES

1. K. Blau and G. S. King, Handbook of Derivatives for Chromatography (Heyden, London, 1977).
2. J. Drozd, J. Chromatog. 113, 303 (1975).
3. M. S. F. Ross, J. Chromatogr. 141, 107 (1977).
4. K. G. Krebs, D. Huesser, and H. Wimmer, in Thin-Layer Chromatography, edited by E. Stahl (Springer-Verlag, New York,
   1969), p. 854.
5. J. G. Kirchner, in Thin-Layer Chromatography, 2nd ed. (John
   Wiley & Sons, New York, 1978), p. 198.
6. D. J. Edwards, in Handbook of Derivatives for Chromatography,
   edited by K. Blau and G. S. King (Heyden, London, 1977), p.
   391.
7. G. E. Means and R. E. Feeney, in Chemical Modification of Proteins (Holden-Day, San Francisco, 1971), p. 118.
8. G. R. Stark, Adv. Prot. Chem. 24, 265 (1970).

9. G. Pataki, Techniques of Thin-Layer Chromatography in Amino
    Acid and Peptide Chemistry (Humphrey Science Publ. Inc., Ann
    Arbor, 1969), p. 126.
10. S. Shaltiel, Biochem. Biophys. Res. Comm. 25, 178 (1967).
11. G. L. Mills and D. Beale, in Chromatographic and Electrophor-
    etic Techniques, edited by I. Smith (Wiley-Interscience, New
    York, 1969), p. 170.
12. J. Rosmus and Z. Deyl, J. Chromatogr. 70, 221 (1972).
13. M. Brenner, A. Niederwieser, and G. Pataki, in Thin-Layer
    Chromatography: A Laboratory Handbook, edited by E. Stahl
    (Springer-Verlag, New York, 1969), p. 756.
14. R. L. Munier and G. Sarrazin, J. Chromatogr. 22, 347 (1966).
15. K.-T. Wang, Y.-I. Lin, and I. S. Y. Wang, Adv. Chromatogr. 11,
    73 (1974).
16. A. Niederwieser, in Chromatography: A Laboratory Handbook of
    Chromatographic and Electrophoretic Methods, 3rd ed. (Van
    Nostrand Reinhold, New York, 1975), p. 393.
17. J. L. Bailey, Techniques in Protein Chemistry (Elsevier,
    Amsterdam, 1967), p. 163.
18. C. Martel and D. J. Phelps, J. Chromatogr. 115, 633 (1975).
19. E. W. Day, Jr., T. Golab, and J. R. Koons, Anal. Chem. 38,
    1053 (1966).
20. A. Zeman and I. P. G. Wirotama, Z. Anal. Chem. 247, 155 (1969).
21. I. C. Cohen, J. Norcup, J. M. A. Ruzicka, and B. B. Wheals,
    J. Chromatogr. 44, 251 (1969).
22. R. P. Shank and M. H. Aprison, Anal. Biochem. 35, 136 (1970).
23. Y. Yamanishi, Y. Kase, T. Miyata, and M. Kataoka, Life Sci. 9,
    409 (1970).
24. R. I. Gregerman and M. A. Kowatch, J. Clin. Endocrinol. 32,
    110 (1971).
25. G. B. Cos, J. Chromatogr. 83, 471 (1973).
26. I. C. Cohen and B. B. Wheals, J. Chromatogr. 43, 233 (1969).
27. I. C. Cohen, J. Norcup, J. H. A. Ruzicka, and B. B. Wheals,
    J. Chromatogr. 49, 215 (1970).
28. J. C. Crawhall and D. F. Elliott, Biochem. J. 61, 264 (1955).
29. A. D. Smith and J. B. Jepson, Anal. Biochem. 18, 36 (1967).
30. D. J. Edwards and K. Blau, Anal. Biochem. 45, 387 (1972).
31. D. J. Edwards, P. Doshi, and I. Hanin, Anal. Biochem. 96, 308
    (1979).
32. R. L. Shriner, R. C. Fuson, and D. Y. Curtin, The Systematic
    Identification of Organic Compounds: A Laboratory Manual
    (Wiley, New York, 1964), p. 253
33. G. A. Byrne, J. Chromatogr. 20, 528 (1965).
34. E. Stahl and H. Jork, in Thin-Layer Chromatography: A Labora-
    tory Handbook, edited by E. Stahl (Springer-Verlag, New York,
    1969), p. 206.

35. E. Denti and M. P. Luboz, J. Chromatogr. 18, 325 (1965).
36. G. Ruffini, J. Chromatogr. 17, 483 (1965).
37. A. Jart and A. J. Bigler, J. Chromatogr. 23, 261 (1966).
38. R. H. Brandenberger and H. Brandenberger, in Handbook of Derivatives for Chromatography, edited by K. Blau and G. S. King (Heyden, London, 1977), p. 234.
39. C. F. Beyer and T. E. Kargl, J. Chromatogr. 65, 435 (1972).
40. L. Treiber and G. W. Oertel, Clin. Chim. Acta 17, 81 (1967).
41. L. Treiber, P. Knapstein, and J. C. Touchstone, J. Chromatogr. 37, 83 (1968).
42. P. Knapstein, L. Trieber, and J. C. Touchstone, Steroids 11, 915 (1968).
43. L. Penzes, P. Menzel, and G. W. Oertel, J. Chromatogr. 44, 190 (1969).
44. J. W. T. Seakins, I. Smith, and M. J. Smith, in Chromatographic Electrophoretic Techniques, edited by I. Smith (William Heinemann, London, 1976), vol. 1, p. 244.
45. A. C. Thompson and P. A. Hedin, J. Chromatogr. 21, 13 (1966).
46. P. Juvic and B. Sundby, J. Chromatogr. 76, 487 (1973).
47. T. Okuyama and K. Satake, J. Biochem. 47, 454 (1968).
48. K. Blau and G. S. King, in Handbook of Derivatives for Chromatography, edited by K. Blau and G. S. King (Heyden, London, 1977), p. 104.
49. D. H. Neiderhiser, R. K. Fuller, L. J. Hejduk, and R. J. Roth, J. Chromatogr. 117, 187 (1976).
50. I. P. G. Wirotama and K. H. Ney, J. Chromatogr. 61, 166 (1971).
51. A. Darbre, in Handbook of Derivatives for Chromatography, edited by K. Blau and G. S. King (Heyden, London, 1977), p. 39.
52. Y. Hirata, T. Takeuchi, and K. Matasumoto, Anal. Chem. 50, 1943 (1978).
53. W. Morozowich and S. L. Douglas, Prostaglandins 10, 19 (1975).
54. J. Y. Chang and E. H. Creaser, J. Chromatogr. 116, 215 (1976).
55. J. Y. Chang and E. H. Creaser, J. Chromatogr. 132, 303 (1977).
56. J. Y. Chang and E. H. Creaser, J. Chromatogr. 135, 245 (1977).
57. S. Datta, S. C. Datta, and R. Sengupta, Biochem. Biophys. Res. Commun. 72, 1296 (1976).
58. J.-K. Lin and J.-Y. Chang, Anal. Chem. 47, 1634 (1975).
59. N. Seiler and L. Demisch, in Handbook of Derivatives for Chromatography, edited by K. Blau and G. S. King (Heyden, London, 1977), p. 346.
60. J. F. Lawrence and R. W. Frei, J. Chromatogr. 98, 253 (1974).
61. N. Seiler, J. Chromatogr. 63, 97 (1971).

62. N. Seiler, in Research Methods in Neurochemistry, edited by
    N. Marks and R. Rodnight (Plenum Press, New York, 1975), vol.
    3, p. 409.
63. N. Seiler and M. Wiechmann, in Progress in Thin-Layer Chroma-
    tography and Related Methods, edited by A. Niederwieser and
    G. Pataki (Humphrey Science Publishers, Ann Arbor, 1970),
    vol. 1, p. 94.
64. H. Tsuzuki, K. Kitani, K. Imai, and Z. Tamura, Chem. Pharm.
    Bull. 20, 1931 (1972).
65. G. Briel and V. Neuhoff, Hoppe-Seyler's Z. Physiol. Chem. 353,
    540 (1972).
66. J. Airhart, S. Sibiga, H. Sanders, and E. A. Khairallah, Anal.
    Biochem. 53, 132 (1973).
67. R. L. Munier and A. M. Drapier, Chromatographi 5, 306 (1972).
68. A. A. Boulton, J. Chromatogr. 63, 141 (1971).
69. O. Heby and G. Andersson, J. Chromatogr. 145, 73 (1978).
70. B. E. Leonard and N. N. Osborne, in Research Methods in Neuro-
    chemistry, edited by N. Marks and R. Rodnight (Plenum Press,
    New York, 1978), p. 443.
71. O. Durden, in Research Methods in Neurochemistry, edited by
    N. Marks and R. Rodnight (Plenum Press, New York, 1978), p.
    205.
72. J. D. MacNeil and R. W. Frei, J. Chromatogr. Sci. 13, 279
    (1975).
73. N. Seiler, Methods of Biochem. Anal. 18, 259 (1970).
74. R. Chayen, R. Dvir, S. Gould, and A. Harell, Anal. Biochem.
    42, 283 (1971).
75. N. Seiler and B. Knodgen, J. Chromatogr. 97, 286 (1974).
76. D. Birkett, N. C. Price, G. K. Radda, and A. G. Salmon, FEBS
    Lett. 6, 346, 1970.
77. Regis Lab Notes, No. 10, Regis Chemical Co., 1971.
78. R. S. Fager, C. B. Kutina, and E. W. Abrahamson, Anal. Bio-
    chem. 53, 290 (1973).
79. H.-J. Klimisch and L. Stadler, J. Chromatogr. 90, 141 (1974).
80. J. C. Hudson and W. P. Price, J. Chromatogr. 117, 449 (1976).
81. H. Kawauchi, K. Tuzimura, H. Maeda, and N. Ishida, J. Biochem.
    66, 783 (1969).
82. S. Stein, P. Bohle, K. Imai, J. Stone, and S. Udenfriend,
    Fluorescence News 7, 9 (1973).
83. H. Nakamura and J. J. Pisano, J. Chromatogr. 154, 51 (1978).
84. W. Dunges, Anal. Chem. 49, 442 (1977).
85. E. Heftmann, Chromatography (Reinhold Publ. Corp., New York,
    1967).
86. M. Jellinek, J. Chromatogr. 69, 402 (1972).
87. Y. Kase, M. Kataoka, and T. Miyata, Life Sci. 6, 2427 (1967).

88. R. R. Linko, H. Kallio, and K. Rainio, J. Chromatogr. 155,
    191 (1978).
89. K.-T. Wang, J. M. K. Huang, I. S. Y. Wang, J. Chromatogr. 22,
    362 (1966).
90. J. Hakl, J. Chromatogr. 61, 183 (1971).
91. R. B. Freedman and G. K. Radda, Biochem. J. 108, 383 (1968).
92. K.-T. Wang and I. S. Y. Wang, J. Chromatogr. 24, 460 (1966).

CHAPTER 6

Newer Instrumentation for TLC

Louis J. Glunz

Instrumentation and special techniques for thin layer chromatography (TLC) seek improvements of the following types:

1. Improvement in resolution.
2. Greater sensitivity, especially in the detection of trace components.
3. Increased sample loading capacity.
4. Speed in development.
5. Reproducibility.

Before examining these techniques, some of the TLC characteristics which will have an effect on these sought-after improvements must be reviewed.

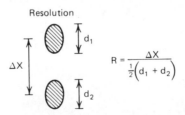

Figure 1.   Definition of resolution.

In Figure 1, resolution in a TLC system is defined by the center-to-center separation of two components divided by the average diameter of the two spots. Resolution is expressed by the equation:

$$R = \frac{\Delta X}{\frac{1}{2}(d_1 + d_2)}$$

The two components are just separated when $R = 1$.

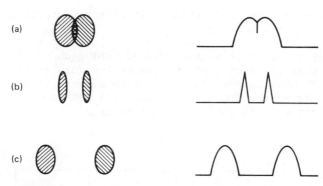

(a)

(b)

(c)

Figure 2.   Effects of solvent selectivity.

In examining this equation, it becomes apparent that resolution can be improved by either increasing the center-to-center separation (ΔX) or by decreasing the average diameter of the spots.

In order to increase the center-to-center separation of the spots, one must increase the selectivity of the system, that is, increase the difference in the migration rates of the two components.

In order to reduce the average diameter of the spots, one must have a more efficient system.

Figure 2 shows three sets of spots and their corresponding densitometer scans. Figure 2(a) shows two components which are not resolved; Figure 2(b) shows these two components separated by increasing the efficiency of the system, without increasing the selectivity; and Figure 2(c) shows the two components separated by increasing the selectivity of the system and not improving the efficiency. Obviously, a combined improvement in both selectivity and efficiency would bring about an even greater improvement in resolution.

Whatever method is used in detection of the components on a TLC plate, sensitivity is related to the number of molecules per unit area. Figure 3 shows three spots of decreasing diameter.

Sensitivity

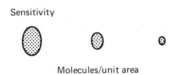

Molecules/unit area

Figure 3.   Concentration as it effects detectibility.

In each of these spots we have placed twelve dots. It is quite
apparent that the spot that has the twelve dots in the smallest
area is the most intense and easiest to detect. Therefore, any
improvements in the TLC process that will make separated zones
more compact will improve the sensitivity.

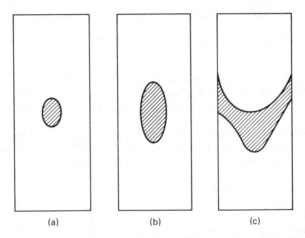

Figure 4.   Illustration of the effect of sample loading.

Figure 4 shows three chromatograms, each with an increased
sample loading. As the sample loading is increased, the following
problems are common: (1) tailing, as shown on plate B, and (2)
edge effects combined with tailing, as shown on plate C. Tailing
and edge effects have adverse effects on resolution, and in many
cases will completely cover close-running trace components. Any
improvements in the TLC process that will permit heavier loading

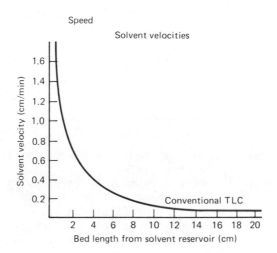

Figure 5. Relationship of solvent velocity and distance of mobile phase from solvent reservoir.

without causing these adverse loading effects will greatly improve the detection of trace components in a mixture.

In considering speed of the chromatographic process, one must bear in mind that the velocity of a solvent drops off as the solvent moves higher on the plate. Figure 5 shows the relationship between solvent velocity and the distance of the solvent front from the solvent reservoir. It becomes obvious that if the desired separations can be accomplished by using only the lower part of the plate, the speed of development can be greatly increased.

Some of the major factors effecting the reproducibility of a separation from plate to plate are listed in Table I.

Since the amount of moisture adsorbed on the TLC plate generally has a marked effect on the separation, the relative humidity to which the plate and solvent are exposed, both before and during the development, must be controlled.

In the TLC process, the problems associated with height and nonlinear flow, especially when working in large TLC tanks, are well known. In order to have reproducible development, a reproducible degree of solvent composition and saturation of the tank is required.

Temperature will also effect the development and the corresponding R_f's of the components. This is especially true when elevated temperatures are used in the TLC process. In these

TABLE I. FACTORS AFFECTING REPRODUCIBILITY

---

1. Moisture
2. Chamber saturation
3. Temperature
4. Solvent demixing

---

cases, the temperature must be reproducibly controlled if the developments are to be reproducible.

Because solvent demixing in many cases plays an important role in the TLC development, care must be taken to insure that this process is reproducible. When using mixed solvents, bear in mind that a change in composition resulting from evaporation of one of the components will cause a corresponding change in the chromatographic process.

The principles involved in instrumentation and special techniques for TLC usually lead to some improvement in one or more of the above-described characteristics.

In cylindrical TLC, a rod or tube is coated with a TLC layer. The sample is spotted at the bottom of the cylinder and the cylinder is developed in a chromatography chamber. Figure 6 shows a developed cylinder; the sample components tend to form bands. Since there are no edges, there are no edge effects. Thus, cylindrical TLC generally permits some heavier sample loading and is especially useful to preparative TLC. Tailing, of course, will continue to occur for some overloaded samples.

One of the more interesting applications of cylindrical TLC is the Chromarod (Figure 7). Chromarods are 0.9 × 150 mm glass rods with a chemically stable powdered glass/silica gel coating. Samples of 1 µl are applied on up to 10 Chromarods supported in a specially designed glass or metal frame. The frame is then positioned vertically in a lined chromatography tank and developed in the normal manner.

The advantage of this system is the flame ionization detection system employed in the companion detector: the Iatroscan TH-10. The schematic representation of the detector is shown in Figure 8.

After the solvent is removed by oven drying for 3 to 5 minutes, the rods are individually transferred to the detector. The rods are automatically passed through the hydrogen flame, generating data that resemble information produced by a gas-liquid chromatograph. The flame also frees the rods of sample and reactivates them for reuse. (Refer to Chapter 36.)

Cylindrical TLC

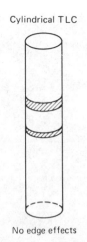

No edge effects

Figure 6.   Developed cylinder of sorbent.

The system has been shown to be sensitive, quantitative, rapid, and reproducible. It is especially suited to systems where the TLC procedure has been worked out. The instrument is manufactured by Iatron Laboratories, Japan and marketed by Newman-Howell Associates, U.K.

Figure 9 shows a distortion of cylindrical TLC that leads to conical TLC and finally to radial TLC. Note the direction of the solvent advance in each of these systems. As the solvent advances, the radius increases and, because the solvent must expand into a steadily increasing volume as it travels through the layer, its forward velocity decreases from the reservoir to the solvent front. A negative solvent gradient is formed. This is unlike conventional TLC where the solvent velocity is generally the same across the entire plate.

Figure 10 shows how a negative solvent gradient effects the development of a spot on a TLC plate. It can be seen that the lower part of the spot is always traveling at a greater velocity than the front of the spot. The result is zone concentration and a narrower spot.

A chromatograph for radial development, the U-Chamber, is manufactured by Camag, Switzerland (Figure 11). In this instrument, an HPTLC plate is placed in a ring holder and secured in an inverted position over a solvent-feeding capillary. The HPTLC layer faces the capillary. A pump drives the solvent through the capillary and onto the plate at a constant,

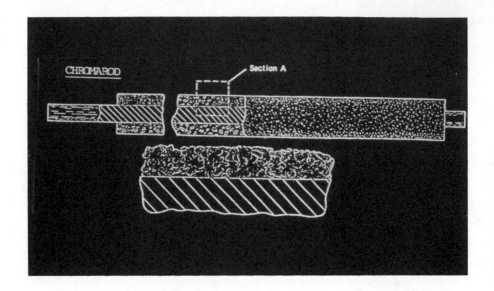

Figure 7.   The chromarod application of cylindrical TLC.

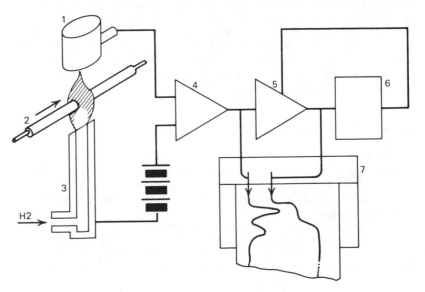

Figure 8.   Schematic of the Iatroscan TH-10 detector.

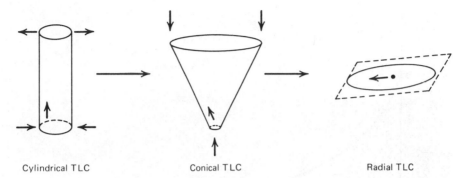

Cylindrical TLC   Conical TLC   Radial TLC

Figure 9.   Effects of various chromatoshapes.

preselected rate.   Once on the plate, the solvent spreads out in
a circular fashion.   The mobile phase flow rate may be varied
according to optimum separation parameters.   The rate is absolutely reproducible since it is electronically set.

   The instrument uses a 5 × 5 cm plate.   The samples (1 μl) are
are spotted in a circular fashion using a template so that up to
12 samples can be run on a single plate.   It is also possible to
inject the sample into the mobile phase in order to compare TLC
results with HPLC.   A gravity-fed, manually operated version of
this instrument has been introduced by the V-Tech Corporation,
New Berlin, Wisconsin.

   Another approach to radial TLC is the Selecta Sol system
marketed by Schleicher & Schuell Inc., Keene, New Hampshire (Figure 12).   This system uses a wick-fed solvent delivery system and
permits an investigator to simultaneously run up to 16 different
samples and/or solvents on a single, standard-size 20 × 20 cm TLC
plate.

   The unit, which was initially designed as a rapid method to
select the ideal solvent system for a chromatographic separation,
is now finding use in such TLC applications as toxicological
screening.

   A negative solvent gradient can also be obtained in evaporative TLC.   Figure 13 shows a TLC plate developing in an open chamber.   As the solvent advances on the TLC plate and eventually
moves outside of the chamber, the rate of evaporation increases.
The result is a more rapid flow of solvent at the bottom of the
plate than at the top of the plate.   This results in a negative
solvent gradient and the corresponding zone concentration.

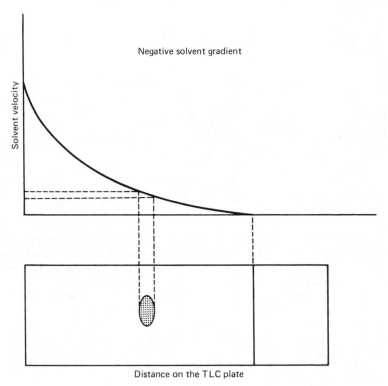

Figure 10.   Results of negative solvent gradients.

Needless to say, however, such a system would not be very reproducible.

The BN-chamber, which is manufactured by Desaga and marketed
in the U.S. by Brinkman, is an apparatus in which evaporative TLC
can be carried out under controlled conditions. The chamber (Figure 14) consists of a TLC-carrier plate and a slide-on cover plate
with ledges. The chamber is small as a sandwich chamber and is
preferably operated horizontally.

Warm, temperature-controlled water is circulated through a
heating block at the end of the chamber to cause controlled evaporation of the solvent from the plate. A cooling block located in
the center of the unit eliminates the condensation of solvent
vapors on the underside of the slide-on cover.

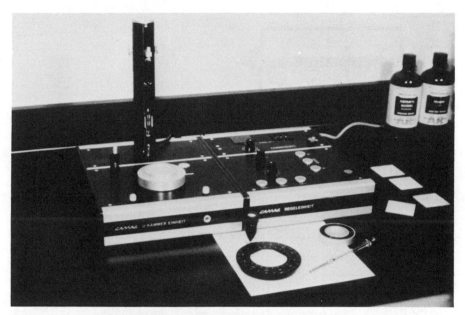

Figure 11(A). CAMAG U-Kammer.

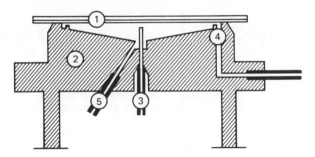

Cross-sectional diagram of CAMAG U–Krammer. The
HPTLC plate (1), measuring 50 x 50 mm, rests with its
layer facing downward on the U–chamber body. (2)
Elution solvent is fed to the center of the plate via a
platinum–iridium capillary (3) of 0.2 mm internal
diameter. Vapor phase, made up externally, may be
passed through the chamber, in through the circular
channel (4) and out through the center bore (5) before,
during, and after chromatographic development. The
direction of gas flow may also be reversed. (Courtesy
of CAMAG Inc., New Berlin, Wis.)

Figure 11(B).

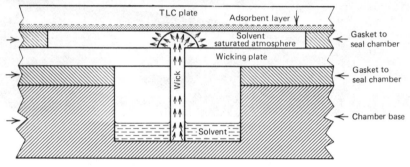

Figure 12.   The Selectra-Sol made by Schleicher and Schuell.

    The BN chamber can also be used for continuous thin-layer
chromatography. In continuous TLC, the solvent is evaporated
constantly at the end of a plate causing a steady flow of solvent
onto the plate.
    A novel approach to continuous development has recently been
introduced by Regis. The short bed/continuous development (SB/CD)
chamber is shown in Figure 15.   The SB/CD chamber consists of a
10.5 × 24 × 4 cm molded glass chamber with a precision ground top,
a cover glass plate and two Teflon wings for sealing the chamber.
Four ridges at the bottom of the chamber permit the TLC plate to
be placed in any of five different positions. In each of these
positions the TLC plate extends outside the chamber. As the solvent advances on the TLC plate, it passes through the lip of the
chamber and evaporates. This causes the solvent to advance onto
the plate at a constant velocity. Depending on the plate position and the level of solvent in the chamber, the bed length--
that is, the distance on the TLC plate from the solvent level to

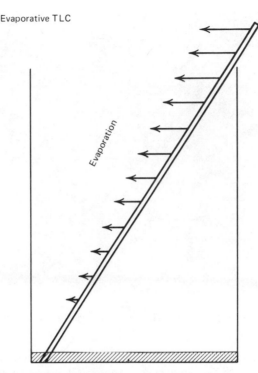

Evaporative TLC

Evaporation

Figure 13.   Effect of development of a TLC in an open chamber,
which results in a negative solvent gradient.

the point of solvent evaporation--can be varied from approximately 2 to 7 cm.

Figure 16 shows the approximate solvent velocity on the TLC
plate for each of the SB/CD positions. The shorter the bed
length, the higher the solvent velocity. Each of these steady-
state solvent velocities is considerably higher than the velocity
obtained on a conventional TLC plate. By using these higher solvent velocities, components that ordinarily would not move off
the origin can now be made to move on the plate in a reasonable
period of time. It is thus possible to explore the area of low
solvent strength chromatography.

Figure 17 shows the relationship that might be observed
between solvent strength and R_f for two components. At solvent
strength 0.4 the components have identical R_f's and will not

Figure 14.   The BN chamber made by Desaga.

separate. However, at solvent strength 0.2, the components can
easily be separated. In conventional TLC, the useful R_f range
would generally be between 0.05 and 0.95. Thus, the solvent
strength range that could be explored in this example would be
between approximately 0.28 and 0.45. By using the SB/CD, the
whole low solvent strength range can be conveniently used to
separate components on a TLC plate.

Obviously, the relationship between R_f and solvent strength
is going to be different for any two components. However, unless
the R_f solvent strength relationship for each of these components
is identical over the entire solvent strength range, resolution
of two components, which do not separate in a particular solvent
system, can almost always be improved by reducing the solvent
strength. It has been possible to improve resolution on many
pairs of compounds by simply adding hexane (solvent strength =
0.00) to the conventional solvent system and developing the
plates in the SB/CD chamber.

Another technique that has been used to develop a controlled
gradient on a TLC plate is Vapor-Programmed TLC. In this

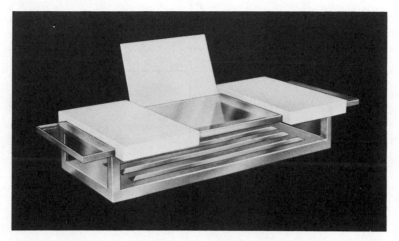

Figure 15. Short bed continuous development chamber introduced by Regis Chemical Company.

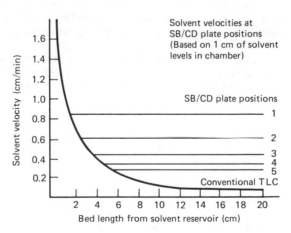

Figure 16. Velocity of solvent flow as effected by angle of the plate in the chamber.

93

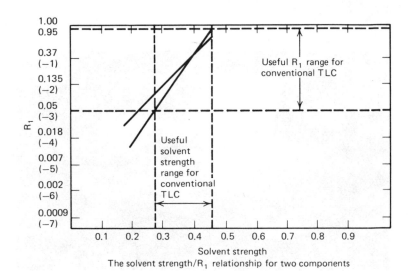

The solvent strength/R_1 relationship for two components

Figure 17.   Effect of solvent strength on velocity.

technique, the TLC plate is developed over a series of troughs,
sometimes referred to as conditioning trays. These troughs can be
arranged either parallel or pendicular to the flow of the mobile
phase. An appropriate liquid is placed in each of the troughs
and a TLC plate is placed over them, exposing the layer to the
different vapors and permitting them to be absorbed onto the dry
layer.
    The mobile phase is delivered to the plate by a paper wick
and then, as it migrates across the plate, it dissolves the
vapors and causes a controlled change in the plate characteristics.
    If these troughs are placed at a right angle to the mobile
phase, a controlled mobile phase gradient can be obtained. If
the troughs are placed parallel to the mobile phase, the effect
that different vapor concentrations (e.g., the amount of moisture)
have on a separation can be systematically studied.
    One instrument utilizing this technique is the CAMAG-KS
chamber (Figure 18). Another vapor programming instrument is the
Desaga VP chamber marketed in the U.S. by Brinkmann.
    Programmed Multiple Development (PMD) is a unique multiple
development technique. PMD is defined as the repeated development of a TLC plate with the same solvent in the same direction
for gradually increasing distances; between developments, the

Figure 18.   CAMAG-KS chamber.

solvent is removed from the thin layer bed by controlled evaporation while the plate remains in contact with the solvent reservoir (Figure 19).   In the PMD developer unit, the thin layer is
automatically cycled through a preset number of developments.   In
each cycle, the solvent is permitted to advance on the plate for
a set period of time and then, through controlled evaporation by
heat from the infrared radiators or by a flow of inert gas, the
solvent front is caused to recede, usually through the spot origin.   In each succeeding development, the mobile phase is
allowed to advance further.   After the last desired cycle, continued controlled evaporation prevents further development.

    Each time the solvent front traverses the sample, both dur-
ing solvent advance and solvent evaporation, spot reconcentration
takes place.   This phenomenon is shown in Figure 20.   The tail
end of the spot, which is developed for longer periods, is swept
into the front of the spot until the typical PMD spot (D in Figure 20) is formed.   As shown in Figure 21, spot reconcentration
leads to improved resolution.   As the spots are more compact,
spot reconcentration also improves sensitivity.

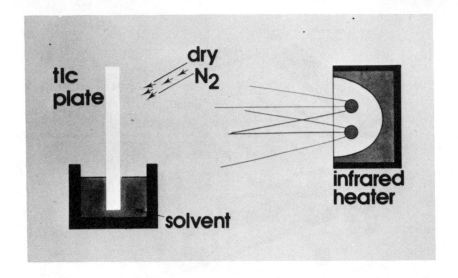

Figure 19. Diagram of essential components of the programmed
multiple development apparatus.

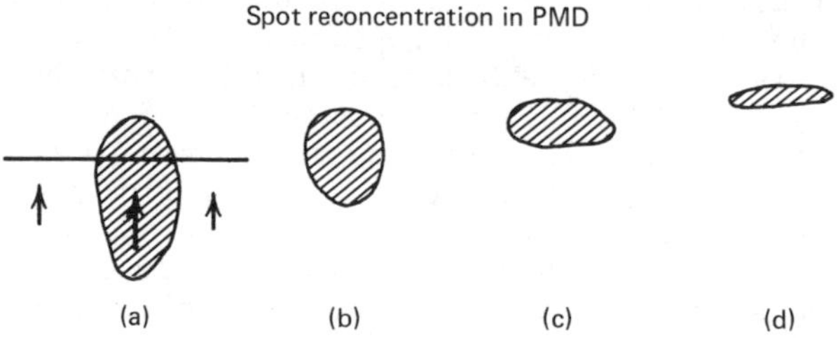

Figure 20. Concentration of the spots effected by programmed
multiple development apparatus.

    The effectiveness of a PMD Development is shown in Figure 22.
This shows densitometer traces for two TLC separations of a complex dye mixture: Sudan Black B. One was done conventionally
and the other two by PMD. The improvements in both resolution
and sensitivity by PMD are obvious.

Spot reconcentration leads to improved resolution
resolution and increased sensitivity.

Figure 21. Improved resolution as a result of programmed multiple development.

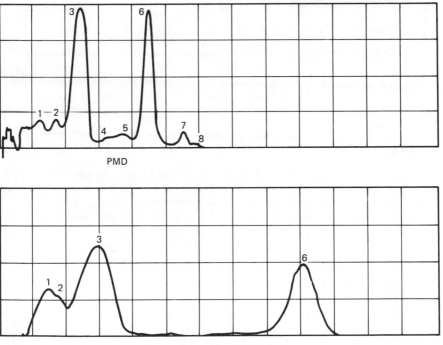

PMD

Conventional TLC

Note the peak definition and intensity in PMD compared to conventional TLC.

Figure 22. Densitometer tracing showing effectiveness of programmed multiple development.

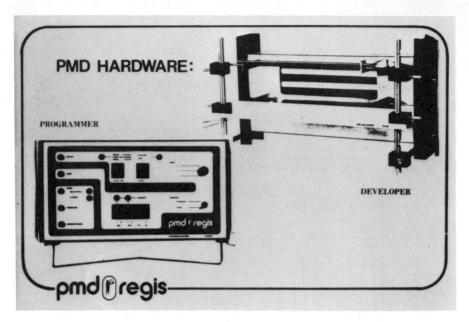

Figure 23.   The PMD apparatus.

    The PMD apparatus that is marketed by the Regis Chemical
Company is shown in Figure 23.

    This has been a brief look at a number of instrumental and
special techniques in TLC. The TLC process is very complex, and
each of these techniques exploits some characteristics of this
process to an advantage. It is important that the user of these
instruments understands their underlying principles in order to
bring all of their advantages to the solution of a problem.

    ACKNOWLEDGMENT

The author would like to thank the V-Tech Corporation, Schleicher
& Schuell, Inc., Brinkmann Instruments, Inc., and Neuman-Howells
Associates for supplying him with technical information and fig-
ures.

CHAPTER 7

Quantitative Densitometry

Melvin E. Getz

Paper and thin layer chromatography are used as qualitative tools for separating and identifying pesticide residues, drugs, biochemical constituents of plants and animals, and environmental contaminants.

There are two popular methods for quantitation. One is to elute or extract the isolated material from the spot and apply colorimetry. An example of this technique is a paper by Stellar and Curry (1), who measured the residues of Cygon [0,0-dimethyl S-(N-methylcarbamoylmethyl)phosphorodithioate)]and its oxygen analog by eluting the two from a silica gel plate and determining the total phosphate by a molybdenum blue procedure.

Another approach was to measure the areas of the spots, since molecules diffuse when they ascend or descend in an adsorbent with interconnecting pores. This diffusement has a relationship to the number of molecules present. Purdy and Truter (2) applied this principle to thin layer separations and showed that there was a linear range over a wide spectrum of concentrations if the square root of the spot area was plotted against the logarithm of the compound concentration.

In situ densitometry offers a simple way of quantitating directly on the chromatogram. If one wished to define quantitative densitometry, it could be "resolving a compound or compounds on a thin layer plate, visualizing the spots and measuring the optical density of each spot directly on the plate. The amounts of material in the unknowns are measured by comparing them to a standard curve from reference standards chromatographed with the same conditions."

Several optical scanning instruments are available to measure in situ spot density with high precision and accuracy. Some
of the initial results were disappointing. As it turned out, it
was not the faulty design of the instruments but the chromatographic techniques and plate quality that were contributing to
the unsatisfactory results.
The factors causing the lack of success were due mainly to
the methods of applying samples and standards, application of the
chromogenic reagents, and the nonuniformity and fragility of the
thin layer adsorbent surfaces. Most of these problems have been
resolved or minimized, so that it is now possible to obtain
results with quantitative TLC comparable to the other chromatographic techniques and densitometry is now accepted (3).

ADSORBENT AND THIN LAYER PLATE

Any material with a capillary structure and pores that can be
coated as a thin layer on some kind of support can be used for
thin layer chromatography. The many TLC publications show silica
gel and alumina to be the most widely used adsorbents. Experience
has shown silica gel to be the best adsorbent for resolving most
pesticide residues. The literature also indicates that it is
widely used for biochemical and drug separations. Apparently it
is the least reactive towards bioactive molecules. Today the
manufacturers of silica gel are producing a high-grade product
for thin layer applications.
Certain specific cases may arise where other adsorbents are
better suited. For example, a certain chromogenic reagent may
only work on alumina, or a specific separation may be possible
with cellulose and reversed phase development. The literature in
combination with the analyst's experience is the best guideline.
Some important surface requirements for quantitative optical
scanning are uniform thickness and layer durability. Uniform
thickness allows the optimum reflection and transmission of light
with a stable base line. The 250 micron thickness that is currently in use has proved quite satisfactory.
The coating equipment available to the laboratory runs the
spectrum from a simple reservoir with a gate that is pushed over
the plates for depositing the adsorbent slurry to an automatic
apparatus that moves the plates under the coating reservoir at
regulated speeds and times. Most companies making precoated
plates use the automatic apparatus. The author knows of at least
one company that uses a computer-operated spray-type coating
machine.
The plates that have been coated by the slurry technique can
sometimes be identified by observing their edges. They will

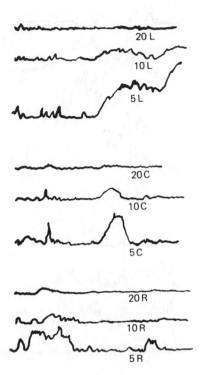

Figure 1. Scans of commercial thin layer plates showing uneven
layers.

exhibit a band of thickness a few millimeters greater than the
rest of the plate. This thickness can cause problems during
development and scanning. Edges that are too thick will slow the
solvent flow and cause a downward turning of the solvent front at
the edges. Spots near the edges will have a different diffusion
and it may not be possible to compare their density to the spot
densities on other parts of the plate.

Scanning the plate is an excellent way of checking the suitability of the thin layer plate. It is scanned at the left side
and right side and at the center. Gross differences can easily
be detected.

Figure 1 shows the blank scans of a commercial precoated
layer prepared by the slurry technique. The left side was
scanned 2 cm from the edge, the right side was scanned 3 cm from
the edge, and then the center of the plate was scanned. The

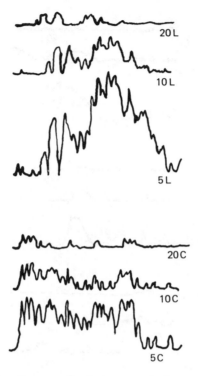

Figure 2.   Scans of plates coated manually.

lower the number the higher the sensitivity.   At the 20 attenua-
tion the base line is smooth.   At the lower attenuations the discrepancies in the plate were detected.

Figure 2 shows the blank scans obtained from a laboratory-
coated plate using the slurry technique.   The center and the left
edge 2 cm from the side were scanned.

The left edge of a plate coated by a spray technique was
scanned 2 cm from the edge.   Figure 3 shows the results.

The peaks obtained from scanning the various plates can be
considered to be optical noise.   This noise can be eliminated or
minimized by electronic means without adversely affecting the
sensitivity.

Several companies make precoats today that are excellent for
in situ applications.

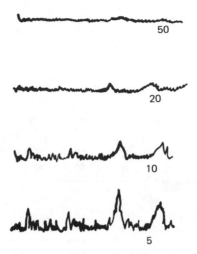

Figure 3.   Scans of a plate coated by a spray technique.

APPLYING SAMPLES AND STANDARDS

The operation of applying a sample to the surface of a thin layer
plate is called spotting. Any type of micropipet or syringe can
be used for qualitative work, but for quantitative work the
spotting technique can contribute the largest error. Fairbairn
(4) explains in detail the difficulties encountered with this
seemingly easy step.

There are two major types of samples--concentrated ones such
as solutions of pharmaceutical products where there are mg/ml
and dilute samples such as environmental ones which may contain
micrograms or less in the whole solution.

Several commercial calibrated micropipets and capillaries
are available that can spot quite accurately. Some of these have
been evaluated and described by Emanuel (5). The concentrated
sample can be quantitatively applied in microliter aliquots to
the plate. However, even under these circumstances, the volumes
and solvents should be the same for unknowns and standards.

An environmental sample usually has to be extracted in order
to isolate a few micrograms of the desired components. This
makes the final solution volume too large to transfer directly to
the plate. Since many of the chromogenic reagents have a sensitivity limit of around 0.1 µg, it would be necessary to spot a
10-g sample in order to get 0.01 ppm sensitivity (0.01 µg/g).

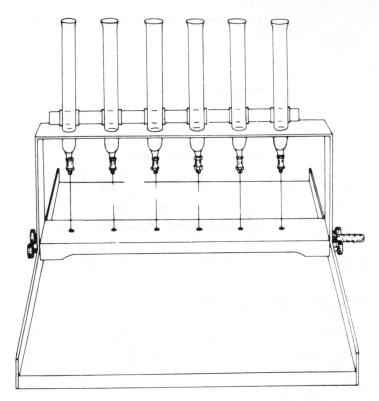

Figure 4.   The Kontes automatic spotting device.

The standard procedure is to concentrate to a volume around
100 µl and then transfer to the chromatogram with the aid of a
microsyringe multimicroliter volumes of solvent and therefore
then transferred to the plate.   During this process the spot
becomes larger and larger.   The operation is very tedious and
time-consuming.   In addition it is not known if the sample has
been completely rinsed out of the tube.
The standards that are spotted in a few microliter volumes
give tight spots at the origin.   When the chromatogram is developed the lateral diffusion of the sample is much greater than that
of the standard.   It is not valid to make quantitative comparisons
for the same reason that it is not possible to directly compare
optical densities from a 10 and 10 cm spectrophotometric cell.

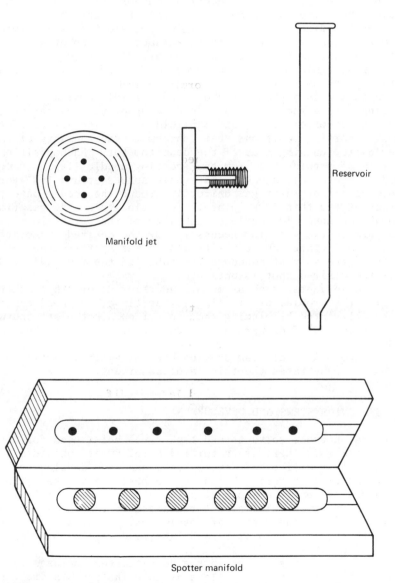

Manifold jet

Reservoir

Spotter manifold

Figure 5. Manifold arrangement of the Kontes spotter.

105

To help alleviate this problem, the author developed an automatic spotter that transfers both samples and standards onto the thin layer plate in an identical manner (6). The solvent, volumes, and evaporation rates are similar so that for practical purposes the diameters of the origin spot are the same. Volumes up to 2 ml can be spotted with the diameter of the origin spot getting no larger than 1 cm in diameter. It has turned out that the 1 cm size minimizes the mushroom effect of lateral diffusion as the R_f values increase.

Figure 4 is a photograph of a commercial version of the spotting device and Figure 5 is a drawing of the spotter manifold, the reservoir tube, and manifold jets. The air manifold has six spotting positions that conform to the geometry of the fiber optic scanning head of the densitometer. Each position consists of a guide hole for the spotting needle ($1\frac{1}{2}$-in. gauge hypodermic needle), and 4 holes evenly spaced around 360° for directing air or inert gas uniformly around the needle onto the surface of the thin layer plate. The velocity of the gas flow through the manifold regulates the size of the origin spot. To predetermine the gas flow necessary for the desired evaporation rate, an old plate can be positioned in the spotter, the desired volume of the solvent added to the tube and the flow adjusted until the desired spot is obtained.

Some solvents, such as methylene chloride or ethyl ether, may give the problem of freezing at the tip of the needle. This can be alleviated by placing the spotter on a hot plate and warming to about 40° C or warming the gas before it enters the manifold.

Figure 6 is a diagram showing how the spotting action works and how it penetrates the thin layer adsorbent.

SOLVENT SYSTEMS AND DEVELOPMENT

There are numerous solvents and tank types being used for developing TLC plates. There are saturated development, unsaturated development, equilibrated development, sandwich development, and variations of each form. Which is the best one to use? The one that gives the optimum separation for a particular problem. The mode of development will affect the shape of the spots, which in turn will affect the degree of resolution. Figures 7 and 8 show the differences in resolution and lateral diffusion between saturated and unsaturated development.

The method of scanning in the densitometer measures the total spot. Other commercially available instruments are capable of having the scanning beam adjusted so that it will cover the whole spot.

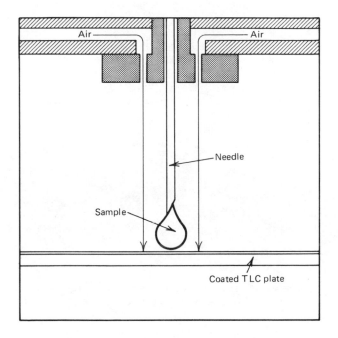

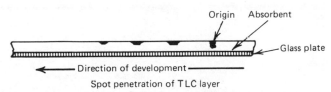

Spot penetration of TLC layer

Figure 6. Action of the spot in penetrating the layer.

Figure 9 shows curves for three different geometric figures with changing diameters. It can be seen that because the total spot is being scanned the curve shape is determined by the geometry of the spot and the area of the spot is the area of the curve of the scan. This phenomenon is important when biological detection reagents, such as enzyme inhibition or bioautography, are used. The areas of the spots are representative of concentration because of zones of inhibition and reaction.

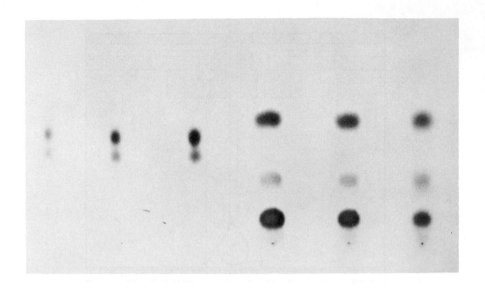

Figure 7. Effect of saturation on spot configuration. Saturated development: 80 ml hexane + 15 ml acetone + 5 ml chloroform.

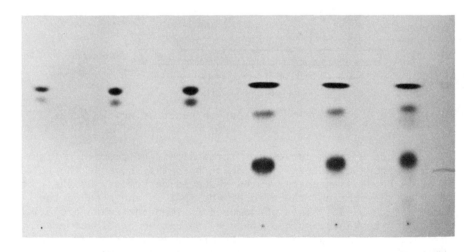

Figure 8. Effect of development in unsaturated tank. Unsaturated development: 80 ml hexane + 15 ml acetone + 5 ml chloroform.

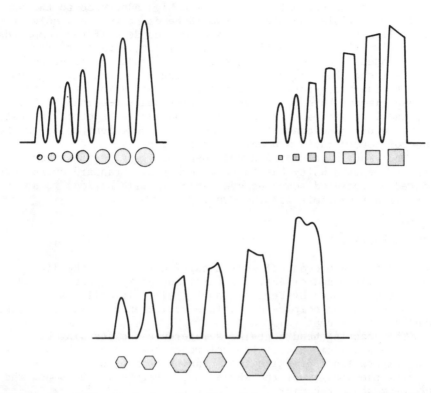

Figure 9.   Effect of changing diameters on scans of separated
spots.

CHROMOGENIC REAGENTS

After the compounds have been satisfactorily resolved on the TLC
plate they have to be visualized.
      There are three major ways to visualize migrated spots.
            1. Formation of color derivative.
            2. Quenching of UV in layers containing phosphor.
            3. Fluorescence, or fluorescence derivative.
      If in situ quantitation is to be performed, the chromogenic
reagent has to be applied in a uniform manner. The conventional
manner of spraying the reagent onto the plate is not always satisfactory. A simple way of applying the reagent uniformly is to
dip the plate into a solution of the reagent. This is another

reason why the adsorbent has to have a high adherence to the support. If possible the reagent should be dissolved in a solvent that will not dissolve the separated compounds. If the compounds have a significant solubility in the solvent, a tailing effect may be induced on the trailing edge of the spot. This becomes quite evident when the spot concentration is several micrograms.

In order to determine the proper concentration of reagent a few experiments have to be run. Reagent concentrations of 0.1, 0.5, and 1.0% are made and used to visualize the spots. In most cases one of these concentrations will be sufficient to cover the range of concentrations under investigation.

The kinetics of color development may vary, so the time effect on spot density has to be determined. Standard curves are plotted by repeated scanning over specific time periods to see if there is a time limit for the scan.

QUANTITATION

The in situ measurement of spot density directly on the TLC plate is simple and direct. In this laboratory we use a scanning instrument invented by Beroza and Hill (7). It utilizes fiber optics and the principles of diffuse reflectance as stated by the Kubelka-Munk concept (8).

The scanning head contains a reference and scanning beam that covers a 1 cm^2 area. The reference beam scans a blank surface next to the spot while the scanning beam simultaneously measures the density of the whole spot. Figure 10 shows how the dual beam scans the plate.

All the scans shown in this chapter were obtained with a commercial version of the reflectance scanner designed by the Kontes Co.

The optical system of the scanner has a branched Y configuration and a single light source. It was found that a monochromator was not needed for most applications and that optical filters were satisfactory. The three colors commonly used were red, green, and blue.

Figures 11 and 12 are plots of mixtures of three dyes resolved on a silica gel plate with hexane-acetone-chloroform (80:15:5) in an unsaturated system. Figure 11 is a plot of a concentration range from 1 to 8 μg with the green filter being used. A is the red dye (Sudan Red), B is the blue dye (indophenol). The yellow spot (Sudan Yellow) cannot be seen.

Figure 12 is the same plate scanned with the blue filter at the light source. This time the red and yellow spot can be seen but not the blue one.

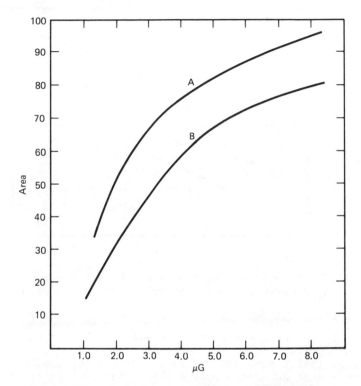

Figure 10.   Dual beam scanning configurations.

When the red filter is used only the blue spot can be seen.
The use of different filters is advantageous when analyzing
actual samples because there can be artifacts and interferences
that chromatograph and can cause background streaking. These
interferences can be eliminated or minimized by choosing the
appropriate colored filter.

CONCLUSIONS

The author in 1971 presented the initial applications of in situ
quantitation for pesticide residue determinations at an international conference (9). Since that time there have been many
applications of this technique in Europe for residue determination.

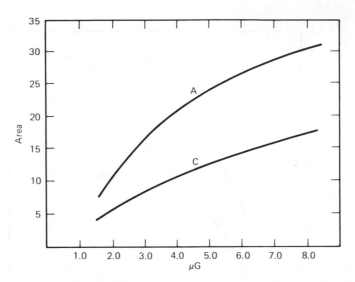

Figure 11. Curves resulting from scans of a Sudan Red and indo-phenol using a green filter. The yellow cannot be seen.

However, in the United States there has only been a minimal amount of application of this method.

Quantitative densitometry is a flexible tool and does have one significant advantage over other techniques; it is highly visual and if the developed colors are stable can be quantitated more than once. The whole path of migration does not have to be scanned if only one spot has to be scanned so that method can be very rapid.

Sherma et al. (10) gave shown that the 1 × 3 in. microslide plates can be as quantitative as the large 20 × 20 plates, which reduces development time from 30-60 minutes down to 3-10 minutes depending upon the developing solvent.

REFERENCES

1. W. A. Stellar and A. N. Curry, J. Offic. Agr. Chem. <u>47</u>, 645 (1964).
2. S. J. Purdy and E. V. Truter, Chem. Ind. 506 (March 1962).
3. J. C. Touchstone and J. Sherma, Eds., <u>Densitometry in Thin Layer Chromatography</u> (Wiley-Interscience, New York, 1979).

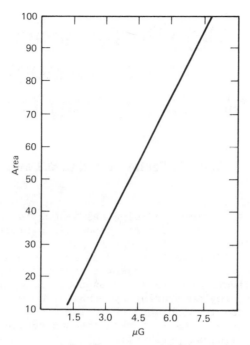

Figure 12. Scan of same plate with a blue filter. Now the Sudan red and Sudan yellow are seen.

4. J. W. Fairbairn, in Quantitative Paper and Thin Layer Chromatography, edited by E. J. Shellard (Academic Press, New York, 1968), pp. 1-15.
5. C. F. Emanuel, Anal. Chem. 45, 1568 (1973).
6. M. E. Getz, J. Offic. Anal. Chem. 54, 982 (1971).
7. M. Beroza, K. R. Hill, and K. H. Norris, Anal. Chem. 40, 1608 (1968).
8. P. Kubelka and F. Z. Munk, Techn. Physik. 12, 543 (1931).
9. M. E. Getz, in Methods in Residue Analysis, edited by A. S. Tahori (Gordon and Breach, New York, 1971), vol. iv, pp. 43-63.
10. J. Sherma, K. E. Klopping, and M. E. Getz, Amer. Lab. 11, 66 (1977).

CHAPTER 8

High Performance
Thin Layer Chromatography
for Monitoring Drug Therapy

David C. Fenimore and C. M. Davis

In recent years clinical medicine has become increasingly aware
of wide variations in patients' responses to treatment with certain drugs. This variability can be particularly troublesome
when the toxic side effects of a drug or a combination of drugs
appear at levels not greatly removed from those necessary for
therapeutic effect. For this reason there is a growing interest
in therapeutic drug monitoring by means of determining drug blood
levels. This can furnish more objective and accurate information
as to the status of a patient than the subjective observations of
the attending clinical personnel.

The quantitation of drugs in whole blood, plasma, or serum
poses a challenging analytical problem. Many of these compounds
are present in extremely small concentrations at therapeutic
levels, and they are often accompanied by metabolic products that
may resemble the parent compound but are devoid of biological
activity. Therefore not only must the analytical procedure be
capable of isolating the compound in question from endogenous
substances in the blood sample, but it must also assure separation from other drugs and drug metabolites, and this must often
be accomplished at concentrations as low as a few parts per billion.

If therapeutic drug monitoring is to have any practical consequences in clinical medicine, economic considerations also
become an important factor. Although many of the most difficult
determinations can be accomplished using highly sophisticated
instrumental methods, the time and expense involved in these procedures restrict such assays to rather limited research studies.
In order to benefit large clinical populations, the analytical
procedures must be performed routinely by other than highly
skilled specialists in instrumentation.

Thin layer chromatography offers many attractive features for the separation of drugs from biological substances, especially when the compounds in question are highly polar and thermally labile, as is often the case with therapeutic agents. Derivatization for the purpose of enhancing stability is usually not required, although certain assays may necessitate the formation of derivatives for the purpose of increasing sensitivity. Recently, however, improvements in instruments used in densitometry for in situ quantitation, together with the introduction of new, high-performance thin layer plates, have done much to eliminate even the need for derivatization to enhance sensitivity of detection. The combination of advances in technique and materials now permits limits of detectability rivaling and in many cases surpassing those associated with column chromatographic systems. The speed, resolution, and sensitivity that can be attained with these improvements appear to justify the term "high-performance thin layer chromatography" (HPTLC) by which this technique is now generally recognized (1).

The drug assays described in this report are a continuation of studies published previously (2-4) concerning therapeutic agents administered for the most part on a chronic basis to patients under treatment for psychiatric disorders. These compounds, which are comprised of the phenothiazine drugs, tricyclic antidepressants, and anticonvulsants are, nevertheless,fairly representative of drugs in general with respect to sample preparation, separation, and quantitation using HPTLC techniques.

PHENOTHIAZINE DRUGS

Numerous phenothiazine derivatives are currently in clinical use as major tranquilizers or antipsychotic agents as well as antiemetics, antipruritic, and antihistaminics. Depending on the nature of substituent groups both on the phenothiazine nucleus and the side chain attached to the heterocyclic nitrogen, the therapeutically effective blood levels for these drugs range from a few ng/ml to several hundred ng/ml (5-9). For the most part, however, therapeutic levels are in the lower portion of this range; therefore the sensitivity of the analytical procedure is an important factor if only small sample volumes are available for assay.

Blood plasma is a convenient sample medium for phenothiazine assays, and the initial extraction and preparation is summarized in Figure 1. Both an internal standard and a "carrier" are added to the blood plasma sample prior to the extraction procedure. These compounds are usually phenothiazines known to be absent in

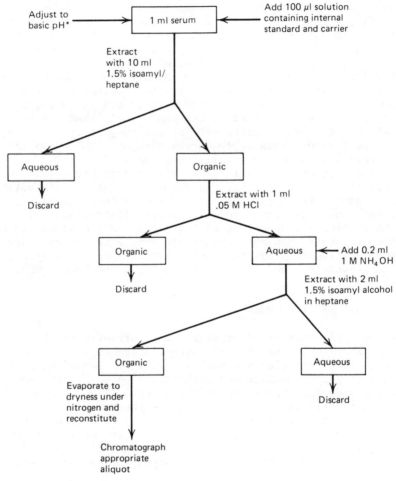

Figure 1. Diagram for extraction for HPTLC determination of phenothiazines and tricyclic antidepressant drugs.

the plasma sample as received. The internal standard serves the usual function of a correction factor for sample loss throughout the procedure; the carrier, which is added in amounts some 20 to 50 times that of the expected concentration of analyte, acts to improve recovery and reproducibility especially at low concentrations of the drug in question (10).

All of the solvents used in the procedure are distilled before use in all-glass systems and stored in glass bottles with poly(tetrafluoroethylene)-sleeved glass stoppers. In addition, all glassware employed in manipulation of samples and extracts is silylated by use of hexamethyldisilizane at elevated temperature and reduced pressure (11).

The chromatographic plates employed throughout these studies are HP-TLC Silica Gel 60 plates (E. Merck, Darmstadt, Germany) that are washed before use in absolute ethanol in the presence of ammonia vapor. Sample extracts, dissolved in an appropriate volume of solvent, are applied to the plates in 200 nl amounts using platinum-iridium pipettes (Camag, New Berlin, Wisconsin), and after spotting the plates are placed in a vacuum desiccator to assure complete removal of the spotting solvent.

The plates are developed in either twin-trough chambers (Camag, New Berlin, Wisconsin) or tanks that permit suspension of the plate above the developing solvent (Desaga, Heidelberg, Germany), either of which can assure thorough equilibration of the surface of the plate with the solvent vapors prior to development. In most instances the mobile phase solvent is allowed to migrate to 4 cm above the origin, and solvent is removed by gently heating the plate in a stream of warm air.

Densitometric measurements of the resolved components are performed with a Zeiss Model KM-3 chromatogram spectrophotometer (Zeiss Instruments, Bad Durkheim, Germany). The phenothiazine drugs display an absorption maximum at 250-260 nm with sufficient extinction to permit detection of less than 1 ng on the HPTLC plate.

Initial studies of the HPTLC assay of blood levels of chloropromazine indicated that ultraviolet absorption measurements furnish reliable values if densitometry is performed immediately following chromatographic development. Phenothiazines will, however, readily auto-oxidize to the respective phenothiazine sulfoxides, especially under the conditions present on the surface of a thin layer plate after the developing solvent has been removed. Figure 2 illustrates the time course of this auto-oxidation process as measured by the decrease in absorption at 254 nm of an isolated chlorpromazine spot and the concurrent increase in absorption at 240 nm, which is the absorption maximum for chlorpromazine sulfoxide. The effect of the auto-oxidation on the accuracy of the

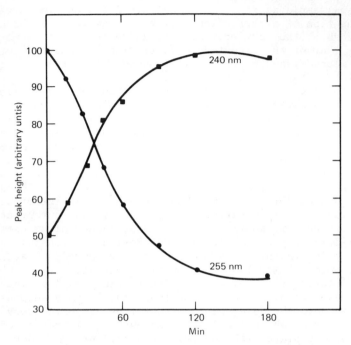

Figure 2.   Auto-oxidation of chlorpromazine (CPZ) to chlorpromazine sulfoxide (CPZ-SO) on a thin layer chromatographic plate as
measured by reflectance densitometry. Absorption maximum for
CPZ is 255 nm and for CPZ-SO in 240 nm.

determination can be minimized by choosing an absorption wavelength that remains relatively constant with time; in this particular case, a wavelength of 250 nm provided sufficient stability to permit reliable measurement.

An alternate approach to the problem of analyte lability is
through conversion by chemical reaction on the plate to form a
more stable product. Reagent sprays formulated from ferric
chloride, perchloric acid, and sulfuric acid or modifications of
this mixture have been employed for visualization of phenothiazines (12,13) but initial examinations of this method indicated
only minor improvement in stability compared to uv densitometry
of the unchanged drugs (4).

One impediment to the use of spray reagents in HPTLC assays
at high sensitivity is the increase in background noise attributed to inhomogeneous deposition of the spray reagent. Vapor

phase reactions avoid this problem to a large extent, and for this reason studies were conducted utilizing nitrogen oxides formed from the reaction of sodium nitrite with hydrochloric acid. This procedure has been applied previously for the visualization and densitometric determination of imipramine and desipramine (14).

After development of the HPTLC plate and removal of developing solvents, the plate is placed in a dry chromatography chamber containing a small beaker with 0.5-1 g $NaNO_2$ to which is added 5 ml 2 M HCl. The plate remains in the chamber for 1 hour at ambient temperature and is then placed in a vacuum desiccator for complete removal of reagent. The plate is then scanned at 380 nm using either reflectance or transmission mode.

An example of the separation and densitometric scan of a mixture of eight phenothiazine drugs is shown in Figure 3. Excellent sensitivity can be achieved with this procedure as illustrated by Figure 4 where chromatograms of 1 ng of chlorpromazine extracted from 1 ml of blood plasma yield peak signals well above background.

The phenothiazine derivatives formed under the above conditions are quite stable and significantly improve the precision of HPTLC assays. A regression of response on concentration of chlorpromazine in blood samples is shown in Figure 5.

TRICYCLIC ANTIDEPRESSANT DRUGS

The tricyclic antidepressant drugs amitriptyline, nortriptyline, imipramine, and desipramine are somewhat similar in chemical structure to the phenothiazines but with the sulfur atom absent from the three-ring nucleus; thus the preparation of plasma samples is essentially the same as that used for the phenothiazines. The initial extraction is carried out at a somewhat lower pH, which appears to improve recovery.

All five drugs can be well separated on a single chromatogram less than 4 cm solvent travel, with two successive developments using hexane-chloroform (70:30) in the presence of ammonia vapor provided by placing a small beaker of concentrated ammonium hydroxide in the developing chamber (Figure 6). Loxapine is employed as the internal standard, and acetophenazine, which remains at the origin, is used as the carrier compound.

The absorption maxima for imipramine and desipramine differ from those of amitriptyline and nortriptyline; therefore the former compounds are measured at 275 nm and the latter at 240 nm. Sensitivity of detection is of the same order of magnitude as observed for uv densitometry of the phenothiazines. All four

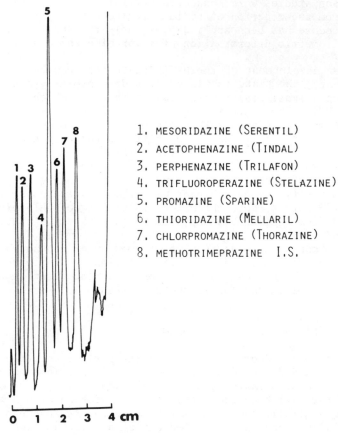

1. MESORIDAZINE (SERENTIL)
2. ACETOPHENAZINE (TINDAL)
3. PERPHENAZINE (TRILAFON)
4. TRIFLUOROPERAZINE (STELAZINE)
5. PROMAZINE (SPARINE)
6. THIORIDAZINE (MELLARIL)
7. CHLORPROMAZINE (THORAZINE)
8. METHOTRIMEPRAZINE I.S.

Figure 3. Separation of a mixture of eight phenothiazine drugs within a 3 cm distance on a HPTLC plate. Developing solvent: toluene-ethyl acetate-NH_4OH (39:20:0.5). 20 ng quantities visualized as the NO adduct and scanned at 380 nm wavelength.

drugs can be determined by uv reflectance densitometry with limits of detectability below 10 ng using wavelengths of 240 nm for amitriptyline and nortriptyline and 275 nm for imipramine and desipramine.

Selectivity in the presence of interfering substances from plasma extracts is enhanced using the color reaction reported by Nagy and Treiber (14) where the plate is exposed to nitrogen

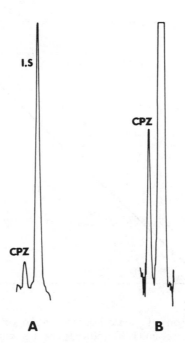

Figure 4. HPTLC chromatogram of 1.0 ng chlorpromazine extracted
from 1.0 ml blood plasma sample. Plate treated with nitrogen
oxide vapors and scanned at 380 nm wavelength. Scan B at 5 ×
amplification of scan A.

oxides generated from sodium nitrite in hydrochloric acid. This
reaction yields yellow addition products with imipramine and
desipramine that are measured at 385 nm (Figure 7). Amitriptyline and nortriptyline do not form colored products, and it would
appear that the presence of a nitrogen atom in the central ring,
as is the case with the phenothiazines (see above), is necessary
for the reaction to occur.

ANTICONVULSANT DRUGS

Drugs used currently for control of epilepsy and other seizure-
producing disorders include phenobarbital, phenytoin, primidone,
and carbamazepine. Therapeutic blood levels for all of these
drugs are normally above 1 µg/ml; consequently, sensitivity of

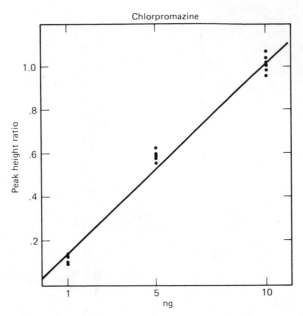

Figure 5. Regression of peak height ratio of chlorpromazine to methotrimeprazine internal standard versus concentration of chlorpromazine in blood plasma at low levels. Six replications of complete procedure at three concentrations.

detection is not a problem in designing assays for these compounds.

Macro thin layer chromatography is an effective technique in this area of drug analysis and, indeed, the first analyses of samples from patients receiving multiple drug therapy was performed by TLC (15). There have been recent contributions to this utilizing in situ densitometry that are illustrative of the applicability of TLC analyses to clinical populations (16,17).

Assay of anticonvulsants in blood samples by HPTLC evolved from these prior studies not so much because there was a need for increased sensitivity, although this may be a factor where sample volumes are limited, as is often the case with pediatric patients, but because development time is substantially less than that required by conventional TLC.

All four of the above-mentioned drugs may be assayed individually or in combination in blood plasma by the extraction procedure shown schematically in Figure 8. A single internal

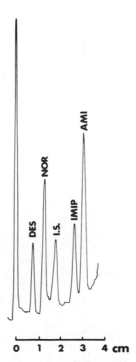

Figure 6. HPTLC scan of desipramine (DES), nortriptyline (NOR), imipramine (IMIP), and amitriptyline (AMI) extracted from blood plasma, containing 60 ng/ml of each drug. Developed twice in hexane-chloroform (70:30) in a NH_4OH atmosphere and scanned at 240 nm. Internal standard (I.S.) is loxapine.

standard, p-tolybarbital (Aldrich Chemical Co., Milwaukee, Wisconsin), is added to the blood plasma, but no carrier compound is required because of the relatively large blood levels of the analytes at therapeutic concentrations. The final volume of the extract is brought to 50 µl, of which only 200 nl is spotted on the plate, indicating that with appropriate microvolume apparatus it should be possible to utilize only a few microliters of original sample with a proportionately large aliquot to achieve satisfactory quantitation.

All four drugs are not separated and determined with a single development of the plate; however, each development time is sufficiently short, about 5 minutes, that multiple development is not a detraction. The wavelengths for densitometric measurements

Figure 7.   HPTLC scan of imipramine (IMIP) and desipramine (DES)
from blood plasma sample containing 50 ng/ml. Scanned at 385 nm
after exposure to nitrogen oxides.   Internal standard (I.S.) is
mepazine.

also differ, as would be expected upon inspection of the in situ
absorption spectra of the drugs (Figure 9).   The procedure therefore begins with isolation of carbamazepine using chloroform as
the developing solvent with a densitometric determination at 285
nm wavelength (Figure 10).   Two developments in chloroform-
isopropanol-ammonium hydroxide (80:19:1.6) then follow, which

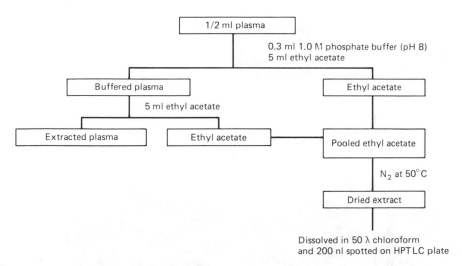

Figure 8.   Extraction diagram for anticonvulsant drugs from blood plasma.

moves the carbamazepine toward the solvent front and effectively separates primidone, phenytoin, and phenobarbital.   The primidone is determined at 215 nm (Figure 11) and the remaining two drugs and internal standard at 230 nm.   Regression curves of known amounts of drugs extracted from plasma are shown in Figures 12 and 13.

   A clinical evaluation of this method, the results of which will be published elsewhere, indicates that the accuracy and precision of the HPTLC method is comparable to that attained with gas chromatographic procedures.   The analysis time, however, is markedly shorter, which should favor this or similar HPTLC determinations in clinical use.

   DISCUSSION

Although numerous analytical procedures have been applied successfully to the determination of drugs in samples of biological origin, most of the methods, particularly those relying on column chromatographic separations, are impractical in situations where a wide variety of compounds must be assayed in a minimum period of time.   Most analysts are aware that the continual changing of chromatographic columns and operating parameters is not conducive

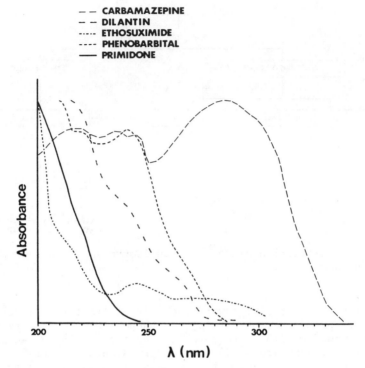

Figure 9. In situ uv spectra of anticonvulsant drugs on HPTLC plate.

to maintaining optimal conditions for separation and quantitative reliability. Thin layer chromatography, on the other hand, usually requires no more than a change of solvent in a developing chamber to permit isolation and subsequent measurement of nearly any drug used in clinical medicine, provided, of course, that the requisite sensitivity can be attained.

Although HPTLC does not represent a radical departure from accepted and proved TLC practices, the improvement in overall performance with respect to sensitivity and analysis time does place HPTLC in a very advantageous position for the types of assays encountered in therapeutic drug monitoring.

The assays described in this chapter must be considered as tentative and subject to considerable improvement with use, if for no other reason than the method and its associated techniques

Figure 10. HPTLC separation of carbamazepine (Ca) from plasma
scanned at 280 nm. Plate developed in chloroform.

which are relatively new and still undergoing continual revision.
The degree of initial success with which HPTLC has been applied
to the present determinations, however, would seem to indicate a
favorable future for this versatile analytical method.

REFERENCES

1. A. Zlatkis and R. E. Kaiser, Editors, HPTLC High-Performance
   Thin-Layer Chromatography (Elsevier, Amsterdam, 1977).
2. D. C. Fenimore, C. J. Meyer, C. M. Davis, F. Hsu, and A.
   Zlatkis, J. Chromatog. 142, 399 (1977).
3. D. C. Fenimore and C. M. Davis, J. High Resol. Chromatog.
   Chromatog. Commun., p. 105 (1978).
4. D. C. Fenimore, C. M. Davis, and C. J. Meyer, Clin. Chem. 24,
   1386 (1978).

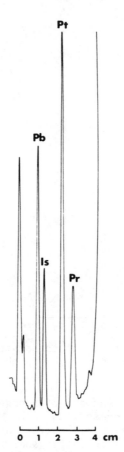

Figure 11. HPTLC separation of phenobarbital (Pb), phenytoin
(Pt), and primidone (Pr) from plasma extract. Developed in
chloroform-isopropanol-NH₄OH (80:19:1.6) in a NH₄OH atmosphere
twice after initial isolation of carbamazepine (Figure 10), which
is now moved into the solvent front. Internal Standard (I.S.) in
p-tolylbarbital.

5. Curry, S. H., in The Phenothiazines and Structurally Related
   Drugs, edited by I. S. Forrest, C. J. Carr, and E. Usdin
   (Raven Press, New York, 1974), p. 335.
6. M. Franklin, D. H. Wiles, and D. J. Harvey, Clin. Chem. 24, 41
   (1978).

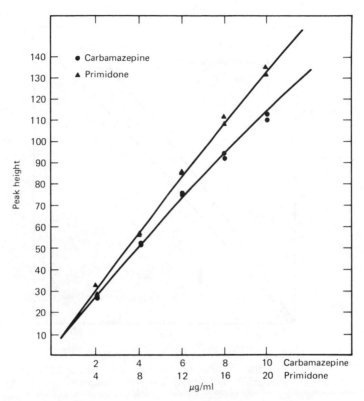

Figure 12.   Regression of carbamazepine and primidone peak
heights versus concentration in blood plasma.

7. F. A. J. Vanderheeren and R. G. Muusze, Eur. J. Clin. Pharmacol. 11, 135 (1977).
8. E. Martensson and B.-E. Ross, Eur. J. Clin. Pharmacol. 6, 8
   (1973).
9. E. Martensson, N. Gosta, and R. Axelsson, Curr. Ther. Res.
   18, 687 (1975).
10. C. M. Davis, C. J. Meyer, and D. C. Fenimore, Clin. Chim.
    Acta 78, 71 (1977).
11. D. C. Fenimore, C. M. Davis, J. H. Whitford, and C. A. Har-
    rington, Anal. Chem. 48, 2289 (1976).
12. I. S. Forrest and E. Usdin, in Psychotherapeutic Drugs,
    edited by E. Usdin and I. S. Forrest (Marcel Dekker, Inc.,
    New York, 1977), p. 699.

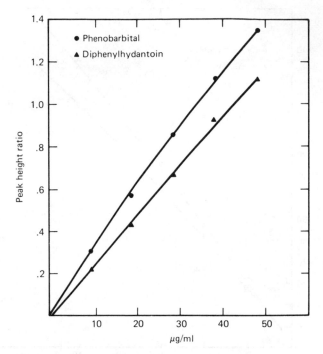

Figure 13.   Regressions of phenobarbital and phenytoin peak
height ratios to internal standard versus concentrations in blood
plasma.

13. T.-L. Chan and S. Gershon, in Quantitative Thin-Layer Chroma-
    tography, edited by J. C. Touchstone (John Wiley & Sons, New
    York, 1973), p. 253.
14. A. Nagy and L. Treiber, J. Pharm. Pharmacol. 25, 599 (1973).
15. H. J. Kupferberg, in Antiepileptic Drugs: Quantitative
    Analysis and Interpretation, edited by C. E. Pippenger and
    H. Kutt (Raven Press, New York, 1978), p. 1.
16. N. Wad and H. Rosenmund, J. Chromatog. 146, 167 (1978).
17. N. Wad, E. Hanifl, and H. Rosenmund, J. Chromatog. 143, 89
    (1977).

CHAPTER 9

Microtechnique Thin Layer Chromatography

George A. Maylin

Thin layer chromatography (TLC) is one of the more commonly used separation techniques in analytical chemistry. Major reasons for its widespread use are the following: it is relatively simple and inexpensive; a wide variety of chemicals can be analyzed; it is suitable for submicrogram analysis as well as preparative separations; separated components can be collected as pure compounds for additional analysis; and a large number of samples can be analyzed simultaneously. Recent developments including high-performance thin layer chromatography (HPTLC), circular thin layer chromatography (CTLC), and densitometers have greatly improved the sensitivity and quantitative aspects of conventional TLC.

This paper describes the use of an additional method, micro-technique thin layer chromatography (MTTLC), to improve the efficiency of conventional TLC. Microtechnique thin layer chromatography (MTTLC) is the application of conventional TLC techniques on a reduced miniature scale. Samples are applied as 2-mm diameter spots and are developed for 5 cm or less in unsaturated tanks.

Sample cleanup of a complex mixture by means of selective extraction is a necessary preliminary step for TLC, just as it is with any chromatographic method. Selective extraction will remove the desired compound or class of compounds from the interfering matrix. In many cases, the better extraction is one that provides a "cleaner extract," as compared to a "richer but dirtier extract." For example, an extraction with petroleum ether that removes only 50% of the compound may be more useful than an extraction with chloroform that removes 80% of the desired compound. In all cases, the optimum extraction procedure must be developed. Care must be taken to prevent loss of material by

oxidation, by photolysis, by volatization, or by irreversible
adsorption to active sites on glass. Concentration should be
performed under vacuum or by means of a gentle stream of nitrogen
at 60° C in subdued light. Irreversible adsorption to glass can
be prevented by silanizing glass or using plastic containers.
Many reactive compounds can be recovered efficiently by adding
trace amounts of alcohol to the solvent.

Because diffusion is a major factor in limiting the sensitivity and resolution in TLC (1), it is imperative to minimize spot
size and development time or distance. This is known from theory
and experimentally confirmed. Resolution is enhanced by the
selection of a solvent system that will provide good chromatographic characteristics for the components in question. As is
the case with selective extraction, trial and error is often
required to separate the compounds of interest from interfering
background material.

Visualization of components is also an important aspect
of MTTLC. Because there is no satisfactory universal detection
system, quenching, native fluorescence, induced fluorescence,
fluorigenic labeling, and chromatogenic reagents are used. By
using these techniques in combination and in series, it is possible to detect 1-20 ng of compound in a characteristic manner on a
TLC plate.

Microtechnique thin layer chromatography provides approximately a tenfold increase in sensitivity over many conventional
TLC procedures. Judicious use of selective extraction and sample
concentration, selection of chromatographic conditions that minimize diffusion and maximize resolution, and visualization
reagents are required.

The usefulness of MTTLC is shown in Table I and Figure 1.
Following the intramuscular administration of radiolabeled meperidine to horses in dosage ranges from 0.05 to 0.2 mg/kg body
weight, the drug and metabolites were isolated from urine using
liquid-liquid extraction techniques. Conjugated metabolites were
cleared by enzyme hydrolysis. All compounds were extracted from
alkaline solutions into chloroform. The metabolites were
detected and quantitated using a combination of ^{14}C-radiolabeled
drug, MTTLC, gas chromatography, and mass spectrometry. Structures for the drug and metabolites isolated in several species
are shown in Figure 1. Despite the similarity of structures, a
conventional TLC solvent system separated the metabolites effectively, as shown in Table I. Concentrations of meperidine N-
oxide and p-hydroxymeperidine ranged from 10 ng/ml to approximately 200 ng/ml. These concentrations were readily detected by
visualization with Dragendorff reagent followed by 25% aqueous
cupric chloride and 5% aqueous sodium nitrite oversprays. Due to

1.

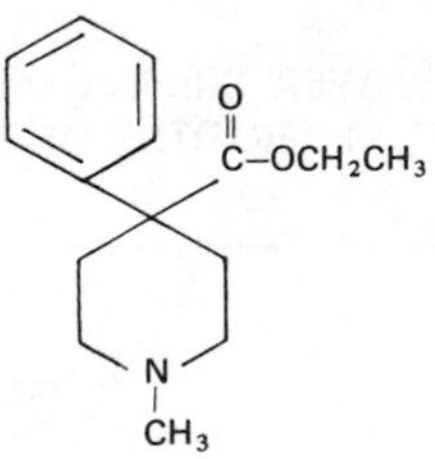

2.

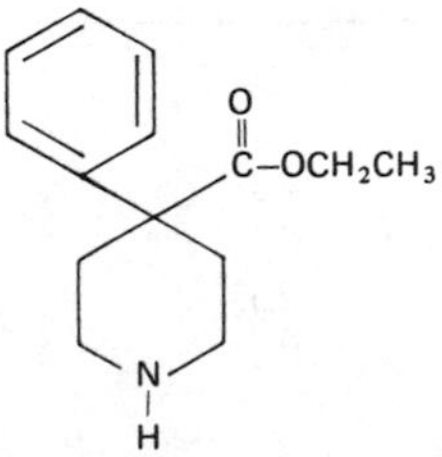

3.

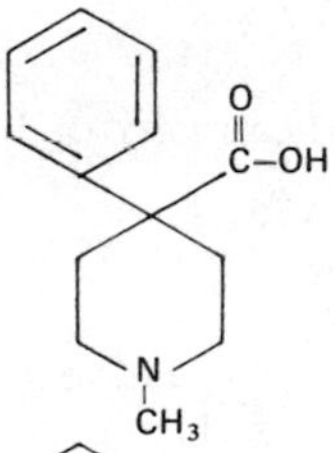

4.

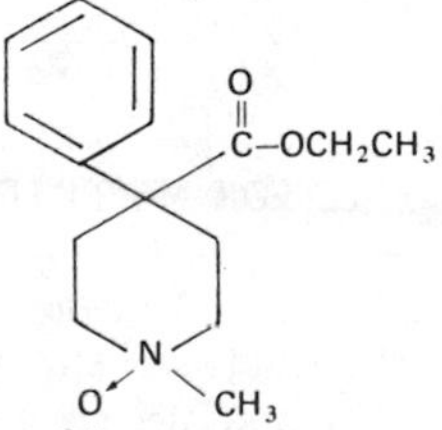

5.

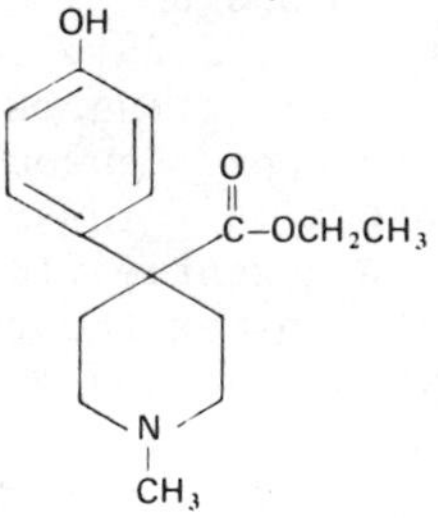

Figure 1. Structures of meperidine and related compounds.
1. Meperidine; 2. Normeperidine; 3. Meperidinic acid; 4. Meperidine N-oxide; 5. p-Hydroxymeperidine.

TABLE I.   THIN LAYER CHROMATOGRAPHIC CHARACTERISTICS
OF MEPERIDINE AND RELATED COMPOUNDS

Compounds	R_f[a]	% Radioactivity[b]	
		Unconjugated	Conjugated
Meperidine	.60	0.1	0
Meperidine N-oxide	.40	90	0
p-Hydroxymeperidine	.65	0.1	9.9
Normeperidine	.20	0	0
Meperidinic acid	.05	0	0

[a] Solvent system--methanol/ammonium hydroxide: 98.5/1.5.

[b] 8-hour pooled urine sample.

the thermal lability of the N-oxide, it could only be identified
by mass spectrometry using solid probe insertion after separation
and isolation by MTTLC.

ANALYTICAL PROCEDURE FOR MEPERIDINE AND METABOLITES

Add 9.0 ml of urine to a 16 × 125 mm screw top test tube fitted
with a Teflon lined cap. Adjust the pH to 9.5 - 10.0 with
approximately 2 ml of 1 M sodium carbonate. Add 4.0 ml of
dichloromethane-2-propanol (75:25). Cap the tube and mix on a
rotorack for 5 minutes. Centrifuge to separate the layers.
Aspirate and discard the upper aqueous phase. Transfer 3.0 ml of
the organic extract into a 13 × 100 mm test tube and concentrate
to dryness at 65° C with a gentle flow of N_2. Add 50 μl of ethyl
acetate to the residue. Vortex the tube to wash down the sides.
Spot 10 μl of the ethyl acetate on a conventional, silica gel G,
TLC plate. With the use of warm air from a hair dryer, apply the
sample as a small spot, less than 2 mm in diameter. Use of
1.0-mm-i.d. glass capillary spotters drawn to a very fine tip
facilitates spotting. Develop the plate 4 cm in methanol-
ammonium hydroxide (98.5:1.5). Visualize the plate by spraying
with Dragendorf reagent followed by 25% aqueous cupric chloride

and 5% aqueous sodium nitrite.   Maximum sensitivity requires three-spray sequence.

REFERENCE

1. A. Zlatkis and R. E. Kaiser, Eds., Thin-Layer Chromatography, Journal of Chromatography Library, vol. 9 (Elsevier Scientific Company, New York, 1977).

CHAPTER 10

Purification of
Beef Extract Mutagen

**Suba Nair, Joyce J. Henry,
and Gregory Cuca**

The application of short-term tests for the detection of chemical
mutagens/carcinogens has considerably enhanced our efforts to
study the incidence and role of such substances in environmental
carcinogenesis. These new methodological developments are based
on the mutagenic effects of such substances and reflect the
theory that malignancy is triggered by one or more somatic muta-
tional events. Although the theoretical link between carcinogenicity and mutagenicity is still the subject of considerable
debate, it is well established that there is a significant correlation between mutagenicity of a substance in these test systems
and its potential carcinogenicity. An important realization
resulting from these studies is that many potential carcinogens
are neither carcinogenically nor mutagenically active in themselves, but need to be activated or converted by metabolism into
their reactive forms as ultimate carcinogens. In most cases such
activation is by microsomal mixed-function oxidases. This new
information has made an important contribution to the development
of in vitro tests where the mammalian enzymes can be incorporated
into the test system and the effect of the products of these enzymatic conversions can be more easily analyzed.
   Systems have now been developed in which suitable strains of
microorganisms are exposed to presumptive carcinogens both in the
presence and absence of microsomal enzyme systems. These methods
are rapid and inexpensive compared to animal bioassays and can be
carried out in the large numbers that are needed to analyze complex environmental samples. Many such short-term assays based
mainly on point mutations in the test organisms have been
described in recent years. Of these the Salmonella-microsome
assay developed by Ames (1) is widely used for screening chemicals and is the most reliable, with close to 90% correlation

between mutagenicity and carcinogenicity (2,3). Combined with
conventional physicochemical fractionation procedures, this
screening test could be used for isolation of carcinogenic substances. When suitably calibrated, it is possible to estimate
their concentration in samples that are mixtures of unknown compounds derived from the environment.

The large number of chemicals that are known to induce malignancy in experimental animals also include possible cancer-
causing or promoting agents found in our diet. Some of these,
like safrole and cycasin, are natural constituents of plants,
whereas others, like aflatoxins, are found in foods contaminated
with fungi (4). Nitrosamines, which are carcinogenic in 20 animal species, may occur in foods that have been preserved by addition of nitrite (6). The occurrence of the well-known carcinogen
benzo(a)pyrene in charcoal-broiled meat has also been clearly
established (6). Malonaldehyde, which may arise from decomposition of peroxidized fatty acids in meat, is mutagenic in the Ames
system and tumorigenic in mice (7). More recently, a new class
of mutagens originating from the action of heat on some constituents of meat have been reported (8,9).

The Salmonella-microsome assay employed in determining the
mutagenicity of these compounds uses strains of Salmonella
typhimurium with a histidine-negative genome, incapable of endogenous synthesis of this essential growth requirement. These
strains are reverted to a histidine-positive genome by the action
of the mutagens on the histidine operon. The frequency of this
reverse mutation is easily measured by counting the colonies that
appear in a medium lacking histidine. The system consists of an
agar-based synthetic medium, bacterial inoculum, and a microsomal
preparation necessary to convert some of the test compounds to
their active form. If a substance capable of inducing mutation
is added to such a plate, the spontaneous rate of mutation is
considerably enhanced.

A number of laboratories (10,11) have consistently observed
that the presence of microsomes in the medium produced an
increase in the background rate of mutation in TA1538, one of the
bacterial strains used, even in the absence of mutagenic agents.
For example, the average values for 200 plates were 13 revertant
colonies per plate when only the bacterial inoculum was present
with 22 colonies per plate in the presence of microsomes. This
phenomenon was present only in strains that are sensitive to
frameshift mutagens. The effect did not interfere with the reliability of the method, because mutagens usually produced significantly higher numbers of mutant colonies. Nevertheless, when the
original method was modified by growing the strain in nutrient
broth in the presence of microsomes and plating an aliquot on

histidine-free medium, the number of revertant colonies was
increased by an order of magnitude, making further analysis feasible. Subsequently the putative substance was extracted from
nutrient broth and shown to be a constituent of beef extract
found in many bacterial nutrients. The result of further studies
using thin layer chromatography in the isolation of this mutagenic constituent in beef extract is described in this chapter.
The feasibility of this methodology in tracing the source of such
substances and studying the nature of their origin is discussed.

MATERIALS AND METHODS

  Extraction

"Bacto" nutrient broth and beef extract used in these studies
were made by the Difco Company. Sample preparations were mainly
by extraction with organic solvents from aqueous solutions of
nutrient broth, beef extract, beef broth, and cooked beef. Preliminary extractions from nutrient broth were as follows: solutions containing 10% dry nutrient broth in 500 ml of distilled
water adjusted to pH 7 were extracted three times with a benzene-
2-propanol (80:20) mixture. Later, for extractions of beef
extract, beef broth, and cooked beef, this procedure was modified
as follows.
  A 500 g sample was dissolved in 1000 ml of water and acidified with hydrochloric acid to pH 2.0. The solution was saturated with ammonium sulfate to precipitate protein. The supernatant was filtered through glass wool and extracted with methylene chloride. The aqueous phase was adjusted to pH 10 with
ammonium hydroxide and extracted three times with equal amounts
of methylene chloride. The pooled extract (from pH 10) was dried
with anhydrous sodium sulfate and evaporated on a rotary evaporator at 37° C. The residue was taken up in chloroform-methanol
(90:10) and used for further analyses. For mutagenesis, desired
amounts of chloroform-methanol samples were evaporated and resuspended in dimethyl sulfoxide.
  Lean ground beef patties (100 g, dry weight portions) were
cooked in an electrically heated hamburger-cooking appliance
(plate temperature 200° C), homogenized in twice their volume of
water, filtered, and extracted. Beef broth was prepared by heating 500 g of ground meat in 1000 ml of water for 20 minutes at
40° C and then boiling it for 10 minutes and filtering to remove
fat and solids.

MUTAGENESIS

Mutagenesis assays were carried out according to the general procedures reported by Ames (12). The tester strain used in these
experiments was TA1538. Strains TA1535, TA1537, TA100, and TA98
were also used on occasion to test the effect of mutagens on
these strains. All platings were done in duplicates and average
number of colonies recorded. Three controls are done concurrently with each experiment: (1) bacteria alone to determine the
spontaneous reversion rate; (2) bacteria in the presence of microsomes; and (3) bacteria and microsomes along with a mutagen that
is known to be activated by microsomal enzymes.

THIN LAYER CHROMATOGRAPHY

Preliminary purification of samples of methylene chloride
extracts was done on a "Sep-Pak" silica gel column (Waters Co.)
using a stepwise gradient elution with hexane-acetone. The mutagenic fraction was subjected to repeated fractionation in different solvent systems (see discussion). Mutagenesis assay was used
at each stage as a means of following the process. Thin layer
chromatography was performed on Gelman silica gel impregnated
glass fiber paper. The developed chromatograms were cut into
successive 1-cm zones, extracted with chloroform-methanol (90:10),
and an aliquot tested for mutagenicity. The number of mutant
colonies produced by each zone describes the chromatographic distribution of the mutagenic material. The chromatogram zone showing the highest mutagenic activity is referred to as the "peak."

RESULTS AND DISCUSSION

Several studies were undertaken to determine the basis for the
apparent mutagenic effect of microsomes on strain TA1538. These
studies showed that mutagenic effects reside in components that
can be extracted by organic solvents from Bacto nutrient broth
and that this substance is converted to its active metabolite by
the microsomal enzyme system (11). The dose-response curve
(Figure 1) confirms these data. It shows that increased amounts
of the extract give a linear response with regard to the induction of mutation in TA 1538 only in the presence of microsomes.
A survey of a number of Difco bacterial nutrients suggested that
the number of revertant colonies produced is roughly proportional
to the beef extract content of the nutrient broth. When beef
extract (Difco) was subjected to the same extraction procedures

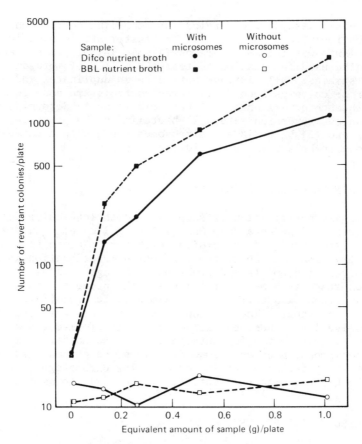

Figure 1. Dose-response curve of benzene-2-propanol extracts of nutrient broth on <u>S</u>. <u>typhimurium</u> TA1538.

as nutrient broth, much higher amounts of mutagenic material could be obtained. Figure 2 shows the dose-response of this sample on all five strains. The mutagenic agent is equally active on strains TA1538 and TA98, about one-fourth as active on TA1537 (TA98 is derived from the original TA1538), but inactive on TA1538 and TA100, strains in which base-pair substitutions are known to occur.

Beef extract is normally prepared by boiling down beef broth to 20% or less of its original volume. It is a common ingredient

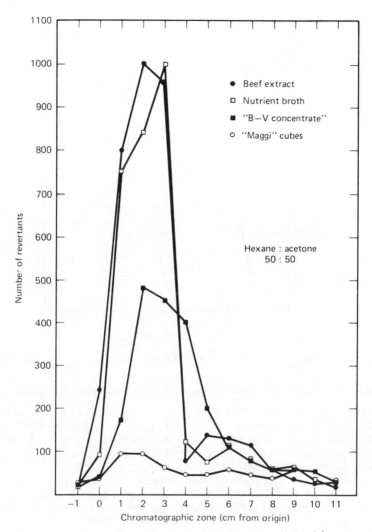

Figure 2. Dose-response curves of methylene chloride extracts of Bacto beef extract on strains of S. typhimurium.

of bacterial nutrients and is commercially available. It is also present in a number of food preparations, such as bouillon cubes and gravy bases. Beef broth itself, when tested, does not show

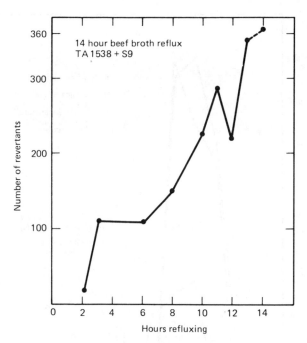

Figure 3.   Effect of time of heating of beef on revertant formation.

any detectable amounts of mutagen, whereas beef extract prepared
from it by boiling does. Accordingly, the conversion process has
been studied by testing beef broth for mutagenicity at 30-minute
intervals during boiling. The results show (Figure 3) that mutagens are formed during the heating-evaporation that occurs in the
conversion of beef broth to beef extract. A similar dose-
response was obtained with extracts of cooked ground beef (Figure
4). Although uncooked samples were inactive, all the cooked samples showed substantial levels of mutagenic activity that
increased with cooking time. All the samples were inactive in
the absence of microsomes. The activity was limited to the outer
layers of the beef patty, indicating that the critical factors
determining the formation of mutagens are temperature and time of
cooking.
    It was found during preliminary studies that the mutagenic
material extracted in such a manner could be localized on a thin

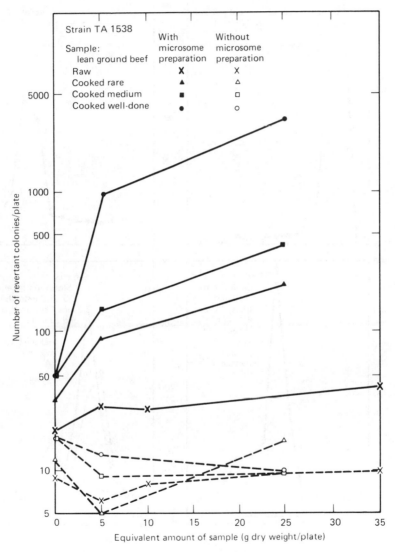

Figure 4. Dose-response curves of methylene chloride extracts of uncooked and cooked ground beef on strain TA1538.

143

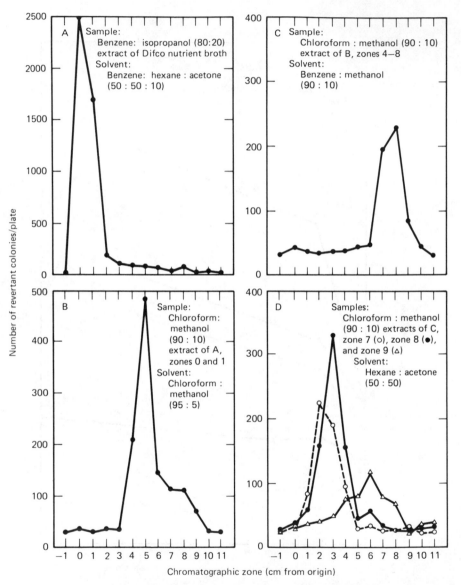

Figure 5. Sequential thin layer chromatographic fractionation of mutagenic material from Difco nutrient broth.

layer chromatogram. Based on this approach, a chromatographic
procedure was devised (Figure 5) to fractionate the mutagenic
materials in the nutrient broth. The results show that when an
initial benzene-2-propanol extract is chromatographed with
benzene-hexane-acetone (50:50:10) as solvent, all of the mutagenic material is retained at the origin with the nonpolar impurities moving toward the solvent front. This zone is extracted
and rechromatographed using chloroform-methanol (95:5) (Figure
5B). Intense mutagenic activity was centered in the middle (R_f
0.5). When the active zone was rechromatographed in benzene-
methanol (90:10), a single peak centered at R_f = 0.8 was obtained
(Figure 5C). Finally, when this zone was re-extracted and chromatographed in hexane-acetone (50:50), mutagens could be resolved
into two components--a major peak with R_f = 0.3 and a minor peak
with R_f = 0.6 (Figure 5D).

A series of similar chromatographic analyses of Difco beef
extract and Difco nutrient broth were carried out with hexane:
acetone and with benzene-methanol as solvents. The results
showed that the chromatographic mobility of the mutagen from both
beef extract and nutrient broth were similar in benzene-methanol
(95:5). R_f values for both mutagens were the same in the hexane:
acetone system also. Thus the mutagen present in the bacterial
nutrient is the same as that found in the beef extract.

Two commercial preparations (Maggi Beef Bouillon Cubes and
BV Broth and Sauce Concentrate) that, according to their labels
contain beef extract, have been tested using methods comparable
to those described for beef extract. From dose-response curves
against various Salmonella strains and from chromatographic
analysis, it is evident that these preparations contain mutagens
with the same characteristics as those found in beef extracts.

These data at first suggested a possible relation between
our observations and earlier evidence that benzo(a)pyrene is
formed in meat when cooked on an open flame. Nagao et al. (8)
have reported that when certain proteins and amino acids are
pyrolyzed at temperatures of 300° C to 600° C mutagens that are
active in the Salmonella test system were formed. A comparative
analysis of these mutagens was undertaken using methylene chloride extracts of Difco beef extract, cooked beef, beef with
benzo(a)pyrene added after cooking, and a mixture of pyrolyzed
amino acids and histone. These materials were chromatographed in
suitable solvent systems under identical conditions. The number
of revertant colonies obtained from chromatographic zones represents the amount of mutagens present. Figure 6 shows the identical behavior of mutagens from Difco beef extract and cooked beef,
both giving major peaks at R_f 0.5 in a benzene-methanol (95:5)
solvent system. Figure 7(a) represents the results of a similar

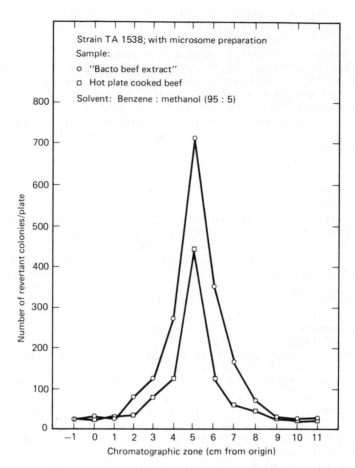

Figure 6. Thin layer chromatographic fractionation of methylene chloride extracts of Bacto beef extract and cooked ground beef.

experiment with methylene chloride extracts of cooked beef, cooked with benzo(a)pyrene (25 μg/kg meat, wet weight) and beef extract chromatographed in hexane. As expected, all of the mutagenic activity associated with beef extract and cooked beef remained at the origin, whereas the sample to which benzo(a)pyrene was added exhibited an additional peak at R_f = 0.85. Figure 7(b) represents a similar comparative chromatographic analysis (using hexane-acetone [50:50] as the solvent) of methylene

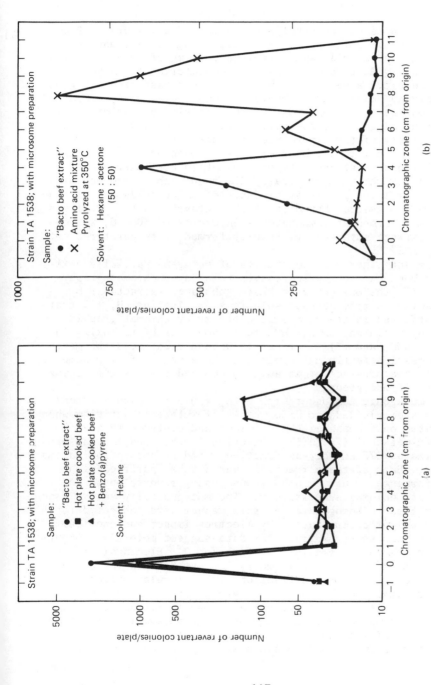

Figure 7. Comparative analysis of chromatographic behavior of methylene chloride extracts. (a) Bacto beef extract, cooked beef and benzo(a)pyrene. (b) Bacto beef extract, pyrolysis products of beef, amino acids, and calf thymus histone.

147

chloride extracts of the pyrolysis product of a mixture of 9 mg
of each of the 18 amino acids that, according to Matsumoto et al.,
yield mutagenic material when pyrolyzed, 30 mg of pyrolyzed histone, and 0.2 g Bacto beef extract. The material from beef
extract exhibited a peak at R_f = 0.3. In contrast, the mutagens
from pyrolyzed amino acids exhibited a major peak at R_f = 0.8 and
a minor peak at R_f = 0.6. Most of the activity of the material
from histone migrated to the solvent front. There was no activity from either of these pyrolyzed products at the region in
which the peak due to the beef extract mutagen occurred. These
results indicate that the mutagen present in beef extract and
cooked beef is chromatographically distinguishable from benzo(a)-
pyrene and from the pyrolysis products of amino acids and histone.
It must be emphasized that most of the mutagens in the pyrolysis
products are formed at temperatures in excess of 300° C, whereas
the beef extract mutagens are readily formed at temperatures not
exceeding 105° C.

      These and other related studies of mutagens partially puri-
fied from beef extract by successive thin layer chromatography
show that (1) the mutagen is a basic substance extractable by
organic solvents from aqueous solutions at alkaline pH, (2) that
it is unaffected in its mutagenic activity or chromatographic
behavior by refluxing in 6 N HCl for 6 hours or by refluxing in
methanolic KOH, and (3) on treatment with nitrous acid the material becomes inherently mutagenic (in the absence of microsomes),
suggesting the presence of an amino group and the possible formation of a nitroso group.

      The mutagenic component with an R_f = 0.3 in hexane-acetone
solvent system yielded two major peaks of mutagenic activity when
fractionated on a Sephadex LH-20 column and eluted with hexane-
chloroform-methanol (75:20:5). Fraction 1 exhibited a uv absorption maxima at 275 nm whereas fraction 2 had an absorption maxima
at 264 nm. Each of these fractions was further purified with
high-performance liquid chromatography using reverse phase and
normal phase columns successively. The mutagenically active fractions purified to homogeneity in this manner were subjected to
molecular weight determination by electron impact and chemical
ionization mass spectrometry. The data suggest molecular weights
of 228 for fraction 1 and 198 for fraction 2. Elemental analysis
by high-resolution mass spectrometry has established the chemical
formula as $C_{11}H_{10}N_4$ for the fraction with molecular weight of 198.
Studies to determine the functional groupings and molecular
structure of this fraction by nuclear magnetic resonance and
infrared spectroscopy are in progress. The mutagenic activity of
the pure form of fraction 2 is extremely high compared to other
mutagens tested in the Salmonella system, yielding approximately

50,000 revertants/µg--the highest reported so far in this system.
On the basis of this estimate the mutagen has a concentration of
200 ng/g in beef extract and   cooked beef contains 1 ng mutagen/
g (wet weight).

  These studies show that this technique can be a very useful
means of monitoring the role of cooking practices on the formation of mutagens. Given the observed correlation between mutagenicity in the Ames system and carcinogenicity in test animals,
these substances must be regarded as likely to act as carcinogens
when tested in laboratory animals.

  Employing the methodology reported here--bioassay supplemented with chromatographic analysis--it was possible to detect a
number of mutagenic constituents in air particulate samples, two
of which were identified as benzo(a)pyrene and benzo(e)pyrene
(13). Mutagens were detected also in the aqueous effluents of
petrochemical plants in the Houston Ship Channel. Work presently
in progress indicates that this methodology can be used successfully to fractionate the mutagens observed in the urine of smokers. In the light of these findings, it is suggested that these
procedures can form a basis for further analysis of the role of
environmental mutagens and their relation to the incidence of
cancer.

REFERENCES

1. B. N. Ames, W. E. Durston, E. Yamasaki, and L. D. Lee, Proc.
   Nat. Acad. Sci. USA 70, 2281 (1973).
2. B. Commoner, "Reliability of Bacterial Mutagenesis Techniques
   to Distinguish Carcinogenic and Noncarcinogenic Chemicals."
   Final Report to U.S. Environmental Protection Agency. EPA
   600 (1-76-022, Government Printing Office, Washington, D. C.,
   April 1976).
3. J. McCann, E. Choi, E. Yamasaki, B. N. Ames, Proc. Nat. Acad.
   Sci. 72, 5135 (1975).
4. L. Stoloff, in Mycotoxins (Pathotox, Park Forest, Soulk, Ill.,
   1977), p. 701.
5. R. Elespuru and W. Lijinsky, Food Cosmotology 11, 807 (1973).
6. T. Thorsteinsson and G. Thodarson, Cancer 21, 390 (1968).
7. R. J. Shanberger, T. L. Andreone, and C. E. Willis, J. Nat.
   Cancer Inst. 53, 1771 (1974).
8. N. Nagao, M. Honda, Y. Seino, T. Yahagi, and T. Sugimura,
   Cancer Lett. 2, 221 (1977; and T. Sugimura, in Origins of Human
   Cancer,(Cold Spring Harbor, New York, 1977).
9. B. Commoner, A. J. Vithayathil, P. Dolora, S. Nair, P. Madyatha,
   and G. C. Cuca, Science 201,913 (1978).

10. D. W. Nebert and J. S. Felton, "Structure, Function and Interaction," in Cytochromes (Plenum, New York), pp. 127-149.
11. A. J. Vithayathil, B. Commoner, S. Nair, and P. Madyastha, J. Tox. Env. Health 4, 189 (1978).
12. B. N. Ames, J. McCann, and E. Yamasaki, Mut. Res. 31, 347 (1975).
13. B. Commoner, P. Madyastha, A. Bronsdon, and A. J. Vithayathil, J. Tox. Env. Health 4, 59 (1978).

CHAPTER 11

Quantitation of Boron at ppb-Levels

Joseph C. Touchstone, M. F. Dobbins,
M. L. Mallinger, and J. Strauss

Boron is an important element in agriculture and industry. It is a metallurgical additive, a component of fertilizers, and is present in cleansing agents. More significant is its importance in plant metabolism. The range of deficiency and toxicity in the ecology is very narrow.

The quantitative impact of boron in these many facets has not been completely evaluated due to lack of suitable analytical methodology. A number of quantitative methods have been reported, but these are either difficult to perform or require expensive equipment. Gladney et al. (1) recently reported the use of thermal neutron-prompt gamma ray spectrometry for determination of boron. This method permits determination directly in the sample. Several colorimetric methods have been described using curcumin (2,3), carminic acid (4), methylene blue (5), and azomethine H (6). Some of these complexing reagents were used for extraction of the boron into organic solvents for flame atomic absorption measurements (7). Direct atomic absorption (7) has also been utilized but, compared to most other elements, the sensitivity of atomic absorbtion of boron is quite low, at the 20-30 parts per million level. Hollow cathode emission has also been attempted but instabilities in the cathode emission make the work difficult (8). Recently, Lambert et al. (9) described a procedure using β-diketones for forming complexes with borate. This method is sensitive at the ppm level.

The present report describes the use of thin layer chromatography, followed by color development with azomethine H and spectrodensitometry, in the quantitation of boron at the 50 parts per billion level. Visibility is much lower than this limit but reproducibility becomes a problem. The method has been used for determination of boron in soil and water samples.

MATERIALS AND METHODS

A Schoeffel Model 3000 spectrodensitometer was used throughout
this work.

All glassware was boron free (soft glass) or plastic (polypropylene). Each piece was washed with dilute hydrochloric acid
before use and then rinsed with deionized water. (Note that all
water used in this experiment was deionized in a non-Pyrex environment. Thus the presence of boron in the Pyrex glassware would
not contaminate the samples.) Plastic that was used to hold samples was used once and then discarded.

The chromatography tanks and paper linings were also washed
with dilute hydrochloric acid and rinsed with deionized water.

An Analytical Instrument Specialties "TLC Multispotter" was
used to apply the sample aliquots.

Drummond disposable glass capillary pipets were used to
apply the standard aliquots.

REAGENTS AND STANDARDS

Azomethine H reagent was purchased from Pierce Chemical Co. (Rock-
ford, Ill.). An amount suitable for spraying one chromatogram
was prepared by dissolving 0.1 g of Azomethine H and 0.1 g of L-
ascorbic acid in 10 ml of deionized water. Fresh reagent was
prepared daily and stored in a refrigerator.

A stock solution of 0.1 g/l of boron was prepared by dissolving 0.5715 g of analytical reagent grade boric acid in deionized
water and diluting to 1 liter. By suitable dilution, standards
for the concentration range of 0.01-0.1 g/l were prepared. These
solutions gave concentrations at the nanogram level when applied
to the chromatogram.

Cellulose MN 300 thin layer plates, 20 × 20 cm, were purchased from Analtech, Newark, Del. These were washed in acetone-
water (1:1) in a developing tank with ascending flow. The solvent was allowed to overflow the top of the plates during an overnight development.

All solvents were reagent grade and redistilled. They were
stored in acid washed "soft glass" bottles.

SAMPLE PREPARATION

The soil sample was placed in a crucible and dried in an oven at
125° C overnight. It was then finely ground and screened through
series 30 mesh to remove extraneous material such as leaves,

stones, and small twigs. To a 5-g sample was added 5 ml of boiling deionized water. After tentative agitation to degas the sample it was placed in a capped polystyrene centrifuge tube and
agitated for several minutes on a Vortex Genie. The liquid was
separated in a Buchner funnel and the remaining soil was reextracted with deionized water. This second extract was added to
the first. The extract volume was decreased by evaporation in
non-Pyrex glassware on a heating mantle. The volume was then
adjusted by pH 2 by addition of HCl, then to 1 ml by addition of
deionized water. The extract was then filtered through an acid-
washed 40-micron millipore filter fitted to a plastic syringe.

To prepare water samples for chromatography the following
procedure was used. Twenty-five milliliters of water was evaporated under nitrogen (or compressed air) in a water bath not over
60° C to a volume of 0.5 ml. This is acidified to pH 1. Aliquots of this are applied to the origin of the layer (10,25 µl)
using an AIS spotter (Analytical Instrument Specialties, Liberty-
ville, Ill.). The speed setting was 2, and the heat setting was
25. A hair dryer was used to blow warm air over the surface of
the starting line to facilitate drying. This resulted in zones
of 1-2 mm in diameter of the applied sample. If the starting
area of the sample was too large, resolution was poor.

SAMPLE APPLICATION AND SEPARATION

Sample aliquot sizes ranging from 5 to 25 µl, along with a standard of 100 ng of boron, were applied to alternate lanes of the
chromatogram (scored into 10 mm lanes) via the Multispotter, with
heat setting 25 and speed setting 4. A hair dryer was used to
blow warm air over the surface of the layer during sample application.

After drying, the plate was developed in a mobile phase of
butanone-deionized water-ethylene glycol (85:13:2), until the solvent front reached 1-2 cm from the top of the plate. This took
30 minutes. The chromatogram was then removed from the tank and
allowed to dry under an air hood for 30 minutes.

DETERMINATION AND QUANTITATION

The chromatogram, when dry, was then sprayed with the azomethine
H reagent until it acquired an even lemon color. This color will
fade slowly until only the boron zone is still yellow. The spot,
with an R_f of 0.43, is easily visible after 30 minutes, and can
also be visualized as a dark spot under long wave ultraviolet

light. After 45 minutes the sample was scanned in the densitometer. The resultant curves were then compared to the standard
boron curve. The scans were made at 400 nm, and the spot was
clear and the peak easily quantitatible at an o.d. of 1.0.

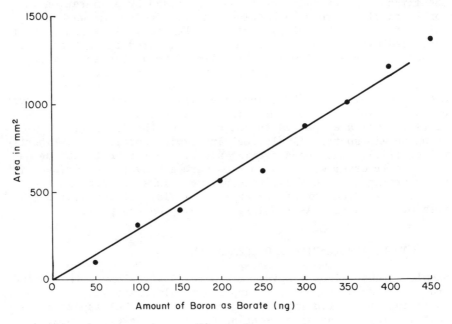

Figure 1

RESULTS AND DISCUSSION

Standard Curve

The standard curve (Figure 1) was set up using standards of 50 to
450 ng of boron (boric acid). The chromatogram was scanned at
400 nm and o.d. of 1.0.

Specificity of the Method

Though highly specific, azomethine H is affected by some ions
that are found in soils and water. John et al. (6) found ionic
concentrations higher than 5,000 ppm fluoride, 3,000 ppm aluminum,

2,500 ppm cupric, and 200 ppm ferric would interfere with the
accuracy of the azomethine method. The specificity of the solvent system used in this work was tested using $Fe_2(SO_4)_3$ as the
source of ferric ions, $CuCl_2$ and $CuSO_4$ as the source of cupric
ions, $AlCl_3$ as the source of aluminum ions, and KF as the fluoride ion source. All chromatograms yielded an R_f of 0.0 for each
of these ions. The mobile phase appears highly specific; throughout the experiment only boron yielded an R_f higher than 0.20.
The specificity of the system is thus a great advantage when working with ion rich soils or water.

Confirmation

For confirmation of the nature of the separated boron zones, the
chromatograms can be oversprayed with a saturated solution of
curcumin in ethanol. After waiting a few minutes the chromatogram is then sprayed with 4 N HCl. The boron spots show a scarlet color on an orange/yellow background after 30 minutes. For
rapid confirmation the chromatogram can be placed in an oven at
125° C immediately after the HCl spray. The scarlet appears in
just a few minutes, though on a much darker, almost brown background. Because curcumin reacts with several substances further
confirmation can be obtained by next exposing the plate to ammonium hydroxide fumes. The instant change from scarlet to purple,
coupled with a rapid fading back to scarlet, is a positive confirmation of the presence of boron. Figure 2 shows the results
of scans of a chromatogram of water extracts prepared as
described. **(See page 156).**
These results indicate that with proper usage this method
for the determination of boron is both reproducible and sensitive. The method is presently being used in various ecological
studies. It should further be noted that the use of the AIS
automatic spotter is highly advantageous because small starting
zones are obtained, an operation difficult to perform manually.

REFERENCES

1. E. S. Gladney, E. T. Jurney, and D. B. Curtis, Anal. Chem. 48,
   2139 (1976).
2. M. R. Hayes and J. Metcalfe, Analyst 87, 956 (1962).
3. A. A. Nemodruk and Z. K. Karalova, Analytical Chemistry of
   Boron (D. Davey & Co., New York, 1965), p. 50.
4. C. Z. Evans and J. S. Machargue, J. Assoc. Off. Agric. Chem.
   30, 308 (1947).
5. P. Lanza and P. L. Buldini, Anal. Chim. Acta 70, 341 (1974).

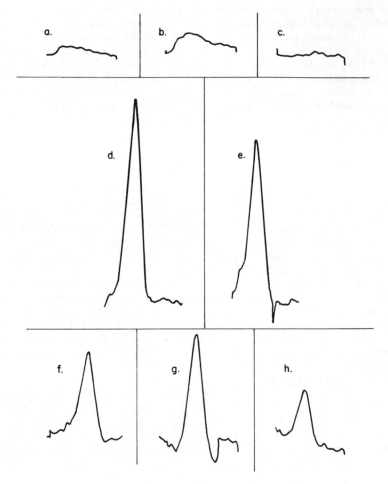

Figure 2.   Boron in water.   a. Container, deionized water, and plastic test tube used throughout the procedure; b. Container, suburban tap water, and plastic test tube; c. deionized water and plastic test tube.   The conditions listed above represent controls used for testing the validity of the procedure.   d. Curve obtained with 100 ng of boron (borate); e. Tap water from rural town (109 ng); f. River water at high tide (63 ng); g. Creek water above tide line (92 ng); h. Creek at low tide (38 ng).   These results represent the scans of chromatograms of extracts of water from the stated sources; the figures in parentheses represent the amount in 5 µl and aliquots from the 250 µl solution.

6. M. K. John, H. H. Chuck, and J. H. Newfeld, Anal. Lett. 8, 559 (1975).
7. W. W. Harrison and N. J. Prakash, Anal. Chim. Acta 49, 15 (1970).
8. E. H. Doughtry and W. W. Harrison, Anal. Chim. Acta 67, 253 (1973).
9. J. L. Lambert, J. V. Paukstelis, and R. A. Bruckdorfer, Anal. Chem. 50, 820 (1978).

CHAPTER 12

Thin Layer Chromatography of Peptides and Related Compounds

Danute E. Nitecki

Peptides are composed of repeated amide units with one saturated carbon atom, carrying the side chain of the amino acid, between the amide bonds. In animals 21 major amino acids are found in proteins; there are considerably more occurring in bacteria and plants and very large numbers of synthetic amino acids. The side chains vary from saturated hydrocarbon chains to polar and positively or negatively charged groups. There is no restriction on the size--peptides exist from dipeptides of molecular weight of about 200 daltons to proteins of molecular weight in tens of thousands. Thus the variations in size, configuration, conformation, charge, hydrophobicity, solubility properties, etc., are practically endless and it is almost presumptuous to speak of TLC of peptides; nevertheless, some generalizations can be attempted.

The field of peptide TLC can be rather sharply divided into chromatography of protected peptides and that of free peptides.

Protected peptides is the field of the synthetic chemist. Here, the amino functions are usually acylated with urethane-type reversibly removable groups. Carboxylic acids are frequently in the form of aliphatic or aromatic esters, amides, or hydrazides. The purity of the protected synthetic intermediates is absolutely essential and TLC is definitely the most important analytical and, occasionally, preparative tool (1,2). Small protected peptides (up to hexapeptide or thereabouts) are usually soluble in relatively nonpolar solvents such as chloroform or ethyl acetate. Correspondingly, the TLC solvent systems tend to be a combination of a rather nonpolar organic solvent with a few percent of acetic acid or methanol. The most generally useful mobile phase at small peptide stage was found to be di-isopropyl ether-chloroform-acetic acid, 6:3:1; the next useful was some permutation of chloroform-methanol-acetic acid mixture.

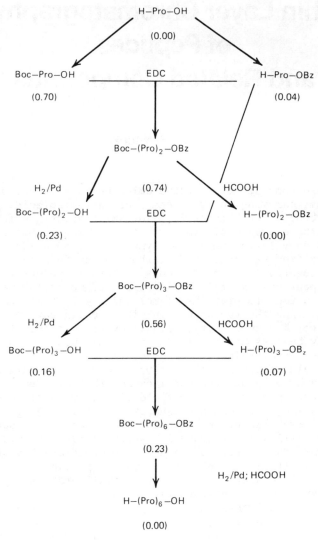

Figure 1. Synthetic pathway of proline hexapeptide and R_f values of intermediates (Silica Gel 60 on glass plates, E. Merck, Germany; CHCl$_3$-HOAc, 9:1; EDC is N-ethyl-N'-dimethylaminopropyl carbodiimide hydrochloride.)

Figure 1 illustrates a practical use of TLC in synthesis of a proline hexapeptide (3), Boc-(Pro)$_6$-OBz (4). R_f values were obtained in CHCl$_3$-HOAc, 9:1, on silica gel plates. With the exception of acidolytic removal of the N-Boc group yielding the not too stable amino acid or peptide esters, the progress of all the other reactions was easily followed by TLC in a single mobile phase. This also illustrates the very large differences in R_f values produced upon exposing one polar group in the protected peptide. For example, the R_f of Boc-(Pro)$_2$-OBz was 0.74, whereas that of Boc-(Pro)$_2$-OH was 0.23; in this system the free amino dipeptide ester H-(Pro)$_2$-OBz does not migrate at all. These differences explain why TLC is such a useful tool in the synthesis of peptides.

Selective cleavage of protective groups is an ever present problem in the synthesis. It can also be conveniently followed by TLC. We are now in the process of preparing N-2,4-dinitrophenyl-6-aminocaproyl-N-ε-2,4-dinitrophenyl-L-lysyl-L-tyrosine (3), (DNP-sac-ε-DNP-lys-tyr-OH) by following sequence of reactions: N-α-Boc-N-ε-DNP-Lys-OH + H-Tyr-O-t-Bu $\xrightarrow{\text{carbodi-imide}}$ N-α-Boc-ε-DNP-Lys-Tyr-O-t-Bu. The protected dipeptide gave an R_f of 0.94 in CHCl$_3$-MeOH-HOAc, 70:30:5, on silica gel layers. At this stage it was necessary to remove the N-α-Boc protective group in order to proceed to couple N-DNP-6-aminocaproic acid onto the α-amino group of lysine. At the same time it was desirable, for solubility reasons, to retain the O-t-butyl ester on tyrosine. Formic acid is the mildest of reagents used for acidolytic deprotection of tert-butyl groups (5). It was found that within 1 hour all of the N-α-Boc group was removed and a ninhydrin positive spot, H-(ε-DNP)-Lys-Tyr-O-t-Bu, R_f 0.76, appeared. Within 24 hours formic acid converted this material into H-(ε-DNP)-Lys-Tyr-OH, with an R_f of 0.45, in the same mobile phase.

We have encountered again the remarkable difference in the rate of acidolysis in formic acid between tertiary butyl ester and tertiary butyl ether of tyrosine. Treatment of N-maleimido-6-aminocaproyl-L-tyrosine-O-t-butyl ester-O-t-butyl ether, R_f 0.84 (in CHCl$_3$-HOAc, 9:1) and 0.67 (in diisopropyl ether-CHCl$_3$-HOAc, 6:3:1), with formic acid produced new material within 30 minutes, of R_f 0.57 and 0.55. This material was positive to diazonium spray; therefore, it contains free phenolic hydroxyl in tyrosine and must be N-maleimido-sac-Tyr-O-t-Bu ester. After 24 hours in formic acid, this was converted into a new derivative, which gave R_f 0.08 and 0.09 in the two mobile phases, respectively. This material was ninhydrin negative, and still contained maleimido group, as was shown by positive sulfhydryl spray (6).

Thus, in this instance also, the two tertiary butyl protecting groups could be cleaved selectively.

TABLE I. R_f VALUES FOR SYNTHETIC PEPTIDES
IN $CHCl_3$-MeOH-H_2O

Z(OMe)-Thr-Pro-OH	0.40
Z(OMe)-Gln-Thr-Pro-OH	0.24
Z(OMe)-Ser-Gln-Thr-Pro-OH	0.17
Z(OMe)-Tyr-Gly-Gly-Phe-Met(O)-OH	0.16
Z(OMe)-Tyr-Gly-Gly-Phe-Met(O)-OMe	0.71
Z(OMe)-Tyr-Gly-Gly-Phe-Met(O)-NHNH$_2$	0.38

The R_f values of protected peptides tend to decrease with increasing size, unless polar functions are removed. This is illustrated in Table I (7).

To illustrate the behavior of small protected amino acid derivatives on silica gel, N-α-Boc-N-ϵ-fluoresceinyl-lysine (3), which contains free carboxylic acid, was employed. The following R_f values, in increasing order, were obtained: 0.05 ($CHCl_3$-HOAc, 9:1); 0.15 (di-isopropyl ether-$CHCl_3$-HOAc, 6:3:1); 0.49 (di-isopropyl ether-EtOAc-HOAc, 6:3:1); 0.68 (di-isopropyl ether-acetone-HOAc, 6:3:1); 0.81 (ether-ETOAc-HOAC, 6:3:1); 0.93 (ether-acetone-HOAC, 6:3:1); 0.95 ($CHCl_3$-MeOH-HOAc, 70:30:5).

A considerably larger protected hexapeptide, Boc-Gly-Phe-Ser-Pro-Phe-Arg(NO$_2$)-OMe (8), showed following migrations in increasing order: R_f 0.0 (di-isopropyl ether-$CHCl_3$-HOAc, 6:3:1); 0.14 (EtOAc-MeOH-HOAc, 100:3:3); 0.16 ($CHCl_3$-acetone-HOAc, 50:50:1); 0.24 (ether-acetone-HOAC, 6:3:1); 0.35 (EtOAc-MeOH-HOAc, 50:3:3); 0.70 (upper phase of n-BuOH-H_2O, 5:2:5); 0.80 (acetone-MeOH, 100:3); 0.87 (acetone-HOAc, 100:3); 0.95 ($CHCl_3$-MeOH-HOAc, 70:30:5). Obviously, the R_f values follow a general increase of polarity of the components in the solvent mixture.

A simple and very ingenious system of preparative purification of protected peptides in large amounts (up to 5 g) has been described (9,10); it utilized preparative silica gel columns, isocratic or gradient solvent conditions, and less than 150 psi back pressure. Solvents consisted of chloroform, alcohols, acetic acid, and mixtures thereof. TLC of crude materials served as a guide in choosing an effective mobile phase and a practical working relationship between TLC and column chromatography was described. This system should be very useful in the synthesis of large protected peptides.

As the synthetic protected peptides increase in size, their solubility shifts from solvents such as ethyl acetate, acetonitrile, or chloroform to more polar domains. In practice, large peptide derivatives of eight or more amino acids are frequently soluble only in solvents such as DMF, DMSO, hexamethyl phosphoric triamide, tetramethyl urea, etc. As these high boiling and viscous solvents are rather useless in TLC, the choice mixtures for chromatography tend to contain increasingly large proportions of alcohols (methanol, propanol and butanols), acetic acid, pyridine and water.

As one leaves the domain of synthetic protected peptides and considers free unprotected (natural or synthetic) peptides, one characteristic is immediately evident. A great many charged groups must be considered--at least the two termini, amino and carboxyl, and usually multiple positively or negatively charged functions in the side chains. These peptides are completely insoluble in most of the organic solvents and may, at best, be somewhat soluble in such solvents as DMF or DMSO; generally, however, water is required to solubilize free peptides. The TLC mobile phases are almost invariably mixtures containing water, alcohols, acetic acid, pyridine, or ammonia; the most frequently used mixture contains n-butanol, water, and either acetic acid or pyridine or both. For reasons that are not clear, n-butanol exerts an almost "magic" touch to peptide TLC--the separated zones in mobile phases containing large proportions of n-butanol are round, well resolved, and show minimal tailing or diffusion.

A series of homo-oligopeptides of glycine, $(Gly)_{1-6}$, and alanine, $(Ala)_{1-6}$, were investigated by TLC (11). Silica gel plates gave insufficient separation with the four mobile phases used (phenol-0.1% NH_3, 10:1; n-butanol-HOAc-H_2O, 4:1:5, upper phase; n-butanol-pyridine H_2O, 6:4:3; collidine-pyridine H_2O, 2:1:1). Cellulose plates gave good separation with the phenol-0.1 N NH_3 as mobile phase. The R_f values in the glycine series were the following: Gly, 0.22; Gly_2, 0.21; Gly_3, 0.26; Gly_4, 0.32; Gly_5, 0.50; Gly_6, 0.63. The alanine series had the following values: Ala, 0.44; Ala_2, 0.56; Ala_3, 0.69; Ala_4, 0.80; Ala_5, 0.94; Ala_6 did not migrate at all. Thus, in both series, the R_f value increased with increasing molecular weight, except glycine and hexaalanine. This probably indicates that as the molecules increase in size they retain less of the charged zwitterionic character and therefore migrate faster. Thus the relative charge distribution and not the size is the determining factor.

Some very fine resolutions can be achieved even with very polar, highly charged compounds on silica gel. For example, it was possible to separate the three azo coupling products of histidine with p-diazonium arsanylate, namely 2-azobenzene-p-arsanylate (illustrated), 4-azobenzene-p-arsanylate, and the 2,4-disubstituted product (3). The R_f values of histidine and Boc-histidine derivatives are shown in Table II.

$$NH_2-CH-COOH$$

$$CH_2$$

N = N — ⬡ — AsO_3H_2

TABLE II. R_f VALUES FOR HISTIDINE AZO DERIVATIVES

	In CHCl$_3$-MeOH- conc. NH$_4$OH 15:20:10	In n-BuOH-HOAc-H$_2$O 12:3:5
2-azo	0.34	
4-azo	0.24	0
bis-azo	0.05	0
N-α-Boc-2-azo	0.52	0.33
N-α-Boc-4-azo	0.44	0.18
N-α-Boc-bis-azo	0.12	0

TLC with cellulose is frequently used for investigation of free peptides. We have compared R_f values of human angiotensin II, an octapeptide H-Asp-Arg-Val-Tyr-Ile-His-Pro-Phe-OH, on silica gel and cellulose plates. In n-BuOH-HOAc-H$_2$O, 4:1:1, the R_f on silica gel was 0.07, on cellulose, 0.42; in n-BuOH-pyridine-HOAc-H$_2$O, 15:10:3:12, the R_f on silica was 0.46, on cellulose, 0.50; in pyridine-HOAc-H$_2$O, 50:30:15, the R_f on silica was 0.94,

on cellulose 0.83. It is not possible to draw conclusions from
such data, although peptides generally tend to migrate a little
faster on cellulose than on silica. Unfortunately, in our hands,
free large peptides have a pronounced tendency to streak and tail
more on cellulose than on silica gel plates, thus obscuring the
presence of unseparated impurities. Both media are used for free
peptides, although direct comparisons for the same compounds and
the same mobile phases are lacking. Table III illustrates this
point in the synthesis of peptide fragments of apolipoproteins
A-I and A-II (12). It is worth noting that peptide II, which has
one basic amino acid, 3-arginine, substituted in the place of
neutral 3-alanine of peptide I, has an R_f value of 0.45 as compared to 0.76 of peptide I.

Table IV illustrates the influence of substitution of polar
amino acids on R_f values of hexa- to deca-peptide on silica gel.
These peptides were synthesized as amino-terminal extension analogs of methionine-enkephalin (Met-Enk, Tyr-Gly-Gly-Phe-Met),
which, in turn, are fragments 60-65 to 56-65 of β-lipoprotein.
The influence of charge is very obvious here, and it is curious
that peptide Asp-Lys-Arg-(Met-Enk) has about the same (or even
higher in one solvent) R_f value as the shorter peptide, Lys-Arg-
(Met-Enk). It is likely that addition of one aspartic acid residue brings about a partially neutralizing effect on the peptide.

Relatively large peptides will migrate on silica gel plates.
This is illustrated in synthesis of opiate peptides of human
β-endorphin analogs of various chain lengths (14). The native
β-endorphin peptide of 31 amino acids, β-EP(1-31), is Tyr-Gly-
Gly-Phe-Met-Thr-Ser-Glu-Lys-Ser-Gln-Thr-Pro-Leu-Val-Thr-Leu-Phe-
Lys-Asn-Ala-Ile-Ile-Lys-Asn-Ala-Tyr-Lys-Lys-Gly-Glu. The synthetic peptides were found to exhibit R_f values on silica gel as
shown in Table V. This table illustrates well a problem encountered with larger peptides. A very polar mobile phase is
required to move these peptides on silica. In the process, fine
resolution is lost, as can be seen in Table V. The five peptides, differing in size from 15 to 20 amino acids, all migrate
with R_f values of 0.49-0.56 in BPAW and 0.12-0.31 in BAW solvent.
The TLC cellulose plates do not represent any improvement. This
may be the reason why many laboratories prefer to use cellulose
sheets (paper) for fingerprinting or, as it is sometimes called,
peptide mapping (15). In the process of protein fingerprinting,
the protein is digested by an enzyme, usually trypsin or chymo-·
trypsin, which frequently results in many tens of peptides of
varying size. The mixture is then processed in one direction by
high-voltage electrophoresis and in another direction by chromatography. In this manner, separations of 20-30 or more peptides
can be achieved. The obvious advantage of paper sheets over

TABLE III. R_f VALUES FOR LARGE SYNTHETIC BASIC PEPTIDES
ON CELLULOSE AND SILICA GEL

Peptides [a]	Medium	Mobile Phase [b]	R_f
I	Cellulose	BPW 1:1:1	0.45
II	Cellulose	BPW 1:1:1	0.76
	Silica	PAW 50:30:15	0.66
III	Cellulose	BPAW 15:10:3:12	0.68
	Cellulose	BAW 4:1:1	0.39
	Silica	PAW 50:30:15	0.35
IV	Silica	BPAW 15:10:3:12	0.41
	Silica	BAW 4:1:1	0.032
V	Cellulose	BAW 4:1:1	0.11
	Silica	BPAW 15:10:3:12	0.35
	Silica	PAW 50:30:15	0.69

[a] Peptides: I, His-Val-Asp-Ala-Leu-Arg-Thr-His-Leu-Ala-
Pro; II, His-Val-Arg-Ala-Leu-Arg-Thr-His-Leu-Ala-Pro
(same as I except 3-Arg substituted for 3-Ala); III,
Leu-Gly-Glu-Glu-Met-Arg-Asp-Arg-Ala-Arg-Ala-His-Val-
Asp-Ala-Leu-Arg-Thr-His-Leu-Ala-Pro; IV, acetyl-Glu-Met-
Glu-Leu-Tyr-Arg-Gln-Lys-Val-Glu-Pro-Leu-Arg-Ala-Glu-Leu-
Gln-Glu-Gly-Ala; V, Gly-Lys-Asp-Leu-Met-Glu-Lys-Val-Lys-
Ser.
[b] Solvents: B, n-butanol; P, pyridine; A, acetic acid;
W, water.

cellulose TLC is the size of the field (Whatman 1MM or 3MM sheets
are 46 × 57 cm). The TLC is also found useful in isolation of
naturally occurring unknown peptides.

A porcine intestinal peptide with somatostatin-like immuno-
reactivity was isolated with the help of TLC on silica gel plates
using n-butanol-pyridine-acetic acid-water (15:10:3:12). Somatostatin, a 14-amino acid peptide, showed an R_f of 0.65, and the
new peptide, composed of 21 amino acids and very probably containing somatostatin extended from its N-terminus by an additional
heptapeptide, showed an R_f value of 0.43 (16).

TABLE IV. R_f VALUES FOR AMINO-TERMINAL EXTENSION ANALOGS
OF METHIONINE-ENKEPALIN ON SILICA

Peptides	Mobile Phases		
	BPA[a]	BAW[b]	BPAW[c]
Arg-(Met-Enk)	0.45	0.45	0.56
Lys-Arg-(Met-Enk)	0.15	0.26	0.43
Asp-Lys-Arg-(Met-Enk)	0.27	0.25	0.39
Lys-Asp-Lys-Arg-(Met-Enk)	0.08	0.13	0.26
Pro-Lys-Asp-Lys-Arg-(Met-Enk)	0.01	0.10	0.24

[a] n- Butanol-0.1 M HOAc-pyridine, 5:11:3, upper phase.

[b] n-Butanol-HOAc-H_2O, 4:1:5, upper phase.

[c] n-Butanol-pyridine-HOAc-H_2O, 6:4:1:9.

TABLE V. R_f VALUES FOR β-ENDORPHIN PEPTIDES ON SILICA

B-EP Peptide	Mobile Phases	
	BPAW[a]	BAW[b]
1-15	0.56	0.31
1-21	0.56	0.21
1-26	0.52	0.24
1-28	0.52	0.22
1-29	0.49	0.12

[a] n-Butanol-pyridine-HOAc-H_2O, 5:5:1:4.

[b] n-Butanol-pyridine-HOAc, 4:1:5, upper phase.

Cyclic peptides form a category by themselves. If they do
not have polar amino or acid side chains, they behave essentially
as protected peptides on TLC, since they do not contain free
charged α-amino or Ω-carboxyl groups. Thus, pentapeptide cyclo
[Val-Orn(Cbz)-Leu-D-Ala-Pro] exhibits an R_f of 0.45 in CHCl$_3$-MeOH,
7:1, and the corresponding dimeric decapeptide, cyclo [Val-Orn
(Cbz)-Leu-D-Ala-Pro]$_2$ showed, in the same solvent, an R_f of 0.60
(17). Similar results were obtained with other cyclic decapep-
tides in such mobile phases as benzene-EtOAc-EtOH, 3:2:1. It is
safe to say that no peptide containing free amino and carboxyl
groups would move in this solvent. Thus again it is not the
amide bonds but the charged zwitterionic groups that slow the
migration of peptides on silica gel.

One of the most interesting demonstrations of the power of
TLC is a resolution of conformational isomers of tentoxin (18).
Tentoxin is a cyclic tetrapeptide cyclo-N-methyl-L-alanyl-L-
leucyl-N-methyl-(Z)-dehydrophenylalanyl-glycyl. The phytotoxic
cyclic tetrapeptide was found to exist in multiple conformations
in solution. Two of these conformers could be isolated by TLC at
4° C in EtOAc-EtOH, 100:5, R_f 0.45 and 0.30, on silica gel plates.
Curiously, the two conformers showed different biological activities in physiological media (19). This is the first demonstra-
tion of a widely accepted hypothesis that separate conformers of
a single molecule can have differing biological activities.

Peptide detection presents its own special problems. Many
detection reagents have been described (1,2), among which are the
following: ninhydrin or isatin reactions for NH-group; Pauly's
diazonium coupling for the phenol of tyrosine and imidazole of
histidine; Sakaguchi's reaction for arginine; specific tryptophan
or sulfhydryl reactions; the α-nitroso-β-naphthol spray for tyrosine; iodine vapor detection, etc. (1,2).

The most useful method for detection of protected peptides
is the chlorination reaction. The NH-group of the amide bond is
converted to N-chloroamide and treatment of this with iodide
liberates iodine, which is the ultimate detected product. It is
obvious that the larger the peptide--the more amide bonds--the
better response to the chlorination reaction. Moreover, substitution in side chains or termini does not interfere with this reaction. The reaction works very satisfactorily on silica, cellulose, and paper chromatograms, but it is not easily controlled in
solution, although attempts have been made to apply it to proteins in solution (20,21). This chromatographic detection reaction consists of first spraying with Clorox (22), t-butyl hypo-
chlorite (23), or exposing the plate to chlorine vapors (24);
destruction of background chlorine follows, usually by means of
ethanol; the final stage is the detection of generated iodine

from KI. The most convenient and sensitive detection reagent
for liberated I_2 was Δ-tolidine. This reagent was found to be
carcinogenic, and a very useful noncarcinogenic replacement,
4,4'-tetramethyldiamino-diphenylmethane, was recently suggested
(25).

A number of new visualization reagents for the amino group
have been developed in recent years; these include o-phthalaldehyde, fluorescamine (27), and o-diacetyl benzene (28). A
detailed comparison of ninhydrin and these reagents was made in
TLC detection methods for histamine from food extracts (29) and
for peptide fingerprints on cellulose and silica gel (30). Fluorescent TLC peptide mapping for protein identification and comparison was achieved in subnanomole range by the use of fluorescamine staining (31,32). A spray based on o-phthalaldehyde was
used for visualizing amino acids in TLC on cellulose; the detectable limit was found to be about 100 pmole per spot (33).

A very extensive investigation of fingerprinting (peptide
mapping) techniques for examining primary structure of proteins
utilizes partial acid hydrolyzates of proteins (34). The hydrolyzates consist of small peptides and amino acids. The mixtures
were separated by two-dimensional TLC on cellulose plates, and
developed by various visualization reactions, making it possible
to identify many of the spots. This technique seems to be very
promising for comparison of normal and modified protein (dis-
eased proteins or genetic variants). The solvents used in the
two dimensions were iso-propanol-butanone-1 M HCl, 60:15:25, and
2-methyl-2-butanol-butanone-acetone-methanol-(0.88) ammonia,
25:20:35:5:20. Samples of 75-300 ng of protein were used and, in
the case of collagen, 61 disparate spots could be distinguished.
Use of ^{14}C-dansylation techniques (35) followed by two-dimensional TLC on polyamide sheets allowed simultaneous identification and quantitation of 22 amino acids in nanoliter samples of
renal tubule fluids; less than 1 pmole of an amino acid could be
measured reproducibly. Novel solvent systems for TLC of dansyl
derivatives of amino acids was described recently (36). Tryptophan related amines and peptides were investigated by TLC on
silica gel. Pre- and post-derivatization with fluorescamine
directly on the plate was used with 5-50 pmole as limits of
detection for dipeptides (37).

A very ingenious chemical assay method applicable to poly-
peptides containing C-terminal amide was recently developed (38).
It was applied for isolation of gastrointestinal hormones. The
method is based on obtaining proteolytic enzyme digests of various fractions of peptide extracts. The mixture is then analyzed
by dansylation and TLC. The C-terminal dansyl-amino acid-amide
is separated easily from dansyl amino acids.

Solubility properties of peptides should be summarized here since they are closely related to TLC. Generalizations are only too frequently untrue; still, these should provide some perspective to the reader not familiar with the field. Protected small peptides (up to hexapeptide or thereabouts) are usually soluble in solvents of dielectric constant of 5 or above (40), such as chloroform (4.8), ethyl acetate (6.0), tetrahydrofuran (7.6), monoglyme (7.2), or dichloromethane (9.1). Ether (4.3), iso-propyl ether (3.9), p-dioxane (2.2), benzene (2.3), or toluene (2.4) very seldom dissolve protected peptides; with exceedingly rare exceptions, the peptides are not soluble in hydrocarbon solvents such as hexane (1.9). These peptides, conversely, are not soluble in water (80 at 20° C). As the protected peptides get larger in the process of synthesis, the solubility requires more polar solvents, making necessary for TLC such solvents as pyridine (12.3), acetone (20.7), butanols (12-18), methanol (32.8), acetic acid (6.15), and formic acid (47.9). The above, useful for TLC, are not generally used for synthetic steps; there, functionally inert solvents such as dimethyl formamide (37.6) and dimethyl sulfoxide (47.0) come into prominence, with occasional use of tetramethyl urea (23.0) and hexamethyl phosphoric triamide (30). Unfortunately, the last four are not useful for TLC as all have very high boiling points (from 153° for DMF to 233° for HMPA), and it is difficult to evaporate them off the plates in order to proceed with visualizing sprays.

Free (underivatized) amino acids and peptides contain minimally two polar functions: α-amino group and Ω-carboxylic acid. The free amino acids are zwitter ions and tend to be soluble in aqueous solutions only. Larger oligopeptides that do not contain charged side chains are occasionally more soluble in such solvents as methanol, DMF, or DMSO than in water. Since most peptides, however, are made of an heterogeneous array of amino acids containing charged basic or acidic groups, hydroxyls, sulfhydryls, amides, heterocycles, etc., in the side chains, the molecules tend to be very polar and soluble only in solvent mixtures containing at least some water. For TLC purposes, solvent mixtures composed of alcohols (methanol to butanols), acetic or formic acids, pyridine, ammonia, and at least a few percent of water are most frequently employed.

It should be pointed out that the dielectric constant of a given solvent is, at best, only an indication of the behavior of this solvent on silica or cellulose plate. The interaction of the solvent and the matrix of the plate creates the true adsorbing surface that the migrating molecule "sees." This interaction is not always parallel to the dielectric constant and, for this reason, the Hildebrand solvent scale value provides more accurate

estimates of polarity (41). For example, the dielectric constant
of acetic acid is 6.15 and that of methanol is 32.8. Yet, it has
been this author's experience that both of these exert approximately similar effects on migration of small peptides on silica
gel plates. A small protected tripeptide would yield an R_f of
0.3 in chloroform and 0.9 in 5% of either acetic acid or methanol
in chloroform. Consultation of Hildebrand scale of solvent polarity indicates that acetic acid has a larger solvent strength E°
value than methanol.

Large peptides with many polar functions in the side chains
are very strongly adsorbed by silica gel and necessitate solvents
of very high polarity, such as aqueous solutions of acetic acid,
pyridine, and alcohol mixtures. In such a process, so much solvent polarity is needed to dissolve and move these peptides that
the fine resolution is frequently completely abrogated. This may
explain the limited success of conventional TLC in chromatography
of large free peptides. For such compounds, reverse phase TLC
contains some promise (39). The method describes plates coated
with silica gel 60 that has been treated with dodecyltrichloro-
silane, thus deactivating the silica gel surface. The elution
was performed with a methanol-H_2O-trifluoroacetic acid mixture,
64:35:1. A very elegant separation of human calcitonin (a 32
amino acid peptide), calcitonin (11-32), (a 22 amino acid peptide), and calcitonin sulfoxide was achieved. This method is
especially intriguing as it apparently can be correlated with
column HPLC methods. The availability of commercial reverse
phase TLC plates should facilitate this development.

This work was supported by USPHS Grant A1-05664.

REFERENCES

1. G. Pataki, Techniques of Thin-Layer Chromatography in Amino
   Acid and Peptide Chemistry. Ann Arbor Science, 1968.
2. E. Wunsch, Houben-Weyl Methoden der Organischen Chemie, edited
   by E. Müller, in Synthese von Peptiden (Georg Thieme Verlag,
   Stuttgart, 1974), vol. 15, Part II, p. 462.
3. D. E. Nitecki, unpublished observations.
4. Abbreviations used: Boc-, N-tertiary-butyloxycarbonyl-; -OBz,
   -benzyl ester; -O-t-Bu, -tertiary butyl ester; DNP-, 2,4-
   dinitrophenyl-; -sac-, -6-aminocaproyl-, Cbz,N-carbobenzyloxy-;
   Z(OMe)-, p-methoxy-carbobenzyloxy-. Other abbreviations used
   as suggested by IUPAC-IUB Commission on Biochemical Nomenclature, Arch. Biochem. Biophys. 150, 1 (1972).
5. B. Halpern and D. E. Nitecki, Tetrahedron Lett. 1967, 3031.

6. O. Keller and J. Rudinger, Helv. Chim. Acta 58, 531 (1975).
7. M. Kubota, T. Hirayama, O. Nagase, and H. Yajima, Chem. Pharm. Bull. Tokyo 26, 2139 (1978).
8. A gift from Bachem, Inc., Torrance, Cal.
9. T. F. Gabriel, J. Michalewsky, and J. Meienhofer, J. Chromatogr. 129, 287 (1976).
10. T. F. Gabriel, M. H. Jimenez, A. M. Felix, J. Michalewsky, and J. Meienhofer, Int. J. Peptide Protein Res. 9, 129 (1977).
11. G. Oshima, H. Shimabukuro, and K. Nagasawa, J. Chromatogr. 152, 579 (1978).
12. D. J. Kroon and E. T. Kaiser, J. Org. Chem. 43, 2107 (1978).
13. N. Ling, S. Minick, and R. Guillemin, Biochem. Biophys. Res. Commun. 83, 565 (1978).
14. H.-W. Yeung, D. Yamashiro, W.-C. Chang, and C. H. Li, Int. J. Peptide Protein Res. 12, 42 (1978).
15. J. C. Bennett, Methods in Enzymology, edited by C. H. W. Hirs (Academic Press, New York, 1967), vol. 11, p. 330.
16. L. Pradayrol, J. A. Chayvialle, M. Carlquist, and V. Mutt, Biochem. Biophys. Res. Commun. 82, 701 (1978).
17. M. Kawai and U. Nagai, Biopolymers 17, 1549 (1978).
18. D. H. Rich and P. K. Bhatnagar, J. Am. Chem. Soc. 100, 2218 (1978).
19. D. H. Rich, P. K. Bhatnagar, R. D. Jasensky, J. A. Steele, T. F. Uchtyl, and R. D. Durbin, Bioorg. Chem. 7, 207 (1978).
20. A. Matsushima, S. Yamazaki, and Y. Inada, Biochim. Biophys. Acta 271, 243 (1972).
21. S. Yamazaki, A. Matsushima, T. Hosoi, and Y. Inada, J. Biochem. Tokyo 73, 195 (1973).
22. D. E. Nitecki and J. W. Goodman, Biochemistry 5, 665 (1966).
23. R. H. Mazur, B. W. Ellis, and F. S. Cammarata, J. Biol. Chem. 237, 1619 (1962).
24. F. Reindel and W. Hoppe, Chem. Berichte 87, 1103 (1954).
25. E. Von Arx, M. Faupel, and M. Brugger, J. Chromatogr. 120, 224 (1976).
26. P. A. Shore, Meth. Enzymol. 17B, 842 (1971).
27. S. Udenfriend, S. Stein, P. Bohlen, W. Daurman, W. Leimgruder, and M. Weigele, Science 178, 871 (1972).
28. M. Roth, Anal. Chem. 43, 880 (1971).
29. E. R. Lieber and S. L. Taylor, J. Chromatogr. 160, 227 (1978).
30. E. Schiltz, K. D. Schnackerz, and R. W. Gracy, Anal. Biochem. 79, 33 (1977).
31. R. E. Stephens, Anal. Biochem. 84, 116 (1978).
32. H. Fujiki and G. Zurek, J. Chromatogr. 140, 129 (1977).
33. H. M. Davies and B. J. Miflin, J. Chromatogr. 153, 284 (1978).
34. J. G. Heathcote and S. J. Al-Alawi, J. Chromatogr. 129, 211 (1976).

35. M. Weise and D. E. Oken, J. Chromatogr. 152, 175 (1978).
36. R. M. Metrione, J. Chromatogr. 154, 247 (1978).
37. H. Nakamura and J. J. Pisano, J. Chromatogr. 152, 153 (1978).
38. K. Tatemoto and V. Mutt, Proc. Nat. Acad. Sci. USA 75, 4115 (1978).
39. E. Von Arx and M. Faupel, J. Chromatogr. 154, 68 (1978).
40. J. A. Ruddick and W. A. Bunger, Dielectric constants taken as approximated values for 20-25° C. In Organic Solvents from the Series of Techniques of Chemistry, edited by A. Weissberger (Wiley-Interscience, New York, 1970), vol. 2.
41. G. Zweig and J. Sherma, Eds., CRC Handbook of Chromatography (CRC Press, Cleveland, 1972), vol. 2, p. 38.

CHAPTER 13

Editorial Note:
The Aflatoxin Precursors
and Metabolites

To the uninitiated, the aflatoxins can be a sea of mystery. How
are they formed and metabolized? What relationship do they bear
toward each other?

The aflatoxins seem to represent a specific form of aceto-
genin metabolism in fungi. Figure 1 is constructed from available literature, particularly two articles by Hsieh and co-workers (1,2). This figure is a composite map that shows the energy
levels and kinetic links among the aflatoxins and their precursors and metabolites. The map is undoubtedly incomplete and the
details are probably not understood. Nevertheless, this figure
indicates the general pattern of relationship and suggests
likely targets for their toxic action. The chemical structures
of the aflatoxins are given in Chapter 16 by Hsieh and co-workers
and others.

The left-hand portion of Figure 1 refers to the biosynthetic reactions in fungi. The right-hand portion refers to the
metabolic (hepatic) reactions in animals, which contribute to
both toxigenesis and detoxification. The upper portion of Figure
1 contains the more reduced compounds, both precursors and detox-
ified products. The lower portion contains the more oxidized
compounds, including the actively toxic metabolites. The toxi-
genic sequence leads from upper left to lower right, whereas the
detoxification sequence leads from upper left to upper right.
Some of the individual steps are undoubtedly reversible.

Aflatoxin assays are usually limited to AF B_1 and B_2 and to
AF G_1 and G_2. Recently, sterigmatocystin has been assayed,
although its relationship to the aflatoxins has not always been
made clear.

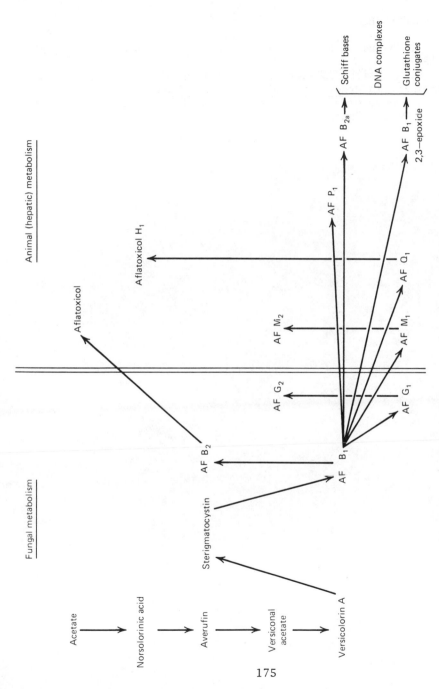

Figure 1. Metabolic relationship among aflatoxins and their precursors and metabolites.

175

REFERENCES

1. R. Singh and D. P. H. Hsieh, Arch. Biochem. Biophys. <u>178</u>, 275
   (1977).
2. D. P. H. Hsieh, Z. A. Wong, J. J. Wong, C. Michas, and B. H.
   Ruebner, in <u>Mycotoxins in Human and Animal Health</u>, edited by
   J. V. Rodericks, C. W. Hesseltine, and M. A. Mehlman (Pathotox,
   Park Forest South, Ill., 1977), p. 37.

CHAPTER 14

Analysis of Submicro Amounts of Aflatoxins by Thin Layer Chromatography

H. Michael Stahr

The professional responsibility of a toxicologist goes beyond diagnosis from the symptoms. He must by legal commitment ascertain the etiological agent of sickness and death. To this end, the toxicologist is a practicing analyst who searches through a morass for traces of toxic substances.

The etiological agent might be determined by a physiological test to obtain an "instant replay" of symptoms. That method is usually impractical, and it may have limited diagnostic value. A nonphysiological means of analysis is more useful and relevant. Chemistry and physics are more dependable aids in making a diagnosis.

Because of its high toxicity, the etiological agent usually exists at low concentrations. In some cases its continued existence in tissues and body fluids is short-lived. Acting rapidly, the toxic substance must be separated from inert or complicating materials in order to obtain a substance-to-junk ratio that is favorable for analysis. A sensitive and specific test method must be developed for the toxic substance. Where metabolism has occurred, characteristic degradation pathways or products must be elucidated in order to determine the action of the toxic substance.

Reinforcement of the diagnosis from symptoms with chemical analysis must be followed by reinforcement of the analytical conclusions with confirmatory tests based on other analytical principles. A thorough, systematic analysis is required to make a diagnosis. Many general TLC references are available. Micro means 10^{-6} (mole) gram molecular weight of a substance and submicro is defined as less than micro. Usually we think of micro as microgram, and submicro as less than microgram.

The following text gives detailed methodology for the analysis of mycotoxins and discusses the importance of using confirmatory tests such as TLC-mass spectrometry.

MATERIALS AND METHODS

Materials and Apparatus

1. Extraction solvents: acetonitrile, Eastman white label, distilled in glass; water - millipore, 'Milli Q' or equivalent (90:10).
2. Development solvents: 85/10/5; chloroform, acetone, propanol-2.
3. Thin-layer plates: silica gel "G," 0.5-mm thick plates, spread with a Desaga spreader (40 g silica gel to 95 ml ethanol) or Merck silica plates, silica N-HR or G, glass or aluminum backed available from Brinkmann Instruments, Inc., 110 River Road, Des Plaines, Ill.; HPTLC and reverse phase TLC plates from Whatman, Inc.
4. Defatting solvent: 2,2,4-trimethylpentane (ACS grade) or petroleum ether distilled in glass.
5. Spotting solvent: chloroform, acetonitrile (98/2).
6. Decolorizing reagents: 15% ferric chloride (15% $FeCl_3 \cdot 6 H_2O$; 4% aqueous sodium hydroxide.
7. Extraction apparatus: Waring-type blender, explosion proofed.
8. Gel preparation apparatus: 50 ml buret equipped with Teflon[R] stopcock; pH meter, Corning model 12-B or equivalent.
9. Standards: aflatoxins, 0.25-.5 µg methanol solution, stock solution dilutions of 25 mg/ml, 1 mg/ml or B_1, B_2, G_1, G_2; Zearalenone, 1 mg/ml methanol; T-2 toxin and diacetoxyscirpenol, 1 µg/ml in methanol, ochratoxin, 1 µg/ml chloroform; sterigmatocystin, 1 µg/ml chloroform; rubratoxin, 1 µg/ml methanol.
10. p-Anisaldehyde spray reagent: 85 ml methanol, 10 ml glacial acetic acid, 5 ml concentrated sulfuric acid, 0.5 mg p-anisaldehyde.
11. Buffer pH 4.0: commercial buffer, Fisher Scientific or equiva-equivalent.

Apparatus. AMINCO Spectrofluorometer; Bausch & Lomb Ultraviolet Visible Spectrometer; Perkin Elmer Infrared Spectrometer (239); Beckman Infrared Spectrometer (IR-4); Finnigen (4000) (GC/MS) Gas Chromatograph-Mass Spectrometer.

MYCOTOXIN

Preparation of Sample Extract for Mycotoxins Other Than
Rubratoxin or Ochratoxin (1)

1. Weigh out 50 g of feed into a Waring blender. Add 200 ml of
   extraction solvent (9/1 acetonitrile - HOH v/v) and blend at
   high speed for 1-2 minutes.
2. Filter 100 ml through filter paper into graduated cylinder and
   transfer into 500 ml separatory funnel fitted with Teflon stopcock.
3. Defat sample by extracting twice with 50 ml of iso-octane or
   pet ether. Discard iso-octane.

Preparation of Gel

1. Place 100 ml of water (Millipore "Q" or equivalent) and 10 ml
   of the 15% ferric chloride solution in a 250 ml beaker.
2. With 4% sodium hydroxide, and using a pH meter, adjust pH to
   4.6.

Decolorization of the Sample Extract

1. Into the separatory funnel containing the sample extract, add
   the gel and shake briefly (1/2 minute).
2. After shaking, allow the gel to settle.
3. Filter 100 ml of the decolorized sample through folded filter
   paper and transfer the filtrate into the 250 ml separatory
   funnel along with 50 ml of distilled water.

Extraction of Mycotoxins

1. Add 50 ml of chloroform to the filtrate and shake.
2. After the layers have separated, drain the chloroform layer
   into a 250-ml Erlenmeyer flask and repeat chloroform extraction. Note: Add additional $CHCl_3$, 50 ml. Use Na_2SO_4 anhydrous to break emulsions. Each sample is handled individually
   so contact time is minimized. The separatory funnel may be
   reused if rinsed thoroughly.
3. Evaporate one-half of the chloroform on a steam bath under
   nitrogen or in vacuo. Do not overdry.
4. Add 0.1-0.4 g of powdered charcoal (Darco G-60) to half of the
   sample. Filter as soon as possible. Evaporate as in 3.

Gas-Liquid Chromatography

1. Determine T-2 toxin and diacetoxyscirpenol directly using a gas chromatograph equipped with a 6-foot, 3-mm i.d. glass column packed with 10% OV-101, 260° C or 3% OV-1 on Chromsorb Q and a flame ionization detector.
2. Redissolve the evaporated sample extract in 0.1 ml of chloroform and inject 5-8 μl into GLC.
3. At a column temperature of 260° C, retention times for T-2 toxin and diacetoxyscirpenol are 20 and 8 minutes, respectively, on the 3% OV-1 column and 28 and 14 minutes on the OV-1 column.
4. Sensitivity of this method is about 1 ppm (1 μg, depending upon the instrument being used and the sample matrix.

Thin-Layer Chromatography

1. Redissolve the sample extract, either before or after GLC, in 500 μl of the spotting solvent.
2. 10% is applied to a thin layer plate along with appropriate standards.
3. Develop the plate in the preferred solvent to within 1-2 cm of the top. (See Table I.)
4. Observe aflatoxins, zearalenone, ochratoxins, and sterigmatocystin by uv light after TLC development.
5. Table I gives R_f data and expected fluorescent colors, as well as levels of detectability, for the above mycotoxins.

Developing Solvents

1. The mobile phase xylene-isoamyl alcohol-methanol (9:10:1) is quite good for separation of aflatoxins B_1, B_2, G_1, and G_2, but can give interfering bands for other mycotoxins. Ordinarily, the first development is performed in toluene-ethyl acetate-acetone (3:2:1) and zearalenone is analyzed. Then if aflatoxins are well enough resolved the plate is sprayed to visualize T-2 and DAS. If not, continued development is done in chloroform-acetone-propanol (85:10:5).
2. The 3/2/1 system was originally used for separation of tricothecenes (10) but will give good separation for zearalenone and sterigmatocystin, though the aflatoxins are not separated as well in the (85:10:5) system.
3. Addition of 1% formic acid or acetic to the 3/2/1 system will move ochratoxins A and B away from the origin and allow their detection.

TABLE I. TLC DATA FOR MYCOTOXINS ON SILICA GEL G-HR

Mycotoxin	R_f (3/2/1)	R_f (3/2/1 + 1% acid)[a]	Fluorescent Color	Lower Detectability Level Silica Gel G	HPT
Aflatoxin B_1	0.48		Blue	1 - 5 ng	0.1 ng
Aflatoxin B_2	0.45		Blue	1 - 5 ng	0.1 ng
Aflatoxin G_1	0.43		Green	1 - 5 ng	0.1 ng
Aflatoxin G_2	0.36		Green	1 - 5 ng	0.1 ng
Zearalenone	0.69		Bluegreen	0.2 - 0.5 μg	0.1 - 0.2 μg
Sterigmatocystin	0.80		Red or yellow[b]	10 - 20 ng / 10 - 20 ng	0.5 - 1 ng / 0.5 - 1 ng
Ochratoxin A		0.64	Blue	10 - 20 ng	0.1 - 0.5 ng
Ochratoxin B		0.48	Blue	10 - 20 ng	0.1 - 0.5 ng

[a] Toluene - ethyl acetate- acetone (3:2:1) + 1% acetic acid.
[b] Yellow fluorescence after spray with aqueous 20% potassium hydroxide.

4. Additional developments, twice or more, with 3/2/1 are possible. One additional development in 85/10/5 is possible and sometimes desirable. HPTLC or reverse phase plates are used if separations are still not complete with two developments.

Quantitation of Zearalenone

1. For quantitation of zearalenone, scrape the fluorescent spot at the same R_f as the zearalenone standard from the TLC plate and elute from the silica gel with methanol.
2. Prepare standards of zearalenone between 1.0 and 10.0 µg/ml in methanol. Dilute the sample accordingly.
3. Using Cary 14 or an equivalent uv spectrophotometer, scan from 250 nm to 310 nm. This gives a characteristic peak.
4. Quantitate by measuring the peak height at the absorbance maximum, 274 nm.
5. Check the purity of zearalenone standard, using the following molar absorptivities: ultraviolet maxima 236 nm (ε = 29,700), 274 nm (ε = 13,909), and 316 nm (ε = 6,020) (11).

Quantitation of Aflatoxins

1. Quantitate aflatoxins in a manner similar to zearalenone.
2. Scrape the fluorescent spot from the TLC plate and elute from the silica gel with methanol.
3. Treat samples and standards in the same manner as described in Secs. 26.004-26.009 (Official Methods, A.O.A.C., 12th ed.).

Screening for T-2 Toxin and Diacetoxyscirpenol

1. Screen, by TLC, for T-2 toxin and diacetoxyscirpenol using the 3/2/1 solvent coupled with p-anisaldehyde spray (3) to give visible spots.

Screening for Sterigmatocystin

The sterigmatocystin plate is spotted as described under aflatoxin. It is developed in 3/2/1 + 1% acid. The sterigmatocystin is visualized with uv light. The TLC plate is sprayed with 20% aqueous potassium hydroxide to increase fluorescent intensity and to change the red color to a more intense yellow for more sensitivity.

Rubratoxin and Ochratoxin Screening in Feeds

Sample Extraction

1. Weigh out 50 g of feed and blend twice with 150 ml of petroleum ether; discard the petroleum ether after filtering each time.
2. Blend the sample with 150 ml of methanol and filter.
3. Discard the feed and evaporate the methanol to dryness under nitrogen.
4. Redissolve the extract with 100 ml of ethyl acetate, wait five minutes, filter, and discard solids.
5. The ethyl acetate is filtered through sand and evaporated to about 0.4 ml under nitrogen.

Thin Layer Chromatography

1. Spot the equivalent of 1-5 g feed on a TLC plate (silica gel G with fluorescent indicator) along with 10 μg of rubratoxin standard.
2. Develop the plate first in 85:15 (V/V) ethyl acetate/acetic acid.
3. In difficult cases, a second development in a different dimension may be required to completely separate the band.
4. Allow development of the plate approximately one-half the way up.
5. (a) Observe rubratoxin, via quenching, under uv light at an R_f of 0.5 in the acid system. (b) Ochratoxin may be observed by its blue fluorescence. Tracking with a standard multiple development may be required for ochratoxin. Use 3/2/1 + acid as above or use the reverse phase separation designated in Table II.

Confirmation

1. Presence of fluorescent compounds of GLC peaks which match standards is only presumptive evidence of mycotoxins and should not be used as confirmation.
2. There are a variety of techniques for derivative formation and reagent sprays used with TLC (3), but use of these methods is not proof of identity. The best confirmation is probably the use of TLC or GLC coupled with mass spectrometry.

DISCUSSION

A toxic substance is a chemical that produces a deleterious effect on a biological organism. All substances have finite toxic

TABLE II. REVERSE PHASE SEPARATION OF MYCOTOXINS

Mycotoxin	R_f (20/30/1) [a]	Fluorescent Color	Lower Detectability Level Silica Gel G
Ochratoxin	0.62	Blue	Same as normal phase
Aflatoxin B_1	0.47	Blue	
Aflatoxin B_2	0.53	Green	
Zearalenone	0.13	Yellow-blue	
T-2 Toxin	0.21	No fluorescence	
DAS	0.36	No fluorescence	
Vomitoxin	0.60	No fluorescence	

[a] Acetonitrile - 0.5% NaCl - acetic acid (20:30:1).

effects. The science of toxicology defines and elaborates these harmful effects. This enables sufficient control to be invoked to allow the necessary uses of chemicals with a minimum degradation of the environment. Defining the biological effects produced by discrete levels of chemicals is basic to the science of toxicology. Once this definition is made, the analytical measurement of kind and level may be made to diagnose intoxication.

The complex matrices that are encountered in analytical toxicology require chromatographic separation of toxic substances and identification by sophisticated means. Techniques such as gas chromatography (GC) - mass spectrometry (MS) are most important to the victim and are rapid compared to biological tests.

Thin layer chromatography has obvious advantages for such applications. The separation process may be designed for the sample. This is the chief advantage of thin layer chromatography. The whole chemical study is done in a very visible way.

Use of selective mobile phases, absorbents, multiple developments, and selective visualization techniques allows separations of class and subclass units; even the most closely allied compounds may be separated and analyzed. Solvent partition, column

chromatography, gel and charcoal decolorization, all are used to
reduce the junk (coextracted and partitioned nontoxic sample
matrix) that can prevent the visualization of chromatographed
toxic substances.

The physical confirmation is the most desirable means to
identify the separated bands on thin layer chromatography (TLC).
Each spectroscopic technique has its inherent advantages and disadvantages. For micro work, infrared (ir) with a conventional
spectrometer and ultraviolet spectrometry will work to submicro
levels of concentration for compounds with conjugated double
bonds. Spectrofluorometry may be used to three orders of magnitude below micro levels to identify and quantitatively analyze
compounds with fluorescence. Nuclear magnetic resonance spectroscopy (w/o FT/nmr) is only applicable to greater than micro
concentrations.

Mass spectrometry, even without computer assistance, can be
used to tens of nanogram levels. Picogram quantities of toxic
substances may be confirmed by these techniques when the best
computer and chromatography are combined in the most sensitive
fashion. Confirmatory information as well as quantitation can
be obtained from these.

Fungal toxins are among the most toxic and carcinogenic of
all known chemicals. Aflatoxins are primary carcinogens. Figure 1 shows some structures of aflatoxins and tissue metabolites.

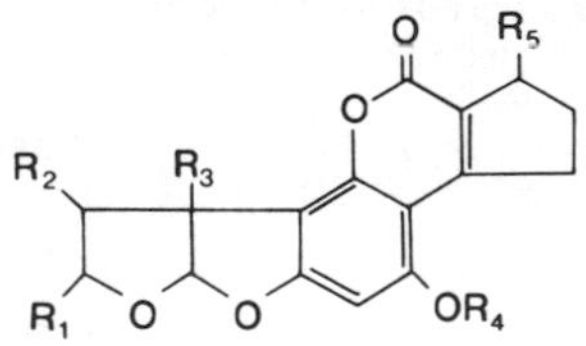

Aflatoxin	R_1	R_2	R_3	R_4	R_5
B$_1$	H	H	H	CH_3	$=O$
B$_2$	H$_2$	H$_2$	H	CH_3	$=O$
B$_{2a}$	OH.H	H$_2$	H	CH_3	$=O$
Aflatoxicol(Ro)	H	H	H	CH_3	$=OH$
M$_1$	H	H	OH	CH_3	$=O$
M$_2$	H$_2$	H$_2$	OH	CH_3	$=O$

Figure 1. The various derivatives of aflatoxin·

Parts per billion analytical sensitivity for these compounds can be obtained directly from thin layer plates.

A metabolism scheme for aflatoxins is shown in Figure 2. It can be seen that the problem of analysis increases when animals

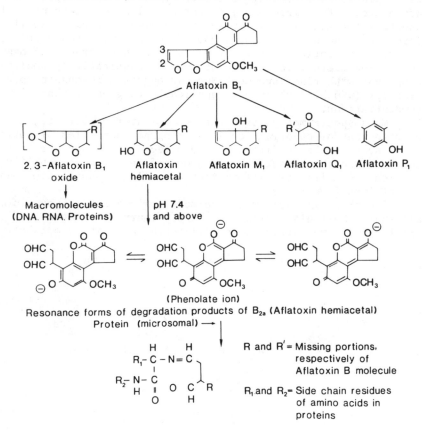

Figure 2.   Scheme for the metabolism of aflatoxin B$_1$.

ingest mycotoxins, and assessments need to be made about the tissue residues and safety of the meat for human consumption. Some other mycotoxins that are routinely analyzed in veterinary medicine are citrinin, ochratoxin, rubratoxins, sterigmatocystin, zearalenone, T-2 toxin, diacetoxyscirpenol, and vomitoxin. Table

TABLE III. MYCOTOXINS

Silica Gel with Chloroform, Acetone, and Propanol (85:10:15)

Compound	R_f	With uv Light/ Long Wavelength
Aflatoxicol	0.75	Blue
Aflatoxin B_1	0.65	Blue
Aflatoxin B_2	0.55	Blue
Aflatoxin G_1	0.45	Green
Aflatoxin G_2	0.35	Green
M_1	0.31	Blue
B_{2a}	0.27	Blue
M_2	0.25	Blue
Zearalenone	0.85	Yellow-green
Sterigmatocystin	0.78	Red

Silica Gel with Toluene-Ethylacetate-Acetone (3:2:1)

Compound	R_f	After Anisaldehyde Spray Visible Light	uv
Vomitoxin (Deoxynivalenol)	0.12	Yellow	Yellow
Diacetoxyscirpenol	0.40	Purple	Yellow-green
T-2 Toxin	0.35	Purple	Yellow-green
Aflatoxin B_1	0.31		Yellow
Aflatoxin G_2	0.21		Yellow
Zearalenone	0.60	Brick red	Yellow-green

Silica Gel with Toluene-Ethylacetate-Acetone Add Formic Acid to 1%

Compound	R_f	Fluorescence/ Long Wavelength uv
Aflatoxin B_1	0.35	Blue
Aflatoxin G_2	0.20	Green
Dicumerol	0.60	Yellow-orange
Ochratoxin	0.65	Blue

TABLE III (continued)

Silica Gel with Ethylacetate-Acetic Acid (85:15)

Compound	R_f	Visible	uv
Rubratoxin B	0.5	Brown (After spraying with 10% H_2SO_4 in Methanol)	Yellow

III shows the R_f's and characteristics of these compounds. Figures 3-5 show mass spectra of compounds which were obtained from aflatoxins separated by thin layer chromatography.

Removal of the toxic substances is possible by mechanical removal of the bands and elution of the toxic substance with methanol. Concentration of the sample in microliter-tipped concentrator tubes permits easy preparation for confirmation and/or quantitation by spectrometric means. Overnight exposure to methanol has been found to be best to quantitatively remove substances from TLC scrapings. Attention to cleanliness of all surfaces contacted is critical to successful spectrometry. Any tract impurity may render nanogram confirmation difficult or impossible. A cleaning scheme is presented in Table IV. This has been successfully used in the Veterinary Diagnostic Laboratory for ten years. TLC plates are routinely washed three times by developing in Nanograde [R] methanol before use as preparatory material for spectrometric work.

The most sensitive detection limits for compounds in any matrix may be achieved by autoradiography. In this process a radioactive atom is incorporated into the molecule to be studied. The course of the compound through biological action and analysis can then be traced by dilution of the radioactivity. An x-ray film placed on top of a TLC plate containing the compounds with the radioactivity in discrete areas produces a film darkening, which indicates the position on the TLC plate and relative concentration of the substance. This information is invaluable in determining the metabolism and excretion products of the substance.

It is similarly invaluable as a development tool to determine the location of compounds in analytical systems using TLC. Using this method, with trace quantities of a labeled compound,

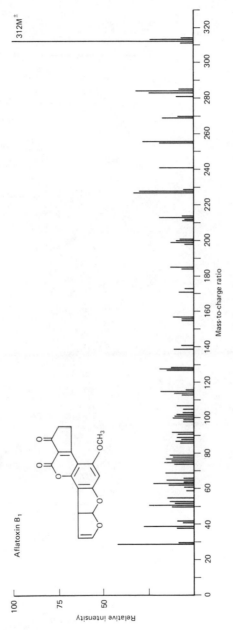

Figure 3. Mass spectra for aflatoxin B_1 after thin layer chromatography.

189

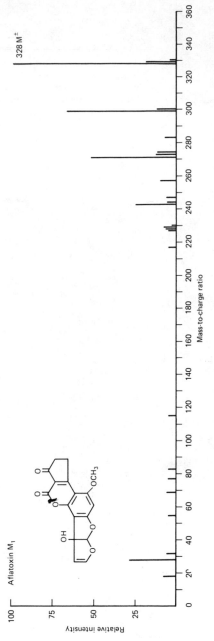

Figure 4. Mass spectra of aflatoxin M_1 after thin layer chromatography.

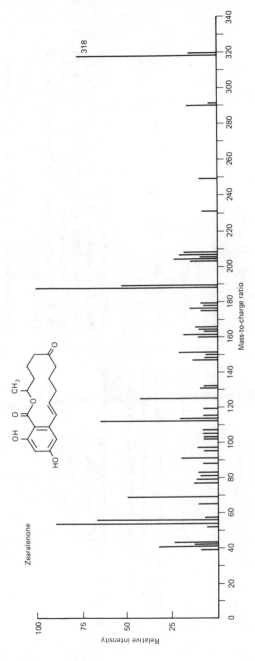

Figure 5. Mass spectra of zearalenone after thin layer chromatography.

TABLE IV.   PROCEDURES RECOMMENDED FOR CLEANING GLASSWARE FOR TOXICOLOGICAL ANALYSIS

Type of Glassware, Purpose	Method of Cleaning
1. Pipets, blenders, and utensils.	Hot detergent.
2. All glassware, including Snyder columns, for general analysis, and for mycotoxin and heavy metal analysis.	Chromic acid ($K_2Cr_2O_7 + H_2SO_4$).
3. All glassware, except Snyder columns, for pesticide analysis.	Muffle oven for 2 hours at 500° C. Place the glassware in the oven at 400° C and raise the temperature to 500° C at a rate of 10°/min. After 2 hours, cool the oven to 400° C at 10°/min. before removing the glassware.
4. Mercury glass.	Hot detergent; store in 50% nitric acid until needed.

After cleaning, the glassware is rinsed with tap water, distilled water, and filtered distilled water, wrapped in aluminum foil, and dried at 110° C. If desired, the glassware may also be rinsed with the organic solvent used for sample extraction. Plastic gloves should not be worn while washing or handling glassware.

192

it is possible to improve separations until even the most difficult separations are achieved. Picogram levels of sensitivity
are achievable with this technique.
    Thin layer chromatography, using proper auxiliary techniques,
can be used in nearly any analysis situation with submicro sensitivity to solve most toxicological problems encountered.

REFERENCES

1. L. Stoloff, S. Nesheim, L. Yin, J. V. Rodricks, N. Stack, and
   A. D. Campbell, J. Assoc. Offic. Anal. Chem. 54, 91 (1971).
2. J. Assoc. Offic. Anal. Chem. 56 (1973). 26.C05-26.C06.
3. L. Stoloff, Clin. Toxicol. 5, 465 (1972).
4. K. Sargent, J. O'Kelly, R. B. A. Carnaghan, and R. Allcroft,
   Vet. Record 73, 1219 (1961).
5. C. O. Ikediobi, J. R. Bamburg, and R. M. Strong, Anal. Biochem. 43, 327 (1971).
6. P. M. Scott, J. W. Lawrence, and W. Van Walbeek, Appl. Microbiol. 20, 839 (1970).
7. R. M. Eppley, J. Assoc. Offic. Anal. Chem. 52, 74 (1968).
8. A. C. Pier, National Animal Disease Center, Ames, Iowa. Pri-
   vate communications.
9. B. J. Wilson and C. H. Wilson, J. Bacteriol. 84, 283 (1962).
10. J. R. Bamburg, N. V. Riggs, and F. M. Strong, Tetrahedron 24,
    3329 (1968).

CHAPTER 15

Factors Affecting the Thin Layer Chromatography of Aflatoxins

Stanley Nesheim

The aflatoxins occur in foods and feeds as a result of mold growth and in products such as milk, meat, and eggs as a result of the ingestion of moldy feed by animals. Some of the aflatoxins are potent carcinogens. Their control requires quantitative methods of analysis that are sensitive to concentrations of micrograms to picograms per kilogram of food or feed.

Since the first analytical methods for aflatoxins were published in 1963 and 1964 (1-3), thin layer chromatography (TLC) has been the only technique capable of detecting and quantitating aflatoxins at these low levels. In the last few years, methods based on high-performance liquid chromatography (HPLC) and radio-immunoassay (RIA) have been developed, but as yet they have not been tested collaboratively. Considering the versatility, simplicity, and economy of the equipment used for the TLC technique, it is doubtful that TLC can be replaced by other techniques for the analysis of aflatoxins in the foreseeable future.

The aflatoxins are well suited for analysis by TLC since most of the compounds fluoresce strongly under longwave uv light. Approximately 0.5 ng/spot can be routinely detected either visually or instrumentally. The TLC technique serves as both a purification and a quantitation step. Before TLC analysis, the afla-toxins are extracted from the sample, usually with an aqueous organic solvent, and the extract is initially purified by one or more techniques, such as solvent partition, heavy metal precipitation, column filtration, or chromatography. These techniques are described in the voluminous literature on aflatoxin methodology. The FDA mycotoxin literature file now contains about 500 references covering this subject. This literature has been reviewed (4-30). Some of the methods have been subjected to interlaboratory testing and may be found among collections of methods such

Solvent system	Adsorbent	Source	Compounds	Reference
Benzene-acetone-EtOAc (16:2:2)	Adsorbosil - 1	Appl. Sci.	M_1, M_2	37
Benzene-EtOAc-AcOH (5:4:1)	Silica Gel G-HR	MN		
	Silica Gel G	Merck	B_1	41
Benzene-EtOAc-MeOH (5:4:1)	Adsorbosil - 1	Appl. Sci.	B_1, M_1, B_{2a}	42
Benzene-EtOH (40:6)	Silica Gel	Not stated	Aflatoxicol	43
Benzene-EtOH-water (46:35:19)	Silica Gel G-HR	MN	4 AT	29 (Secs. 26.015, 26.062), 44-46
	Adsorbosil - 1	Appl. Sci.	4AT, M_1, M_2	37, 47, 48
	Silica Gel G	Merck	4 AT	49
	Silufol 254 UV	Kavalier	4 AT	49
	Bio Sil A	Bio Rad	4 AT	35, 45, 50
	SilicAR TLC-4G	Mal	4 AT	35, 50
	SilicAR TLC-4GF	Mal	4 AT	50
	SilicAR TLC-7G	Mal	4 AT	35, 50
	Sil G-200 (RM)	MN	M_1	51
Ether	Silica Gel G	Merck	4 AT	52
	Silica Gel G-HR	MN		

Table I (continued)

Mobile Phase	Sorbent and Trade Name	Manufacturer	Aflatoxins Separated	Reference(s)
Ether	SilicAR 7G, 7GF, 4GF	Mal		
	Adsorbosil - 1, 5 Adsorbosil - 1 Silica Gel	Appl. Sci. Appl. Sci. Not stated	M_1, M_2 Aflatoxicol	29 (Sec. 26.010), 53 43
Ether-acetone-water (96:3:1)	Silica Gel G-HR	MN	M_1, M_2	37
Ether-benzene-CH_3CN (10:9:1)	Adsorbosil - 1	Appl. Sci.	4 AT	36
Ether-CCl_4-hexane-$CHCl_3$-acetone-MeOH (35:30:15:10:5:5)	Silica Gel G-HR	MN	4 AT	54, 55
Ether-$CHCl_3$-AcOH (2:2:1)	Silica Gel H	Merck	4 AT	56, 57
Ether-hexane-acetone-$CHCl_3$-MeOH (4:4:1:0.5:0.5)	Silica Gel G-HR	MN	4 AT	54, 55
Ether-hexane-CCl_4-$CHCl_3$-acetone-MeOH (35:30:15:10:5:5)	Silica Gel G-HR	MN	4 AT	36, 54, 55

198

Solvent system	Precoats:		Compound	Reference
Ether-MeOH (94:4, 98:2)	Adsorbosil - 1 Bio Sil A Q-5	Appl. Sci. Bio Rad Whatman	4 AT	36
Ether-MeOH-water (96:3:1)	Silica Gel G-HR	MN	4 AT	29 (Sec. 26.010), 58
Ether-MeOH-water (96:3:1)	Silica Gel G-25 HR	MN	4 AT, M_1	59
Ether-MeOH-water (96:4:0.25)	Silica Gel G	Merck	4 AT	60
Ether-MeOH-water (94:4.5:1.5)	Silica Gel G-HR	MN	4 AT	36, 61
EtOAc-CHCl$_3$ (3:1)	Silica Gel	Not stated	Aflatoxicol	43
EtOAc satd. with water EtOAc-CH$_3$CN (1:1) EtOAc-NH$_4$OH (1:1, upper phase)	Not stated Sil G-200 (RM)	Not stated MN	B$_{2a}$, G$_{2a}$ M$_1$	29 (Sec. 26.A17) 51
	Polygram Sil N-HR (RM)	MN	B$_1$, G$_1$	62
CHCl$_3$-AcOH (92:3) (9:1)	Silica Gel G-HR Silica Gel	MN Not stated	4 AT M$_1$, aspertoxin	35 63

199

Table I (continued)

Mobile Phase	Sorbent and Trade Name	Manufacturer	Aflatoxins Separated	Reference(s)
$CHCl_3$-acetone (96:4)	Silica Gel G-HR	MN	4 AT	60, 64
$CHCl_3$-acetone (9:1)	Silica Gel G-HR	MN	4 AT	29 (Sec. 26.002), 44, 54; 55, 61, 65, 66
	Silica Gel	S & S	4 AT	67
	Silica Gel	Not stated	M_1, aspertoxin	63
	Adsorbosil - 1	Appl. Sci.	D_1 acetate	68
	Adsorbosil - 1	Appl. Sci.	B_{2a}, G_{2a}	69
	Adsorbosil - 1	Appl. Sci.	4 AT	29 (Sec. 26.010), 70
	Silufol 254 UV (Silica Gel, RM)	Kavalier	B_1	49
	Alufolien (Silica Gel, RM)	Merck	4 AT	71, 72
	Polygram Sil N-HR	MN	4 AT	62, 73-76
	Not stated		M_1 acetate	29 (Sec. 26.090)
(95:5, 85:15)	Adsorbosil - 1, 5	Appl. Sci.	4 AT	29 (Sec. 26.002, 26.015)
$CHCl_3$-acetone (88:12) (85:15)	Silica Gel G-HR-25	MN	B_1	75, 76
	Silica Gel D-5	Camag	M_1, M_2	77

Solvent system	Adsorbent	Source	Aflatoxins	Ref.
CHCl$_3$-acetone (85:15)	SilicAR 4G, 7G	Mal	4 AT	29 (Secs. 26.002, 26.015), 49
	Silica Gel G	Merck	B$_1$	49
	Silufol 254-UV	Kavalier	B$_1$	49
(4:1)	Silica Gel D-5	Camag	M$_1$, M$_2$, 4AT, B$_2$a, G$_2$a	78
CHCl$_3$-acetone-n-amyl alcohol (80:10:10)	Alufolien (Silica Gel, RM)	Merck	M$_1$	79
	Sil G-200 (RM)	MN	M$_1$	51
CHCl$_3$-acetone-benzene-hexane-petroleum ether (6:1:1:1:1)	Silica Gel G-HR, H	MN	4 AT	32, 51, 54, 55
CHCl$_3$-acetone-MeOH (90:10:2)	Not stated	Not stated	M$_1$	80
CHCl$_3$-acetone-2-propanol (90:6:3)	Adsorbosil - 1	Appl. Sci.	M$_1$	81
(87:10:3)	Adsorbosil - 1	Appl. Sci.	M$_1$, M$_2$, para-siticol, water adducts & acetates; M$_1$, M$_2$, acetates, M$_2$a, B$_2$a, G$_2$a.	37

Table I (continued)

Mobile Phase	Sorbent and Trade Name	Manufacturer	Aflatoxins Separated	Reference(s)
$CHCl_3$-acetone-2-propanol				
(85:10:5)	Silica Gel G-HR	MN	M	29 (Sec. 26.083), 82
	Polygram Sil G	MN	4 AT	83, 84
	Adsorbosil - 1, 5	Appl. Sci.	M_1, M_2	29 (Sec. 26.083), 37, 85
	Adsorbosil - 1	Appl. Sci.	4 AT	86
	SilicAR 4G, 7G	Mal	M_1, M_2	29 (Sec. 26.083)
(85:12.5:2.5)	Adsorbosil - 1	Appl. Sci.	4 AT	87, 88
(85:15:2.5)	Silica Gel	Not stated	Aflatoxicol	43
(82.5:15:2.5)	Silica Gel G-HR	MN	4 AT	89
	Adsorbosil - 1	Appl. Sci.	H_1	91
(80:10:7)	Sil G-200 (RM)	MN	M_1	51
	Adsorbosil - 1	Appl. Sci.	M_1 TFA product	90
(8:1:1)	Adsorbosil - 1	Appl. Sci.	B_{2a}, G_{2a}, M_1	69
$CHCl_3$-acetone-2-propanol-water (88:12:1.5:1)	Adsorbosil - 1	Appl. Sci.	4 AT	29 (Sec. 26.010), 92

Solvent system	Adsorbent	Source	Compounds	References
CHCl$_3$-acetone-water (88:12:1.5)	Adsorbosil - 1	Appl. Sci.	4 AT	29 (Sec. 26.010), 62,93
			B$_1$, Q$_1$, M$_1$, M$_2$, B$_{2a}$, H$_1$	37,94,95
				91
	Chromatogram K 301 V (RM)	Eastman	B$_1$, G$_1$	62, 82
CHCl$_3$-EtOH (93:19)	Silica Gel G	Merck	B$_{2a}$, G$_{2a}$	96
CHCl$_3$-ether-AcOH (17:3:1)	Silica Gel G-HR	MN	B$_1$, G$_1$	97, 98
CHCl$_3$-ether-acetone (5:4:1)	Sil Gel G-HR	MN	4 AT	35
	Silica Gel G	Merck	B$_1$	41
CHCl$_3$-EtOAc (7:3)	Adsorbosil - 1	Appl. Sci.	4 AT	36
CHCl$_3$-n-hexane-acetone (85:20:5)	Silica Gel	S & S	B$_1$, B$_2$	99
(85:20:15)	Adsorbosil - 1	Appl. Sci.	H$_1$	91
(85:20:10)	Adsorbosil - 1	Appl. Sci.	M$_1$, M$_2$	37

Table I (continued)

Mobile Phase	Sorbent and Trade Name	Manufacturer	Aflatoxins Separated	Reference(s)
$CHCl_3$-n-hexane-acetone (85:20:15)	SilicAR TLC-4GF	Mal	4 AT	36, 100
	Adsorbosil - 1	Appl. Sci.	4 AT	36
	Adsorbosil - 1	Appl. Sci.	B_1, Q_1, M_1, B_{2a}	94
$CHCl_3$-hexane-petroleum ether-acetone (5:2:2:1)	Silica Gel G-HR	MN	4 AT	54, 55
$CHCl_3$-MeOH (99:1, 93:7)	Adsorbosil - 1	Appl. Sci.	4 AT	29 (Sec. 26.056), 35, 45
	Silica Gel G-HR	MN	4 AT	29 (Sec. 26.056), 35, 45, 73
	Silica Gel N-HR	MN	4 AT	95, 101, 102
	Polygram Sil N-HR (RM)	MN	B_1, G_1	62, 103
	Bio Sil A	Bio Rad	4 AT	62
	Silica Gel D-5	Camag	4 AT	35, 45, 50, 104
	SilicAR TLC-7G, 4G, 4GF		4 AT	35
	Silica Gel G	Mal	4 AT	35, 50
	Silica Gel G	Merck	4 AT	2, 45, 49, 50, 73, 105
	Silica Gel G	Serva	4 AT	43, 45

Solvent system	Adsorbent	Manufacturer		Reference
CHCl$_3$-MeOH (99:1, 93:7)	Silica Gel C	Acme	B$_1$	106
	Silufol 254-UV (RM) Chromatogram K301V (RM)	Kavalier	B$_1$	47, 49
	Alumina, neutral	Eastman	B$_1$, G$_1$	62, 83
		Woelm	4 AT	105
		Merck	4 AT	105
CHCl$_3$-MeOH-formic acid (190:10:1)	Silica Gel G	Merck	4 AT	7
CHCl$_3$-MeOH-hexane (10:2:1)	Silica Gel	Not stated	B$_1$, Q$_1$	95
CHCl$_3$-MeOH-water (98:1:1)	Adsorbosil - 5	Appl. Sci.	B$_1$, G$_1$	107
(90:1:1)	Adsorbosil - 5	Appl. Sci.	4 AT	108
CHCl$_3$-HOAc-hexane-petroleum ether-benzene-acetone-CH$_3$CN (2:3:1:1:1:1)	Silica Gel G-HR	MN	4 AT, M$_1$	32
CHCl$_3$-alcohol (95:5) n-propyl, isopropyl, n-butyl, t-butyl, n-amyl	Adsorbosil - 1	Appl. Sci.	M$_1$, M$_2$	37
	Silica Gel G-HR	MN	4 AT	37

Table I (continued)

Mobile Phase	Sorbent and Trade Name	Manufacturer	Aflatoxins Separated	Reference(s)
$CHCl_3$-hexane-benzene-acetone-petroleum ether (6:1:1:1:1)	Silica Gel H	MN	B_1, M_1	32, 55
$CHCl_3$-hexane-petroleum ether-acetone (5:2:2:1)	Silica Gel G-HR	MN	4 AT	54
$CHCl_3$-methylene chloride-isoamyl alcohol-formic acid (15:8:3:1)	Adsorbosil - 1	Appl. Sci.	B_1	109
$CHCl_3$-methylisobutyl-ketone (4:1)	Silica Gel G + oxalic acid	Merck	B_1, G_1	110
$CHCl_3$-2-propanol (99:1)	Not stated	Not stated	4 AT	29 (Sec. 26.010)
$CHCl_3$-pyridine (9:1)	Silica Gel	Not stated	M_1, asper-toxin	63
$CHCl_3$-trichloroethane-t-butyl alcohol-formic acid (85:10:4:1)	Adsorbosil - 1	Appl. Sci.	4 AT	36
$CHCl_3$-trichloroethylene-n-amyl alcohol-formic acid (80:15:4:1)	Silica Gel H	Merck	4 AT	29 (Sec. 26.010) 111

206

Solvent system	Adsorbent	Manufacturer	Detection	Reference
CHCl$_3$-trichloroethylene-n-amyl alcohol-formic acid (80:15:4:1)	Sil G-200 (RM)	MN	M$_1$	51
Methylene chloride-trichloroethylene-n-amyl alcohol-formic acid (80:15:4:1)	Silica Gel G-HR + H (1:1)	MN	4 AT	29 (Sec. 26.010), 111
	Silica Gel H	Merck	4 AT	66
Toluene-EtOH (1:1)	Silica Gel G-HR	MN	B$_1$, G$_1$	97
Toluene-EtOAc (8:1)	Silica Gel G-HR	MN	B$_1$, G$_1$	97
	Silica Gel G	Merck	B$_1$, G$_1$	50
	Bio Sil A	Bio Rad	B$_1$, G$_1$	50
	SilicAR TLC-4G, 4GF, 7G	Mal	B$_1$, G$_1$	50
Toluene-EtOAc-AcOH (5:4:1)	Silica Gel G	Merck	B$_1$	41
Toluene-EtOAc-formic acid (6:3:1) (5:4:1)	Silica Gel G-HR	MN	B$_1$, G$_1$	97
	Silica Gel H	Merck	B$_1$	112
Toluene-EtOAc-CHCl$_3$ (2:1:1)	Silica Gel G-HR	MN	B$_1$, G$_1$	97

Table I (continued)

Mobile Phase	Sorbent and Trade Name	Manufacturer	Aflatoxins Separated	Reference(s)
Toluene-EtOAc-CHCl$_3$-formic acid (25:13:10:2)	Sil G-200 (RM)	MN	M$_1$	51
Toluene-isoamyl alcohol-MeOH (90:15:2) (90:32:3)	Silica Gel G Silica Gel G	Merck Merck	4 AT B$_1$, G$_1$	60, 113 114
Xylene-t.-butanol-formic acid (94:5:1)	Silica Gel sprayed with t.-butanol-formic acid-water (10:1:2:5)	MN	4 AT	111
Xylene-isoamyl alcohol-MeOH (90:20:2)	Silica Gel G	Merck	4 AT	113
Water-acetone (9:1)	Polyamide TLC, CHCl$_3$ + MeOH	Woelm	4 AT	115
Water-formic acid (7:3)	Polyamide TLC, CHCl$_3$ + MeOH	Woelm	4 AT	115
Water-2-propanol (3:1)	Polyamide TLC, CHCl$_3$ + MeOH	Woelm	4 AT	115
Water-MeOH (11:9)	Polyamide TLC + CHCl$_3$-MeOH	Woelm	4 AT	115

Water-MeOH-acetone (8:7:5)	Polyamide TLC + CHCl$_3$-MeOH	Woelm	4 AT	115

a Solvents (v/v): AcOH = acetic acid; CH$_3$CN = acetonitrile; CCl$_4$ = carbon tetrachloride; CHCl$_3$ = chloroform; EtOH = ethanol; ether = ethyl ether; EtOAc = ethylacetate; MeOH = methanol; McOAc = methyl acetate.

b (RM) = manufactured TLC plates.

c Manufacturers: Bio Rad = Bio Rad Laboratories; Mal = Mallinckrodt; Merck = E. Merck; Eastman + Eastman Kodak; Appl. Sci. = Applied Science Laboratories; MN = Macherey & Nagel; S & S = Schleicher & Schuell.

d Aflatoxins: 4 AT = B$_1$, B$_2$, G$_1$, and G$_2$. TFA = trifluoroacetic acid.

Single-strength window glass is recommended. Glass of this
thickness is capable of withstanding handling, cleaning, and
heating during repeated use. Thinner glass is too fragile and
thicker glass tends to shatter on being heated to 100° C or above.
The glass must be clean and free of alkaline residues from cleaning agents, which tend to adhere to glass and which, if left,
will cause the aflatoxins to decompose. The alkaline residues
are best removed by rinsing with dilute mineral acid and deionized water.

Until recently, laboratory-prepared plates have been preferred; they were cheaper, were generally cleaner, and yielded
the best resolution. They can be prepared by using commercially
available spreaders such as those manufactured by Desaga, Camag,
Reeve-Angel, or Instrutec. Of these hand-operated spreaders, the
Instrutec is preferred because of a glass plate leveller feature
that compensates for differences in thicknesses among the glass
plates and allows the spreader to move smoothly from plate to
plate.

Over the years, commercially prepared TLC plates have
improved in uniformity of layer thickness and hardness. They can
be obtained with many of the silica gels listed in Table I and
separate the four aflatoxins (B_1, B_2, G_1, and G_2) quite well. On
occasion they require rinsing in a solvent such as methanol and
reactivation before use to remove impurities.

Optimum separation is achieved on plates 0.25-0.5 mm thick
(35), although plates up to 2 mm thick have been used for preparatory TLC isolation of the aflatoxins (51,117). For best resolution and quantitation, 20 × 20 cm plates are used, but 5 × 5 cm
and 10 × 10 cm plates have also been used to advantage (53,118),
particularly for preliminary screening (53). The so-called high-
performance TLC plate made from a narrow-range silica gel of
small particle size is the latest to become available but has not
been evaluated for aflatoxins.

The hand-made or manufactured TLC plates are scored along
edges to minimize edge effect during development, or are scored
in 1-cm-wide channels for use in densitometry and then activated
at 80°-160° C for ca. 1 hour. They are stored in a desiccator to
avoid contamination with atmospheric moisture and pollutants,
particularly acids, bases, and oxidants.

TLC plate coating materials other than silica gels have also
been tried but so far have not had much use. They include paper
(119), alumina (1,105), polyamide (115), formamide-impregnated
diatomaceous earth, or cellulose (33,34) and C_{18} hydrocarbon
bound to TLC silica gel. This system (reverse phase TLC) of separation on an organic stationary phase using aqueous developing

solvents holds promise because it works well for the HPLC of aflatoxins.

REFERENCE STANDARDS

Standard aflatoxins are obtainable from many sources (see Table II). The standard is often a source of error in analysis. The accuracy and purity of the standard or standard solution is the responsibility of the individual analyst. The standard solution must be prepared with the same solvent that will be used for the sample extract. Acetonitrile-benzene (2+98) is used for the more soluble aflatoxins (B_1, B_2, G_1, and G_2) (29, Sec. 26.008; 120, 121), whereas either chloroform (29 Sec. 26.084) or benzene-acetonitrile (9+1) (23) is used for the less soluble hydroxy-lated aflatoxins such as M_1 and M_2. Chloroform has the disadvantages of being too volatile and too easily decomposed (the decomposition products in turn may catalyze changes in the aflatoxins). Methylene chloride has been suggested as an alternate solvent (60), but, although it may be more stable and less toxic, it is even more volatile than chloroform and thus difficult to work with in quantitative analysis.

The standard solutions are calibrated (122-126) for concentration (8-10 µg/ml) by uv (29) and for purity (the absence of extraneous spots) by TLC (29). About 40-50 ng of each toxin is spotted in the test for purity. The calibration is important for (a) detection of errors in preparing the standard, (b) detection of concentration changes from adsorption of the aflatoxins to the glass walls of containers (127), and (c) detection of deterioration or decomposition during storage of the solutions (14,127). The aflatoxins are labile to light or uv irradiation both as solids (128-130) and in solution in such solvents as benzene, chloroform, ethanol, methanol (57,131-136), or water (54). When not used, the standard solutions should be stored in the dark and at low temperature (126).

SPOTTING OF THE TLC PLATE

The extract is dissolved for convenient application of amounts equivalent to 0.01 g of original sample for products expected to contain 5-100 µg/kg of aflatoxin B_1 and extract equivalent to 5 g of sample for products, such as meat, milk, and eggs, that are expected to contain less aflatoxin, i.e., between 0.02-10 µg (B_1 or M_1)/kg. The use of microliter syringes (with square-cut needle) and the application of 1-50 µl of solution per spot has

TABLE II.   SOURCES OF MYCOTOXINS REFERENCE STANDARDS

1. Aldrich Chemical Co., 940 St. Paul Ave., W. Milwaukee, WI 53233--Aflatoxins B_1, B_2, G_1, G_2, Ochratoxin A, Sterigmatocystin, Patulin, Penicillic Acid, Citrinin, Rubratoxin B_1.

2. Applied Science Laboratories, Inc., P.O. Box 440, State College, PA 16801--Aflatoxins B_1, B_2, G_1, G_2, and M_1, Citrinin, Diacetoxyscirpenol, Luteoskyrin, Ochratoxin A, Patulin, Penicillic Acid.

3. Dr. M. Bachman, Commercial Solvents Corp., Terre Haute, IN 47808--Zearalenone.

4. Dr. H. Burmeister, USDA/ARS, 1815 N. University St., Peoria, IL 61404--T-2 Toxin.

5. Calbiochem, 10933 N. Torrey Pines, LaJolla, CA 92037--Aflatoxins B_1, B_2, G_1, G_2, M_1, Ochratoxin A, Sterigmatocystin, Patulin, Rubratoxin B, Diacetoxyscirpenol, Roridin A, T-2 Toxin, Citrinin, Luteoskyrin.

6. Eureka Labs., Inc., 401 N. 16th St., Sacramento, CA 95186--Aflatoxin M_1.

7. ICN-K&K Labs, 121 Express St., Plainview, NY 11803--Aflatoxins B_1, B_2, G_1, G_2, Patulin, Diacetoxyscirpenol.

8. Dr. L. Leistner, Federal Meat Research Institute, 865 Kulmbach, Germany, for European distribution only--Aflatoxin M (Reference standards only).

9. Makor Chemicals Ltd., Box 6570, Jerusalem, Israel--Aflatoxin B_1, B_2, G_1, G_2, ^{14}C-B_1, M_1, M_2, Ochratoxin A, Sterigmatocystin, Patulin, Penicillic Acid, Rubratoxin B, Diacetoxyscirpenol, Roridin A, Verrucarin A, Citrinin Luteoskyrin, T-2 Toxin, HT-2 Toxin, T-2 Tetraol, T-2 Triol, Verrucarol, Zearalenone.

10. Moravek Biochemicals, 15302 E. Proctor Ave., City of Industry, CA 91745--^{14}C and ^3H-labeled Aflatoxin, Patulin.

11. Myco Lab Co., P.O. Box 321, Chesterfield, MO 63017, c/o Mr. Robert Sankey--Aflatoxins B_1, B_2, G_1, G_2, Zearalenone, T-2 Toxin, Diacetoxyscirpenol, Vomitoxin.

Table II (concluded)

12. Dr. A. E. Pohland, Food & Drug Administration, 200 C St.,
    S.W., Washington, DC 20204 (Reference standards only)--Afla-
    toxins B_1, B_2, G_1, G_2, Ochratoxins A, B, C, Sterigmatocystin,
    Patulin, Penicillic Acid, Zearalenone, T-2 Toxin, Citrinin.

13. RFR Corp., 1 Main St., Hope, RI 02831--Aflatoxins B_1, B_2, G_1,
    G_2, Patulin, Penicillic Acid, Sterigmatocystin.

14. Rijksinstitut voor de Volksgezondheid, P.O. Box 1, Bilthoven,
    The Netherlands--Aflatoxins B_1 (Reference standards only).

15. C. Roth, Postfach 1387, 7500 Karlsruhe 1, W. Germany--Afla-
    toxins B_1, B_2, G_1, G_2.

16. Senn Chemicals, CH-8157, Duesseldorf, West Germany--Aflatox-
    ins B_1, B_2, G_1, G_2, Ochratoxin A, Sterigmatocystin.

17. Sigma Chemical Co., P.O. Box 14508, St. Louis, MO 63178--Afla-
    toxins B_1, B_2, G_1, G_2, Citrinin, Diacetoxyscirpenol, Ochra-
    toxin A, Luteoskyrin, Patulin, Penicillic Acid, Rubratoxin B,
    Sterigmatocystin, Zearalenone.

18. Mr. R. M. Stubblefield, USDA/ARS, 1815 N. University St.,
    Peoria, IL 61604--Aflatoxin M_1 and M_2.

19. Supelco, Inc., Supelco Park, Bellefonte, PA 16823--Aflatoxins
    B_1, B_2, G_1, G_2, Citrinin, Diacetoxyscirpenol, Ochratoxin A,
    Luteoskyrin, Patulin, Penicillic Acid, Rubritoxin B, Sterig-
    matocystin, Zearalenone.

20. TCI Tridom Chemical, Inc. (Fluka AG), 255 Oser Ave., Haup-
    pauge, NY 11787--Aflatoxins B_1, B_2, G_1, G_2, Ochratoxin A,
    Sterigmatocystin, Rubratoxin B, Diacetoxyscirpenol.

become standard practice in aflatoxin methodology. Application
of less than 1 μl without special equipment may give rise to
large errors; likewise, the use of large volumes can result in
diffuse spots and increased chance of damaging the silica gel
layer. Automatic spotters, commercially available, can be used
to overcome these problems.

Application of sample should be done rapidly under subdued
light without pricking or marring the surface, in an atmosphere
of less than 60% relative humidity. At higher humidity the silica gel can be protected (a) by heating the plate during spotting,
(b) by covering the silica with a clean glass plate separated
from the TLC plate by applying a narrow strip of masking tape
along both sides (53,137), (c) by spotting under inert gas in a
spotting chamber such as that marketed by Brinkmann, Inc., or (d)
by spotting the plate after it has been equilibrated with a solvent (50). The control of moisture is of utmost importance,
often spelling the difference between success and failure (35,64).
Some moisture in the silica gel improves the resolution and eliminates tailing. Under very dry conditions, it is necessary to add
water to the developing chamber (70) or to the solvent (62,93).
However, too much moisture results in poor resolution, slow development, and excessively high R_f's. Techniques to control humidity also reduce chances that the plate will be contaminated with
airborne contaminants, such as fluorescent lint or pollutants,
that can cause fading of the aflatoxins. The fading phenomenon
is a particularly bad problem in the summer, during periods of
high humidity and air pollution (102,137). The change of aflatoxins to nonfluorescent substances is probably catalyzed by silica
gel and promoted by uv irradiation and acids (128).

For quantitation of the aflatoxins by densitometry, amounts
of each toxin are spotted according to the level expected in the
sample, the level at which analysis is desired, and the relative
fluorescence intensities of the aflatoxins (138,140). The intensity of fluorescence of the individual aflatoxins relative to
each other depends on the type of silica gel, residual solvent on
the TLC plate, and the uv light or densitometer characteristics.
Even when these parameters are controlled, the plate-to-plate
variation in fluorescence intensity dictates that each plate have
its own set of standards. The amounts of the aflatoxins commonly
spotted (in duplicate) are 5 ng each of aflatoxin B_1 and G_1 , 1
ng each of aflatoxins B_2, G_2, M_1, M_2, B_{2a}, G_{2a}, P, Q, H, and aflatoxicol. The TLC plates are usually prescored in 1-cm-wide chan-
nels with commercially available scoring devices. The samples
can also be applied as bands (129).

For visual estimation 1, 2, 3, and 5 μl portions of standard
solution (1, 1, 0.2, 0.2 ng/μl B_1, C_1, B_2, and G_2; respectively)

are used. The same volumes of extract solution are spotted after
the aflatoxin concentration in the sample extract has been adjusted to approximately that of the standard solution, based on an
educated guess or a preliminary TLC analysis. The amount of aflatoxin in the extract is then determined by visually matching the
intensity of fluorescence of the sample spot(s) with that of the
standard spots. To avoid erroneous identification of unknown
spots as aflatoxin, the sample should also be cochromatographed
with superimposed standard aflatoxins (internal standard). This
procedure helps overcome the confusion that may arise from uneven
development across the TLC plate, and from the interaction of
aflatoxin with constituents of the extract, particularly fatty
materials (14).

DEVELOPMENT OF THE CHROMATOGRAM

The TLC should be developed as soon as possible with the chosen
solvent (discussed in the next section). Only one chromatogram
should be developed at a time in a standard chamber (size 25 ×
25 × 10 cm). The chamber should be well insulated to minimize
temperature gradients (not necessary if the chamber is glass) and
sealed to prevent solvent loss. For most solvent combinations
the best resolution is obtained in a chamber not equilibrated
with solvent before the plate is developed. Equilibration, if
used, speeds up the development and gives uniform R_f's across the
plate, but at the expense of poorer resolution. Chambers of very
small volume such as the "sandwich chamber" where little evaporation from the plate surface can take place require special mobile
phases (35,111,141). A development chamber for controlling the
vapor atmosphere is available from Camag, Inc. The conditions
have been worked out for separation of the aflatoxins using this
system (64). Development (10-12 cm) should take 30-90 minutes
depending on the solvent, the silica gel particle size, and activity (moisture content). A fixed volume of freshly mixed solvent
should be used and preferably placed in a solvent trough rather
than in the bottom of the chamber. This procedure saves solvent
and improves reproducibility and resolution of the chromatography
(29, Sec. 26.019).

MOBILE PHASES

In choosing solvents and solvent combinations the analyst can
take advantage of a wide range of solvent selectivities for the
individual aflatoxins and interfering constituents of the extract

and greatly improve the analytical results. Tables I and IV list
many of the mobile phases that have been used. The neutral ace-
tone-chloroform mixture is recommended for testing the performance of silica gels and other conditions used for TLC (29). The
acidic developers, toluene-ethyl acetate-formic and benzene-
acetic acid-methanol, are less affected by silica gel properties.
They give excellent resolution of aflatoxins B_1, B_2, G_1, and G_2
from each other and from certain extract interferences, even
though many acidic and most other materials are moved to higher
R_f's relative to the aflatoxins than they are with acetone-
chloroform. Benzene-ethanol-water and ether-methanol-water are
examples of developers in which the separation may take place by
partition. The moisture content of the silica gel has less of an
effect on separations with aqueous solvents than with mixtures
containing no water.

QUANTITATION

The aflatoxins after separation by TLC can be quantitated both in
situ and in solution after elution from the silica gel. Some of
the approaches taken include the following: The spots are
scraped from the plate, and the aflatoxins are extracted from the
silica gel with a polar solvent such as chloroform-methanol
(3+1 - 2+1) (29). The eluted toxins are then determined by uv
(103,120), fluorometry (96,131,142,143), laser fluorometry (144),
polarography (145), or radiography (94,114). These are time-
consuming procedures and may incur losses of up to 50% of the
aflatoxins (131), particularly the less stable compounds such as
B_1, G_1, and M_1 (12). In addition, the limits of determination
are increased as the recovered toxins are diluted in the large
solvent volumes needed for analysis. A solution to these problems may be a currently available apparatus capable of quantitatively eluting aflatoxins from TLC spots without disturbing the
silica gel (146).
     The estimation of aflatoxins visually or by densitometry by
measuring the intensity of fluorescence of the aflatoxin zones is
the most widely used technique. It can be done equally successfully in several different ways, according to the analyst's pref-
erences. Basically, in visual estimation two techniques have
been used. One is the method of serial dilution of unknown
extracts and standards to the point at which fluorescence is not
detectable (about 0.1 - 0.5 ng for aflatoxins B_1 or G_1 [8,13]).
This point of extinction depends on the intensity of the uv light,
the silica gel, and the residual solvent in the silica gel (47,71,
72,128,147), as well as the compactness of the spots, the quantity

and type of extract interferences, the darkness of the room, and
the visual acuity of the observer.

The second technique for visual estimation involves comparing the fluorescence intensity of extract spots with those of
standard aflatoxin spots. Differences of about 20% are detectable in the range of 0.2 - 10 ng aflatoxin (44). For viewing the
TLC spots a Chromato-Vue cabinet equipped with one or more 15 W,
longwave uv lights is adequate. However, the contrast can be
enhanced using a Transilluminator (Ultraviolet Products) under
the TLC plate and cutoff filters in the viewing window. The eyes
should be allowed to adjust to subdued lights before the fluorescence intensity comparisons are made. If the intensity of the
unknown spot is between those of two of the standard spots it is
interpolated as half the difference of the volume or amounts in
the two spots, i.e., 2.5 if between 2 and 3 μl or ng. If the
sample falls outside the range of the standard spots, it must be
diluted or concentrated and rechromatographed; estimates must
never be made by extrapolation from a series of standard spots.
To increase the precision and objectivity of visual estimation,
the Kodak "Gray Scale" has been used to visually compare the
intensity of aflatoxin B_1, G_1, and M_1 TLC spots (83,84,104).
The technique was applied to standards and not to extracts and
has not been evaluated by comparison with other techniques.

Analysis by visual estimation is subjective; because the
smallest detectable difference between spots is about 20% (44),
the precision obtainable is about 15%. On the other hand, instrumental in situ quantitation is objective and can, at least in
theory, give precision of better than $\pm$ 5%. Densitometry by uv
absorption spectrophotometry has been used (73,99,148), but
because it requires microgram amounts of toxin for detection, it
is not practical for analysis of samples in the μg/kg range.
Laser fluorometry has also been used, and has detected as little
as 0.2 ng of B_1 (149) but is still in the experimental stage.
Fluorodensitometry is the most widely used technique and is gradually supplanting visual estimation. Again, good chromatography
with well-resolved, small spots is essential for good quantitation, even more than for visual estimation, and the same factors
influence both--humidity, silica gel, developing solvent, solvent
residues, layer thickness and uniformity, contaminants, R_f, and
type and mode of operation of instrument. The TLC densitometry
procedures as described even in collaboratively tested or "Official Methods" (29) are very general, more in the form of guidelines than procedures. This tendency is an outgrowth of the variety of procedures used in different laboratories.

One of the basic problems in the instrumental measurement of
the aflatoxin spots is minimizing the effect of extract

interferences, a problem as yet only partially solved. One
approach is that of standard addition, i.e., graded amounts of
standard aflatoxin are added to the unknown extract (44). The
observed responses are plotted against amounts added. The amount
of aflatoxin originally present in the unknown is calculated from
the observed response for the point at which no aflatoxin was
added. This technique is more tedious than others and offers no
compensating advantages. The second approach is to spot the
standards at four concentrations, close to and including the
concentrations of the unknowns, construct a standard curve for
each TLC plate, and make the determination of the unknown from
the standard curve. This would appear to be the safest approach
because the system is tested for linearity for each chromatogram.
The third approach as described in the AOAC Method (29) is the
simplest, most efficient, and most widely used. In this procedure the reproducibility and reliability of the TLC system are
established, the densitometric parameters are optimized, and the
linearity of response with aflatoxin concentration is determined.
The samples and standards are then spotted in duplicate at similar concentrations. In this procedure only an occasional check
on the response linearity is done as an analytical quality control.

In the scanning of the chromatogram the following precau-
tions may help improve the results. Because the amount and
nature of solvent remaining in the silica affects the fluorescence intensity of the aflatoxins, volatile developing solvent
must be rigorously eliminated. For the solvents commonly used,
drying the plate 15 minutes at 25° C or 5 minutes at 40°-80° C is
sufficient. Fading of the aflatoxins can be prevented by covering the chromatogram while still hot with a clean glass plate
separated with narrow strips of masking tape at the edges (53,
137). Although it is advisable to scan the plate as soon after
development as possible, some delay can be tolerated. If delay
is necessary, the chromatograms are best stored in the dark, protected from atmospheric reactants, preferably in a refrigerator.
Storage for as long as a week has shown no problems (128).

The general location of the aflatoxins can be marked to
expedite the scanning and aid in the interpretation of the data.
Great care must be taken in correctly identifying spots as aflatoxins. The R_f's may vary considerably across the plate and even
between neighboring channels. The duration and intensity of uv
and other light exposure must be kept to a minimum to avoid
undesirable changes. Exposure of aflatoxin to light for 4 hours
has resulted in a 40% reduction (128).

The spots should be scanned in the direction parallel to the
direction of development and from low to high background

(increasing base line). If fading of the aflatoxin spots occurs, the standards usually fade faster than the aflatoxin spots in the extracts. The standards should therefore be scanned before the unknowns. The TLC and scanning results should be evaluated as they are being carried out. Duplicate spots should agree within + 5% or else be rescanned or possibly rechromatographed. The spot first scanned should be rescanned and the two values compared should be in agreement. If the response ratios B_1/B_2 and G_1/G_2 decrease, this is usually indicative of decomposition of B_1 and G_1, whereas a change in response for the B compared with that for the G aflatoxins may be due to loss of residual solvent from the silica gel.

As already discussed, many factors can contribute to changes in the fluorescence intensity; one additional concern arises from the use of instruments equipped with high-intensity light sources, such as xenon lamps. The ozone produced by these lamps can also attack the spots (137). The lamp housings on such instruments must be properly vented.

A variety of TLC equipment has been used for aflatoxin analysis and is listed in Table III. Common features include single beam operation, a source of excitation light close to 365 nm, a filter or monochromator, and a uv cutoff filter passing light above 420 nm to a detector capable of detecting the fluorescence of 1 ng of aflatoxin on the layer. The detector can be located so as to measure reflected fluorescence or fluorescence transmitted through the glass. In the so-called flying spot densitometer a very small detector (about 1 mm^2) is used to scan the TLC in a zigzag manner and the signal is integrated. The instrument's detection limit is about the same as that of the conventional scanners. Errors from irregular spots and noisy base lines are reduced.

The instrumental parameters to be optimized include the following: selecting excitation light filters or monochromator setting for maximum response; setting the gain for the maximum signal-to-noise ratio; selecting scanning and recorder speeds; selecting uv cutoff filter or monochromator setting before the detector for maximum response, minimum noise, and ease and/or accuracy in measurement of instrument response; and selecting optimum slit widths and lengths for the best signal-to-noise ratio consistent with adequate resolution of adjacent spots. The output from the instrument can be a strip chart recording on which peak areas can be measured manually or mechanically, or it can be electronically integrated using any of the many types of integrators. These range from systems that are manually started and stopped to automatic, microprocessor-operated systems capable of base line tracking, base line correcting, and data handling.

TABLE III. THIN LAYER CHROMATOPLATE SCANNERS FOR ANALYSIS OF AFLATOXINS

Type	Model	Manufacturer	Reference(s)
Filter fluorometer, reflectance	Uniscan 900	Perkin-Elmer (Nester Faust)	70
Filter fluorometer, transmittance	Model 530	Photovolt Corp.	44, 46, 64, 74, 78, 87, 89, 93, 150
Filter fluorometer, reflectance	Model 111 with TLC scanner	GK Turner Associates	107, 151, 152
Filter fouorometer, reflectance	Model 111 with T-Scanner	GK Turner Associates Camag, Inc.	27
Fluorodensitometer	SD-91	Ozumor	135
Filter fluorometer	Science and Mechanics Darkroom Densitometer	Olden Camera Lens Co.	101
Filter fluorometer	Laboratory constructed		47
Spectrophotofluorometer, reflectance	Aminco-Bowman Model 4-8202 with TLC Scanner 4-8247	American Instrument Corp.	129, 133, 134

220

Instrument	Model	Manufacturer	References
Spectrodensitometer, reflectance or transmittance	SD 3000	Schoeffel Instrument Corp.	37, 53, 68, 81, 85, 86, 94, 99, 137
Spectrophotometer	UV Scan	Carl Zeiss	60, 61, 72, 73, 136, 153
Spectrophotometer, reflectance or transmission, zigzag scanning head (flying spot)	CS-900	Shimadzu	
Filter fluorometer, reflectance	TLD-100	Vitatron Scientific Instruments	59, 66, 154, 155, 156
Fiber optics, reflectance or transmission	K 49500	Kontes	

After the appropriate instrumental parameters are selected,
the linearity of response to concentration must be determined by
preparing a standard curve covering the range of concentrations
of interest. For instruments such as the Schoeffel, peak areas
appear best related to concentration, whereas the Zeiss spectro-
photofluorodensitometer (72) accurately measures concentration by
peak height as well.

MULTIPLE DEVELOPMENT AND TWO-DIMENSIONAL TLC

For many commodities, i.e., eggs, some cheeses, tissue, milk,
figs, spices, toasted cottonseed and corn, mixed feeds, and fish-
meals, one-dimensional TLC is not adequate for separating the
aflatoxins from interfering constituents. It is also not adequate when levels lower than 1 μg/kg are to be detected.
    The simplest technique for improving separation is multiple
development with the same or different solvents. Table IV lists
a number of these systems. An extension of this technique is a
procedure in which the sample and standard are spotted along a
horizontal line midway on a silica gel-coated aluminum sheet.
The sheet is predeveloped with ethyl ether, which moves many
interferences to the top of the chromatogram but leaves the aflatoxins near the origin. The top of the chromatogram is cut off
and the sheet is rotated 180° and developed with a solvent appropriate for aflatoxins (116). An apparatus that automatically
performs multiple development is commercially available. It has
been applied to aflatoxin in a preliminary test (165-167).
    For samples in which the above techniques are inadequate,
the final resort is two-dimensional TLC, a very powerful technique because of the many different solvent options which can be
used. Table IV lists a number of systems that have been successfully applied. The technique is time-consuming, because only one
sample can be applied per TLC plate (178). Usually the standards
are developed in only one direction, but this appears to be satisfactory, as demonstrated by collaborative studies (147,177,178).
Both visual methods and densitometry can be used for estimation
(53,61).
    Two difficulties arise in the fluorodensitometry of two-
dimensional thin layer chromatograms: properly locating the spot
over the sensor, and placing the reference standards on the plate.
Generally, only the sample spot can be developed in two directions; the standards are developed in one direction only. Several
investigators have addressed these problems successfully. One
investigator covered the chromatogram with a clean glass plate
and covered the area around the aflatoxin spot with masking tape

for guidance in scanning (53); another applied several spots
diagonally across the TLC plate to allow development of the plate
in two directions without running the spots together or over each
other (61).

CONFIRMATION OF IDENTITY

Because initial identification of unknowns in a chromatogram is
based on the similarities of their R_f values with those of standard aflatoxins, additional proof of identification is needed.
This need can hardly be overemphasized, particularly for regulatory samples and for commodities with a limited history of analysis or incidence of contamination. Several techniques that have
been used for confirmation of identity are listed in Table V.
Rechromatography with several different solvents such as those
listed in Table I and the use of the spray reagents in Table V
are not specific tests and do not give conclusive positive identification. They are conclusive only when they show that an
unknown is not aflatoxin. The mineral acid sprays are widely
used tests; the acid changes the fluorescence of the aflatoxins
from blue to yellow. If an unknown remains blue fluorescent on
spraying, one can safely conclude that it is not aflatoxin.
    The test for biological activity in fertile chicken eggs
(29,197) has been applied for the past 10 years in several laboratories. In FDA tests of more than 400 samples it has demonstrated 100% correlation with the chemical derivative tests (29).
    The preparation of chemical derivatives of the aflatoxins
with changed chromatographic properties is the simplest method
for positive confirmation of chemical identity. Widely used
tests for aflatoxin B_1, G_1, and M_1 involve the formation of a
water addition product, with trifluoroacetic acid as the catalyst,
or formation of acetates. The derivatives of the unknowns are
then compared by TLC with the derivatives of standard aflatoxins.
By the preferred methods (29) the derivatives are formed directly
on the TLC either before or after it is developed.
    More specific than any of the above techniques is that of
mass spectrometry. Recent improvements have greatly lowered its
detection limits, now approaching nanogram amounts of aflatoxin
B_1 from crude extracts.

CONCLUSION

The number of aflatoxin TLC methods published each year has been
on the decline because methods are now available to handle nearly

TABLE IV. MULTIPLE DEVELOPMENT OR TWO-DIMENSIONAL TLC OF AFLATOXINS

Mobile Phase[a]	Sorbent[b] and Trade Name	Manufacturer[c]	Aflatoxins[d] Separated	Reference(s)
Multiple Development[e]				
1. Benzene-EtOAc-formic acid (80:20:0.5) 2. $CHCl_3$-acetone (9:1)	Silica Gel G-HR	MN	4 AT	157
1. Benzene-hexane (3:1) 2. Benzene-AcOH (9:1) or Toluene-EtOAc-formic acid (5:4:1)	Adsorbosil - 5	Appl. Sci.	4 AT	158
1. $CHCl_3$-MeOH (97:3) 2. Ether	Silica Gel G Silufol 254 UV (RM)	Merck Kavalier	B_1	49
1. Ether 2. $CHCl_3$-acetone (9:1)	Adsorbosil - 1	Appl. Sci.	4 AT	159
or 2. $CHCl_3$-acetone-n-amyl alcohol (80:10:10)	Alufolien (Silica gel, RM)	Merck	M_1, M_2	79
or 2. $CHCl_3$-acetone-2-propanol (85:15:2.5)	Not stated		Aflatoxicol	43

or 2. Cut foil turn 180° $CHCl_3$-acetone-water (88:12:0.2)	Alufolien (Silica gel, RM)	Merck	4 AT	27
or 2. $CHCl_3$-MeOH (97:3)	Silica Gel G	Merck	B_1, G_1	132, 160
or 2. Cut foil turn 180° $CHCl_3$-trichloro-ethylene-n-amyl alcohol-formic acid (80:15:4:1)	Alufolien (Silica gel, RM)	Merck	4 AT	27
or 2. $CHCl_3$-trichloroethylene amyl alcohol-formic acid (80:15:4:1)	Alufolien, Silica Gel 60 F - 254 (RM)	Merck	4 AT, M_1	161
or 2. $CHCl_3$-trichloroethylene t-butanol-formic acid (85:10:4:1)	Silica Gel D 5	Camag	4 AT, M_1	162
or 2. EtOAc-CH_3CN (1:1)	Alufolien (Silica gel, RM)	Merck	M_1, M_2	79
1. Ether twice 2. Benzene twice 3. $CHCl_3$-trichloroethylene-n-amyl alcohol-formic acid (80:15:4:1)	Silica Gel 60 (Alufolien, RM)	Merck	4 AT	116

225

Table IV (continued)

Mobile Phase	Sorbent and Trade Name	Manufacturer	Aflatoxins Separated	Reference(s)
or 2. Toluene-isoamyl alcohol-MeOH (90:15:2)	Not stated		B_1	133
1. Hexane 2. CHCl$_3$-acetone (9:1)	Silica Gel G-HR	MN	B_1, G_1	163
1. MeOAc 2. CHCl$_3$-MeOH (98:2)	Silica Gel G-HR	MN	4 AT	164

Programmed Multiple Development

Mobile Phase	Sorbent and Trade Name	Manufacturer	Aflatoxins Separated	Reference(s)
CHCl$_3$-tetrahydrofuran-EtOAc (18:1:1)	Silica Gel 60 F 254 (RM)	Merck	4 AT	165–167

Two Dimensional Thin Layer Chromatography

Mobile Phase	Sorbent and Trade Name	Manufacturer	Aflatoxins Separated	Reference(s)
1. Benzene-MeOH-AcOH (18:1:1) 2. CHCl$_3$-acetone (9:1)	Adsorbosil - 1	Appl. Sci.	B_1, B_2	98,109
1. CHCl$_3$-acetone (9:1) 2. CHCl$_3$-acetone (9:5)	Silica Gel G-HR	MN	B_1	168
or 2. CHCl$_3$-acetone-2-propanol (86.5:10:3.5)	Not stated		M_1	169
or 2. EtOAc-2-propanol-water (10:2:1)	Silica Gel G-HR	MN	4 AT	170

Solvent system	Adsorbent	Source	Compound	Ref.
or 2. Spray with MeOH-H$_3$PO$_4$ (4:1)	Silica Gel	Not stated	4 AT	171
+ 3. Benzene-AcOH-water (85.5:15:0.5)				
+ 4. Benzene-CHCl$_3$-MeOH-AcOH-water (65:15:8.5:10:1.5)				
or 2. Toluene-EtOAc-formic acid (5:4:1)	Adsorbosil - 1	Appl. Sci.	4 AT	172, 173
or 2. Toluene-EtOAc-formic acid (5:4:1)	Silica Gel (Alufolien, RM)	Merck	B$_1$	174
1. CHCl$_3$-acetone-MeOH (90:10:2)	Silica Gel G-HR (RM)	MN	M$_1$	61, 175
2. Ether-MeOH-water (94:4.5:1.5)				
1. CHCl$_3$-acetone-2-propanol (90:10:1)	Adsorbosil - 1	Appl. Sci.	M$_1$ and M$_1$ TFA reaction product	52
2. CHCl$_3$-acetone-2-propanol (85:10:7)				
1. CHCl$_3$-acetone-2-propanol (80:15:5)	Silica Gel 60 (Alufolien, RM)	Merck	M$_1$	176

227

Table IV (continued)

Mobile Phase	Sorbent and Trade Name	Manufacturer	Aflatoxins Separated	Reference(s)
2. Toluene-EtOAc-formic acid (6:3:1)				
1. Ether				
2. Acetone-CHCl$_3$ (1:9)	Adsorbosil - 1 (non activated)	Appl. Sci.	B$_1$	53
1. Ether	Silica Gel G	Merck	B$_1$, G$_1$	41
1. CHCl$_3$-MeOH (97:3)				
2. CHCl$_3$-acetone (9:1)				
2. CHCl$_3$-acetone (98:2)				
1. Ether-MeOH-water (96:3:1)				
2. CHCl$_3$-acetone (9:1)	Adsorbosil - 1	Appl. Sci.	B$_1$	177
1. Ether-MeOH-water (94:4.5:1.5)				
2. CHCl$_3$-acetone (9:1)	Silica Gel G	Merck	B$_1$, G$_1$	41, 154, 178, 179
1. Ether-MeOH-water (94:4.5:1.5) (sat'd chamber)				
2. CHCl$_3$-acetone-2-propanol (85:10:5)	Silica Gel G (RM)	Merck	M$_1$, M$_2$	180
or 2. CHCl$_3$-acetone-MeOH (90:10:2)	Silica Gel G-HR	MN	B$_1$, M$_1$	181

1 and 2 as for Ref. 181 and in   Silica Gel G-HR   MN B_1, M_1 181
direction 1 again with: $CHCL_3$-
MeOH-HOAc-water (92:8:2:0.8)

1. Toluene-EtOAc-formic acid   Adsorbosil – 1   Appl. Sci.   4 AT   182
   (5:4:1)
2. $CHCl_3$-acetone (9:1)

a, b, c, d See notes a, b, c, d in Table I.

e Under column Mobile Phase numbers 1-5 represent sequence of development in the same
  direction.

f Under column Mobile Phase numbers 1 and 2 indicate development in first and second dimension
  and numbers 3 and 4 repeated development in direction 1.

229

TABLE V. CONFIRMATION OF IDENTITY OF AFLATOXINS

	Aflatoxin	Detection	Reference(s)
Reduction with sodium borohydride	B_1, B_2	TLC	183
Preparation of acetate derivative	B_1, G_1, M_1	TLC	14, 29 (Secs. 26.066-71), 184, 185, 186
Preparation of water adduct	B_1, G_1, M_1	TLC	29 (Sec. 26.A17), 90, 172, 181, 186-189
TLC-mass spectrometry	B_1	Mass spectrometry TLC	51, 190-193
Irradiate with uv on TLC plate and develop	B_1		130
TLC sprays:			
10% HCl	B_1	TLC	106
25% Inorganic acid	4 AT	TLC	189
33% Sulfuric acid	4 AT	TLC	29 (Secs. 26.A15, 26.A17)
20% Sulfuric acid, heat 15 min, 120°	4 AT	TLC	112
2,4-Dinitrophenylhydrazine	4 AT	TLC	106, 194, 195
Antimony chloride	4 AT	TLC	153

230

	Aflatoxin	Detection	Reference(s)
TLC sprays			
Basic bismuth nitrate	B_1	TLC	153
Anisidine	B_1	TLC	196
Iodine	B_1	TLC	106
TLC: Development in several different solvents	4 AT	TLC	29 (Sec. 26.056), 49, 51, 136, 172, 189
Preparatory TLC isolation	B_1	Chick embryo toxicity	29 (Secs. 26.073-26.078), 197

231

any product, and thus the need for new methods has decreased.
TLC methods constitute the bulk of these methods. The precision
and accuracy of the TLC technique can be as good as that of any
spectroscopic method, namely, 1-2%. In aflatoxin analysis, workers have been satisfied with 5-20% because of the large errors
resulting from variability due to sampling heterogeneous materials. TLC will continue to be widely used for many reasons: its
economy and simplicity, its versatility, its usefulness in screening large numbers of diverse materials, its utility for scouting,
methods development, and separation or preparative TLC for use
with other techniques such as HPLC or mass spectrometry.

    This brief review is intended to be a helpful guide in solv-
ing analytical problems--from the selection of a method, solvent,
and adsorbent to performing of the quantitation. It is intended
to be helpful in identifying causes of problems and supplying
remedies for their solution, thus providing better control and
quality of analytical results.

    For the future--adsorbents, instrumentation, and techniques
will continually improve and promise possibilities for better
accuracy, greater precision, lower levels of detection, and the
use of automation.

REFERENCES

1. J. H. Broadbent, J. A. Cornelius, and G. Shone, Analyst 88,
   214 (1963).
2. H. De'Iongh, J. G. van Pelt, W. O. Ord, and C. B. Barrett,
   Vet. Rec. 76, 901 (1964).
3. S. Nesheim, J. Assoc. Off. Anal. Chem. 47, 1010 (1964).
4. A. D. Campbell, INCAP Publ. L-6, 76 (1976).
5. A. D. Campbell, Toxic Infections of Food Origin, edited by
   A. E. Olszyma Marzys (Proceedings of the Inter-American Conference, INCAP, Guatemala, 1974), pp. 76-88.
6. J. de Waart, C. van Zadelhoff, and A. Edelbroek, Alimenta 13,
   35 (1974).
7. L. Fishbein and H. L. Falk, Chromatogr. Rev. 12, 44 (1970).
8. J. G. Heathcote and J. R. Hibbert, Aflatoxins: Chemical and
   Biological Aspects (Elsevier, New York, 1978), pp. 54-82.
9. C. W. Hesseltine and O. L. Shotwell, Pure Appl. Chem. 35, 259
   (1973).
10. W. Horwitz, J. Assoc. Off. Anal. Chem. 59, 238 (1976).
11. B. D. Jones, in Mycotoxic Fungi and Chemistry of Mycotoxins,
    edited by T. D. Wyllie and L. G. Morehouse (Marcel Dekker,
    New York, 1977), pp. 201-237.
12. B. D. Jones, Alimenta 13, 18 (1974).

13. B. D. Jones, Methods of aflatoxin analysis. Trop. Prod. Inst. Rep. No. G70. London, England (1972).
14. F. Kiermeier, Z. Lebensm. Unters. Forsch. 148, 331 (1972).
15. L. O. Mijatov, Acta Vet. (Belgrade) 25, 327 (1976).
16. S. Nesheim, in Trace Organic Analysis: A New Frontier in Analytical Chemistry, edited by S. N. Chesler and H. S. Hertz (National Bureau of Standards Special Publication 519, 1979), pp. 355-372.
17. T. Romer, Feedstuffs 48, 18 (1976).
18. U. Samarajeewa and S. N. Aseculeratne, J. Food Sci. Technol. 12, 27 (1975).
19. P. L. Schuller, W. Horwitz, and L. Stoloff, J. Assoc. Off. Anal. Chem. 59, 1315 (1976).
20. O. L. Shotwell and M. L. Goulden, J. Assoc. Off. Anal. Chem. 60, 83 (1977).
21. L. Stoloff, Clin. Toxicol. 5, 465 (1972).
22. L. Stoloff, in Quantitative Thin Layer Chromatography, edited by J. C. Touchstone (John Wiley-Interscience, New York, 1973), pp. 95-142.
23. L. Stoloff, J. Assoc. Off. Anal. Chem. 62, 356 (1979).
24. T. B. Whitaker, J. W. Dickens, and R. J. Monroe, J. Am. Oil Chem. Soc. 51, 214 (1974).
25. T. B. Whitaker, M. E. Whitten, and R. J. Monroe, J. Am. Oil Chem. Soc. 53, 502 (1976).
26. W. A. Pons, Jr., and L. A. Goldblatt, in Aflatoxin, Scientific Background, Control, and Implications (Academic Press, New York, 1969), pp. 77-105.
27. H. Arnold, in Proceedings of Workshop: Health Hazard of Aflatoxins, Zurich, March 21-22, 1978 (Inst. for Toxicol. of ETH and the University of Zurich, 1978), p. 215-222.
28. P. L. Schuller and H. P. van Egmont, Ref. 27, pp. 176-195.
29. W. Horwitz, Ed., Official Methods of Analysis, 12th ed. (Association of Official Analytical Chemists, Washington, D. C., 1975).
30. R. O. Walker, Ed., Official and Tentative Methods of the American Oil Chemists' Society (American Oil Chemists' Society, Champaign, Ill., 1973).
31. W. C. Schaefer, Ed., Approved Methods of the American Association of Cereal Chemists (American Association of Cereal Chemists, Inc., St. Paul, Minn., 1969).
32. E. L. Strzelecki, Zentralbl. Veterinaermed. Reihe B. 25, 194 (1978).
33. A. Adye and R. I. Mateles, Biochim. Biophys. Acta 86, 418 (1964).
34. W. C. Jacobsen, W. C. Harmeyer, and H. G. Wiseman, J. Dairy Sci. 54, 21 (1971).
35. S. Nesheim, J. Am. Oil Chem. Soc. 46, 335 (1969).

36. D. M. Takahashi, J. Assoc. Off. Anal. Chem. 57, 875 (1974).
37. R. D. Stubblefield, O. L. Shotwell, and G. M. Shannon, J.
    Assoc. Anal. Chem. 55, 762 (1972).
38. E. Varsavsky and S. E. Sommer, Ann. Nutr. Aliment. 31, 539
    (1977).
39. L. Stoloff, S. Nesheim, L. Yin, J. V. Rodricks, M. Stack, and
    A. D. Campbell, J. Assoc. Off. Anal. Chem. 54, 91 (1971).
40. P. M. Scott, J. W. Lawrence, and W. van Walbeek, Appl. Micro-
    biol. 20, 839 (1970).
41. D. I. Kahlau and B. R. Gedek, Zentralbl. Veterinaermed. Reihe
    B 23, 230 (1976).
42. L. Friedman and L. Yin, J. Nat. Cancer Inst. 51, 479 (1973).
43. Z. A. Wong and D. P. H. Hsieh, Science 200, 325 (1978).
44. A. C. Beckwith and L. Stoloff, J. Assoc. Off. Anal. Chem. 51,
    602 (1968).
45. J. Janicki, K. Szebiotko, M. Kokorniak, M. Wiewiorowska, J.
    Chelkowski, and B. Godlewska, Nahrung 18, 27 (1974).
46. R. D. Stubblefield, O. L. Shotwell, C. W. Hesseltine, M. L.
    Smith, and H. H. Hall, Appl. Microbiol. 15, 186 (1967).
47. K. Szebiotko, J. Chelkowski, M. Kokorniak, W. Socalska, and
    M. Wiewiorowska, Nahrung 18, 185 (1974).
48. A. E. Waltking, G. E. Bleffert, M. Chick, and R. Fogerty,
    J. Am. Oil Chem. Soc. 50, 424 (1973).
49. B. Stefaniak, J. Chromatogr. 44, 403 (1969).
50. J. G. Heathcote and J. R. Hibbert, J. Chromatogr. 108, 131
    (1975).
51. F. Kiermeier, M. Miller, G. Weiss, and G. Behringer, Z.
    Lebens. Unters.-Forsch. 166, 274 (1978).
52. M. W. Trucksess, J. Assoc. Off. Anal. Chem. 57, 1220 (1974).
53. M. W. Trucksess, L. Stoloff, W. A. Pons, A. F. Cucullu, L. S.
    Lee, and A. O. Franz, J.Assoc.Off.Anal.Chem. 60, 795 (1977).
54. E. L. Strzelecki, Acta Microbiol. Pol. 4, 155 (1972).
55. E. L. Strzelecki, Pure Appl. Chem. 35, 297 (1973).
56. K. L. Hanna and T. C. Campbell, J. Assoc. Off. Anal. Chem. 51,
    1197 (1968).
57. W. Lijinsky and W. H. Butler, Proc. Soc. Exp. Biol. Med. 123,
    151 (1966).
58. J. Velasco, J. Am. Oil Chem. Soc. 46, 105 (1969).
59. M. Jemmali and T. R. K. Murthy, Z. Lebensm. Unters.-Forsch.
    161, 13 (1976).
60. H. Muller and V. Siepe, Deutsch. Lebensm. Rundsch. 74, 133
    (1978).
61. P. I. Schuller, C. A. H. Verhulsdonk, and W. E. Paulsch,
    Arzneim. Forsch. 20, 1517 (1970).
62. J. Reiss, J. Chromatogr. 49, 301 (1970).
63. J. V. Rodricks, A. D. Campbell, and J. M. Verrett, Nature 217,
    688 (1968).

64. R. A. de Zeeuw and H. S. Lillard, J. Assoc. Off. Anal. Chem. 54, 98 (1971).
65. R. M. Epply, J. Assoc. Off. Anal. Chem. 49, 473 (1966).
66. P. R. Beljaars and J. H. M. Fabry, J. Assoc. Off. Anal. Chem. 55, 775 (1972).
67. L. S. Lee and A. F. Cucullu, J. Agric. Food Chem. 26, 881 (1978).
68. W. A. Pons, Jr., A. F. Cucullu, L. S. Lee, H. J. Janssen, and L. A. Goldblatt, J. Am. Oil Chem. Soc. 49, 124 (1972).
69. P. B. Marsh, M. E. Simpson, and M. W. Trucksess, Appl. Microbiol. 30, 52 (1975).
70. W. A. Pons, J. Assoc. Off. Anal. Chem. 54, 870 (1971).
71. L. S. Oldham, F. W. Oehme, and D. C. Kelley, J. Milk Food Technol. 34, 349 (1971).
72. W. Postel and A. Gorg, Brauwissenschaft 29, 193 (1976).
73. H. Bosenberg and W. Neugebauer, Zentralbl. Bakteriol. Hyg. Abt. 1, Orig. B 155, 70 (1971).
74. J. L. Ayers and R. O. Sinnhuber, J. Am. Oil Chem. Soc. 43, 423 (1966).
75. L. M. Seitz and H. E. Mohr, Cereal Chem. 51, 487 (1974).
76. L. M. Seitz and H. E. Mohr, J. Assoc. Off. Anal. Chem. 59, 106 (1976).
77. I. F. H. Purchase and M. Steyn, J. Assoc. Off. Anal. Chem. 55, 1316 (1972).
78. J. C. Schabort and M. Steyn, Biochem. Pharmacol. 18, 2241 (1969).
79. F. Kiermeier and M. Buchner, Z. Lebensm. Unters.-Forsch. 164, 82 (1977).
80. K. Lemieszek, Zesz. Probl. Postepow. Nauk. Roln. 189, 263 (1977).
81. N. L. Brown, S. Nesheim, M. E. Stack, and G. L. Ware, J. Assoc. Off. Anal. Chem. 56, 1437 (1973).
82. D. S. P. Patterson and B. A. Roberts, Food Cosmet. Toxicol. 13, 541 (1975).
83. J. Reiss, J. Appl. Microbiol. 2, 183 (1975).
84. J. Reiss, Fresenius Z. Anal. Chem. 293, 138 (1978).
85. R. D. Stubblefield, G. M. Shannon, and O. L. Shotwell, J. Assoc. Off. Anal. Chem. 56, 1106 (1973).
86. W. A. Pons, Jr., A. F. Cucullu, and L. S. Lee, J. Assoc. Off. Anal. Chem. 56, 1431 (1973).
87. W. A. Pons, J. Assoc. Off. Anal. Chem. 51, 913 (1968).
88. W. A. Pons, A. F. Cucullu, A. O. Franz, and L. A. Goldblatt, J. Am. Oil Chem. Soc. 45, 694 (1968).
89. W. A. Pons, J. A. Robertson, and L. A. Goldblatt, J. Am. Oil Chem. Soc. 43, 665 (1966).
90. M. W. Trucksess, J. Assoc. Off. Anal. Chem. 59, 722 (1976).

91. A. S. Salhab and D. P. H. Hsieh, Res. Commun. Chem. Pathol. Pharmacol. 10, 419 (1975).
92. J. Karmelic, M. Israel, S. Benado, and C. Leon, J. Assoc. Off. Anal. Chem. 56, 219 (1973).
93. R. D. Stubblefield, G. M. Shannon, and O. L. Shotwell, J. Assoc. Off. Anal. Chem. 52, 669 (1969).
94. D. P. H. Hsieh, J. I. Dalezios, R. I. Krieger, M. S. Masri, and W. F. Haddon, J. Agric. Food Chem. 22, 515 (1974).
95. G. H. Buchi, P. M. Muller, B. D. Roebuck, and G. N. Wogan, Res. Commun. Chem. Pathol. Pharmacol. 8, 585 (1974).
96. K. K. Maggon and S. Gopal, Agric. Biol. Chem. 38, 681 (1974).
97. B. A. Roberts and D. S. P. Patterson, J. Assoc. Off. Anal. Chem. 58, 1178 (1975).
98. L. Yin, A. D. Campbell, and L. Stoloff, J. Assoc. Off. Anal. Chem. 54, 103 (1971).
99. E. H. Reimerdes, Arch. Inst. Pasteur Tunis 54, 335 (1977).
100. J. I. Teng and P. C. Hanzas, J. Assoc. Off. Anal. Chem. 52, 83 (1969).
101. R. E. Peterson, A. Ciegler, and H. H. Hall, J. Chromatogr. 27, 304 (1967).
102. T.-W. Kwon and J. C. Ayres, J. Chromatogr. 31, 420 (1967).
103. J. Reiss, Anal. Biochem. 55, 643 (1973).
104. C. Agthe, W. Lijinsky, and D. Oremus, Food Cosmet. Toxicol. 6, 627 (1968).
105. W. T. Trager, L. Stoloff, and A. D. Campbell, J. Assoc. Off. Anal. Chem. 47, 993 (1964).
106. V. Nagarajan, R. V. Bhat, and P. G. Tulpule, Environ. Physiol. Biochem. 3, 13 (1973).
107. G. C. Aldermann, C. O. Emeh, and E. H. Marth, Z. Lebensm. Unters.-Forsch. 153, 305 (1973).
108. C. N. Shih and E. H. Marth, J. Milk Food Technol. 32, 213 (1969).
109. G. M. Shannon and O. L. Shotwell, J. Assoc. Off. Anal. Chem. 58, 743 (1975).
110. P. S. Steyn, J. Chromatogr. 45, 473 (1969).
111. G. W. Engstrom, J. Chromatogr. 44, 128 (1969).
112. I. Balzer, C. Bogdanic, and S. Pepeljnjak, J. Assoc. Anal. Chem. 61, 584 (1978).
113. T. V. Reddy, L. Viswanathan, and T. A. Venkitasubramanian, Anal. Biochem. 38, 568 (1970).
114. S. N. Khan, K. K. Maggon, and T. A. Venkitasubramanian, Appl. Environ. Microbiol. 36, 270 (1978).
115. A. Friedlander and M. Gonen, Israel J. Chem. 8, 87 (1970).
116. H. Arnold, Fleischwirtschaft 7, 985 (1975).
117. M. Stack and J. V. Rodricks, J. Assoc. Off. Anal. Chem. 54, 1310 (1971).

118. R. M. Eppley, J. Assoc. Off. Anal. Chem. 52, 311 (1969).
119. T. J. Coomes and J. C. Sanders, Analyst 88, 209 (1963).
120. L. Stoloff, A. C. Beckwith, and M. E. Cushmac, J. Assoc. Off. Chem. 51, 65 (1968).
121. J. A. Robertson, W. A. Pons, Jr., and L. A. Goldblatt, J. Assoc. Off. Anal. Chem. 53, 299 (1970).
122. J. V. Rodricks and L. Stoloff, J. Assoc. Off. Anal. Chem. 53, 92 (1970).
123. J. V. Rodricks and L. Stoloff, J. Assoc. Off. Anal. Chem. 53, 96 (1970).
124. J. V. Rodricks and L. Stoloff, Collaborative study of a method for determination of concentration and purity of aflatoxin standards and use of the method for measuring stability of the standards. IUPAC Inf. Bull. Tech. Rep. No. 1 (1971).
125. J. V. Rodricks, J. Assoc. Off. Anal. Chem. 56, 1290 (1973).
126. I. F. H. Purchase and B. Altenkirk, J. Assoc. Off. Anal. Chem. 56, 1115 (1973).
127. J. V. Rodricks, J. Assoc. Off. Anal. Chem. 52, 979 (1969).
128. P. J. Andrellos, A. C. Beckwith, and R. M. Eppley, J. Assoc. Off. Anal. Chem. 50, 346 (1967).
129. L. Toth, F. Tauchmann, and L. Leistner, Fleischwirtschaft 50, 1235 (1970).
130. T. Shanta, V. S. Murthy, and H. A. B. Parpia, J. Food Sci. Technol. 11, 194 (1974).
131. J. Chelkowski, B. Godlewska, M. Kokorniak, K. Szebiotko, and M. Wiewiorowska, Nahrung 18, 19 (1974).
132. J. Nabney and B. F. Nesbitt, Analyst 90, 155 (1965).
133. H. R. Prasanna, S. R. Gupta, L. Viswanathan, and T. A. Venkitasubramanian, Z. Lebensm. Unters. Forsch. 159, 319 (1975).
134. R.-D. Wei and F. S. Chu, J. Assoc. Off. Anal. Chem. 56, 1425 (1973).
135. K. Aibara and S. Yamagishi, in Proceedings First U.S. Japan Conference on Toxic Microorganisms, Washington, D. C. (1970), pp. 211-221.
136. F. Kiermeier and W. Mucke, Z. Lebensm. Unters. Forsch. 152, 18 (1973).
137. S. Nesheim, J. Assoc. Off. Anal. Chem. 54, 1444 (1971).
138. R. D. Stubblefield, G. M. Shannon, and O. L. Shotwell, J. Am. Oil Chem. Soc. 47, 389 (1970).
139. J. A. Robertson, W. A. Pons, and L. A. Goldblatt, J. Agric. Food Chem. 15, 798 (1967).
140. J. A. Robertson and W. A. Pons, J. Assoc. Off. Anal. Chem. 51, 1190 (1968).
141. F. A. El-Nawawy, Wien. Tieraerztl. Monatsschr. 61, 66 (1974).
142. H. R. Prasanna, L. Viswanathan, and T. A. Venkitasubramanian, Indian J. Biochem. Biophys. 9, 119 (1972).

143. L. Toth, F. Tauchmann, and L. Leistner, Fleischwirtschaft 50, 349 (1970).
144. A. B. Bradley and R. N. Zare, J. Am. Chem. Soc. 98, 620 (1976).
145. R. J. Gajan, S. Nesheim, and A. D. Campbell, J. Assoc. Off. Anal. Chem. 47, 27 (1964).
146. H. J. Issaq, E. W. Barr, and W. L. Zielinski, J. Chromatogr. Sci. 13, 597 (1975).
147. P. L. Schuller, C. A. H. Verhulsdonk, and W. E. Paulsch, Zesz. Probl. Postepow. Nauk. Roln. 189, 255 (1977).
148. C.-M. Chang and J. Q. Lynd, Agron. J. 60, 582 (1968).
149. M. R. Berman and R. N. Zare, Anal. Chem. 47, 1200 (1975).
150. A. A. Sekul, F. G. Dollear, and L. P. Codifer, J. Agric. Food Chem. 25, 1314 (1977).
151. C. N. Shih and E. H. Marth, J. Milk Food Technol. 34, 119 (1970).
152. G. S. Torrey and E. H. Marth, Appl. Environ. Microbiol. 32, 376 (1976).
153. R. Lotzsch, L. Leistner, and M. K. Ghosh, Fleischwirtschaft 56, 1773 (1976).
154. P. R. Beljaars, J. C. H. M. Schumans, and P. J. Koken, J. Assoc. Off. Anal. Chem. 58, 263 (1975).
155. K. Polzhofer, Z. Lebensm. Unters. Forsch. 163, 175 (1977).
156. P. R. Beljaars, F. H. M. Fabry, M. M. A. Pickott, and M. J. Peeters, J. Assoc. Off. Anal. Chem. 55, 1310 (1972).
157. B. G. E. Joseffson and T. E. Moller, J. Assoc. Off. Anal. Chem. 60, 1369 (1977).
158. S. N. Hagan and W. H. Tietjen, J. Assoc. Off. Anal. Chem. 58, 620 (1975).
159. J. Dantzman and L. Stoloff, J. Assoc. Off. Anal. Chem. 55, 139 (1972).
160. S. K. C. Obi, Zentralbl. Veterinaermed. Reihe B, 25, 173 (1977).
161. F. Kiermeier and G. Weiss, Z. Lebensm. Untersuch. Forsch. 160, 337 (1976).
162. B. Altenkirk, J. Chromatogr. 65, 456 (1972).
163. R. Lotzsch, F. Tauchmann, and W. Meyer, Fleischwirtschaft 54, 943 (1974).
164. M. Wiley, J. Assoc. Off. Anal. Chem. 49, 1223 (1966).
165. J. A. Perry, T. H. Jupille, and L. J. Glunz, Sep. Purif. Meth. 4, 97 (1975).
166. J. A. Perry and L. J. Glunz, J. Assoc. Off. Anal Chem. 57, 832 (1974).
167. Regis Chemical Co., Regis Programmed Multiple Development News, No. 8, Oct. 1974 (Regis Chemical Co., Morton Grove, Ill., 1974).

168. G. L. Schoenhard, R. O. Sinnhuber, and D. J. Lee, J. Assoc. Off. Anal. Chem. 56, 643 (1973).
169. L. Stoloff, M. Trucksess, N. Hardin, O. Francis, J. R. Hayes, C. E. Polan, and T. C. Campbell, J. Dairy Sci. 58, 1789 (1975).
170. R. E. Peterson and A. Ciegler, J. Chromatogr. 31, 250 (1967).
171. J. I. Suzuki, B. Dainius, and J. H. Kilbuck, J. Food Sci. 38, 949 (1973).
172. P. M. Scott and B. P. C. Kennedy, J. Assoc. Off. Anal. Chem. 56, 1452 (1973).
173. P. M. Scott and B. P. C. Kennedy, Can. Inst. Food Sci. Technol. J. 8, 124 (1975).
174. L. G. M. Th. Tuinstra, C. A. H. Verhulsdonk, J. M. Bronsgeest, and W. E. Paulsch, Neth. J. Agric. Sci. 23, 10 (1975).
175. P. L. Schuller, C. A. H. Verhulsdonk, and W. E. Paulsch, Pure Appl. Chem. 35, 291 (1973).
176. D. S. P. Patterson, E. M. Glancy, and B. A. Roberts, Food Cosmet. Toxicol. 16, 49 (1978).
177. S. Nesheim and M. W. Trucksess, J. Assoc. Off. Anal. Chem. 61, 569 (1978).
178. P. R. Beljaars, C. A. H. Verhulsdonk, W. E. Paulsch, and D. H. Liem, J. Assoc. Off. Anal. Chem. 56, 1444 (1973).
179. P. J. Lardinois, Off. J. Eur. Commun. 1, 102 (1976).
180. L. G. M. Th. Tuinstra and J. M. Bronsgeest, J. Chromatogr. 111, 448 (1975).
181. H. P. Van Egmond, W. E. Paulsch, and P. L. Schuller, J. Assoc. Off. Anal. Chem. 61, 809 (1978).
182. L. Allen, J. Assoc. Off. Anal. Chem. 57, 1398 (1974).
183. S. H. Ashoor and F. S. Chu, J. Assoc. Off. Anal. Chem. 58, 617 (1975).
184. M. E. Stack, A. E. Pohland, J. G. Dantzman, and S. Nesheim, J. Assoc. Off. Anal. Chem. 55, 313 (1972).
185. P. Cauderay, J. Assoc. Off. Anal. Chem. 62, 197 (1979).
186. G. Weiss, M. Miller, and G. Behringer, Milchwissenschaft 33, 409 (1978).
187. M. Messripour and S. Nesheim, Arch. Inst. Pasteur Tunis 54, 363 (1977).
188. C. A. H. Verhulsdonk, P. L. Schuller, and W. E. Paulsch, Zesz. Probl. Postepow. Nauk. Roln. 189, 277 (1977).
189. W. Przybylski, J. Assoc. Off. Anal. Chem. 58, 163 (1975).
190. W. F. Haddon, M. Wiley, and A. C. Waiss, Anal. Chem. 43, 268 (1971).
191. K. Bencze, F. Kiermeier, and M. Miller, Z. Lebensm. Unters.-Forsch. 159, 7 (1975).
192. J. A. Sphon, P. A. Dreyfuss, and H.-R. Schulten, J. Assoc. Off. Anal. Chem. 60, 73 (1977).

193. D. G. I. Kingston, J. Assoc. Off. Anal. Chem. 59, 1016
     (1976).
194. E. V. Crisan, Contrib. Boyce Thompson Inst. 24, 37 (1968).
195. P. J. Wakelyn, Anal. Chim. Acta 69, 481 (1974).
196. W. Muche and F. Kiermeier, Z. Lebensm. Unters. Forsch. 146,
     329 (1971).
197. M. J. Verrett, J.-P. Marliac, and J. McLaughlin, Jr., J.
     Assoc. Off. Anal. Chem. 47, 1003 (1964).

CHAPTER 16

Analysis of Aflatoxin Metabolites

Dennis P. H. Hsieh, M. Y. Fukayama,
D. W. Rice, and J. J. Wong

This chapter is concerned with the thin layer chromatographic (TLC) analysis of fungal and animal metabolites related to aflatoxin B_1 (AFB_1), the major and most potent member of the aflatoxin family. Of particular interest are versicolorin A (VLA), aflatoxicol (AFL), and aflatoxin M_1 (AFM_1) (Figure 1).

VLA is an intermediate in AFB_1 biosynthesis that can be produced in large quantities by mutants of aflatoxigenic strains of Aspergillus parasiticus (1). According to Wong et al. (20), the Ames mutagen assay has revealed its significant mutagenicity to the bacterial tester strain Salmonella typhimurium TA 98. Animal carcinogenicity has also been observed in feeding trials with rainbow trout (personal communication with R. O. Sinnhuber, Oregon State University, Corvallis, Oregon). The biological activity of VLA and its possible concurrence with aflatoxins in contaminated commodities warrant development of analytical methods for surveillance of its occurrence in foodstuffs. AFL is a metabolite of AFB_1 formed in animal species sensitive to aflatoxicosis (3). It is the most mutagenic and carcinogenic of all animal metabolites of AFB_1 isolated to this date (4). The activity of AFL formation from AFB_1 in an animal seems to be correlated with the sensitivity of the animal to aflatoxicosis (5). Analysis of AFL as a biochemical indicator of susceptibility, therefore, is a crucial step in the comparative toxicological study of AFB_1. AFM_1 is a common metabolite of AFB_1 found in all exposed animals (6). It also possesses significant bacterial mutagenicity (4) and animal carcinogenicity (7). The presence of AFM_1 in the milk and edible tissues of animals exposed to AFB_1 is an issue of great concern in food safety (8). An action level of 0.5 ppb in milk has recently been promulgated by the Food and Drug Administration. Methods for analysis of sub-ppb levels of AFM_1 in milk are currently under rapid development.

241

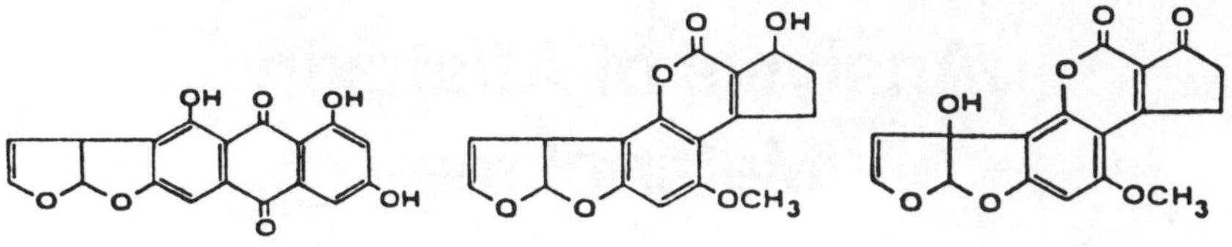

VERSICOLORIN A   AFLATOXICOL   AFLATOXIN M₁

Figure 1.   Structures of versicolorin A, aflatoxicol, and aflatoxin M₁.

TLC coupled with spectrodensitometric quantitation has been
the primary official method of analysis for aflatoxins and
related compounds (9).   Numerous brands of sorbents and an even
greater number of mobile phases allow a variety of choices and
versatile modifications of TLC systems to suit individual situations.   The commercial availability of precoated TLC plates of
acceptable quality greatly increases the speed and reproducibility of TLC analysis.   Some representative sorbents and precoated
TLC plates and various established developing mobile phases are
shown in Tables I and II.

TABLE I.   SELECTED TLC SORBENTS FOR AFLATOXIN ANALYSIS

          Silica Gel
              Machery-Nagel GHR
              Applied Science Adsorbosils-1 or -5
              Mallinckrodt Silica AR 4G or 7G
          Precoated Plates
              E. Merck Silica gel 60
              Whatman Silica gel
              Whatman PLK5 Linear-K (reverse phase)

ANALYSIS OF VLA

Since VLA is an intermediate in the pathway of aflatoxin biosynthesis, its detection and analysis from fungal cultures of contaminated foodstuffs are subject to possible interferences by other
fungal metabolites related to the same pathway, notably sterigmatocystin (ST), averufin (AR), versiconal hemiacetal acetate (VHA),

TABLE II. SELECTED TLC SOLVENT SYSTEMS FOR AFLATOXIN ANALYSIS

Solvent System	Reference
Chloroform-methanol (97+3)	15
Chloroform-acetone (9+1)	16
Chloroform-acetone-water (88+12+1.5)	17
Chloroform-acetone-hexane (85+15+20)	10
Chloroform-acetone-2-propanol (83+15+2.5)	12
Benzene-ethanol-water (46+35+19)	16
Benzene-methanol-acetic acid (24+2+1)	18
Ethyl acetate-2-propanol-water (20+2+1)	19
Ethyl ether-methanol-water (96+3+1)	17
Hexane-acetone-acetic acid (18+2+1)	20
Toluene-ethyl acetate-formic acid (5+4+1)	18

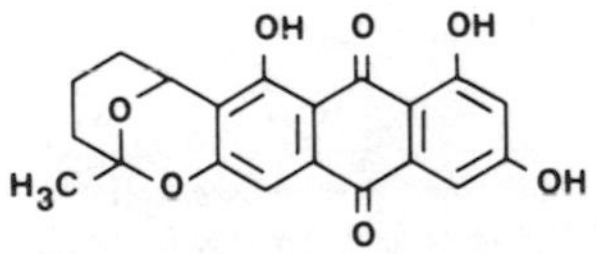

AVERUFIN 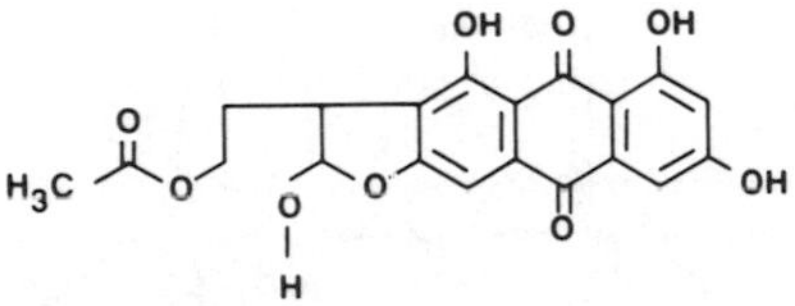   VERSICONAL HEMIACETAL ACETATE

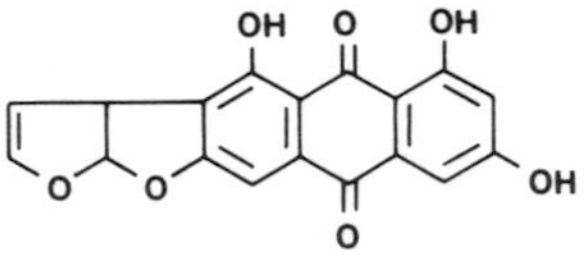

VERSICOLORIN A 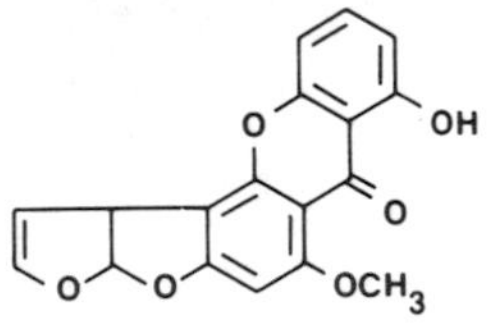   STERIGMATOCYSTIN

Figure 2. Some intermediates in aflatoxin biosynthesis.

243

and aflatoxins (Figure 2). The R_f values and relative densitometric response of these compounds on a silica gel plate devel-
opened in chloroform-acetone-hexane (85:15:20) are shown in
Figure 3. When the densitometer was operated in a transmission
mode using an interference wedge monochromator at a specific
wavelength, the weakly fluorescent orange and brick red pigments,
VHA, VLA, AR, and ST, produced inverse quantitative responses.
Resolution between VLA and AR was poor, and the minimum

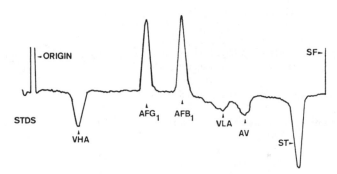

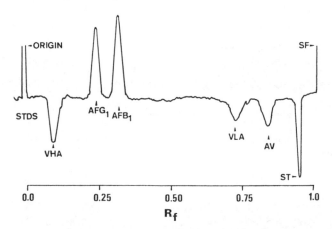

Figure 3. Mobility and spectrodensitometric response of some
fungal metabolites related to aflatoxin biosynthesis on a silica
gel thin layer chromatographic plate developed in chloroform-
acetone-hexane (85 + 15 + 20) (CAH) and dichloromethane-hexane-
methanol-acetic acid (70 + 30 + 0.4 + 5) (DHMA (10).

detectability of VLA was approximately 150 ng per spot as compared to 0.5 ng per spot for AFB_1. A new mobile phase, dichloro-
methane-hexane-methanol-acetic acid (70:30:0.4:5), was found to
significantly improve the resolution between AR and VLA (Figure
3). The presence of a few percent of acetic acid in the solvent
systems has the effect of reducing the tailing of spots, hence
enhancing spot compactness and increasing resolution between these
closely related compounds. Using these two systems, VLA was
detected in the culture of the VLA-producing mutant of A. parasiticus but not in the culture of the aflatoxigenic A. parasiticus.

ANALYSIS OF AFL

AFL is one of the six animal metabolites of AFB_1 that have been
isolated from in vitro biotransformation systems containing liver
enzymes (11). In comparative in vitro metabolism studies, AFL
must be analyzed in the presence of AFQ_1, AFP_1, $AFLH_1$, AFM_1, AFB_2a
and the residual AFB_1 (Figure 4). There has been no single TLC
system that can completely separate all the seven compounds on a
silica gel plate. Therefore, a sequence of developing solvent
systems was employed (Figure 5). The TLC plate was first developed in pure ethyl ether, thereby mobilizing AFL and completely
separating it from other metabolites. After quantitation for AFL,
the same TLC plate was then developed in chloroform-acetone-
2-propanol (85:15:5) according to Pons et al. (12). In this
system AFQ_1 and $AFLH_1$ were completely separated and were easily
measurable, but AFM_1 and AFB_2a were not separated. Separation of
the last two was achieved by a third development in benzene-
ethanol (BE, 40:4) or chloroform-methanol (CM, 19:1). Due to the
relatively low fluorescent intensity of AFL, its minimum detectability was 5 ng per spot as compared to 0.5 ng per spot for AFB_1.

ANALYSIS OF AFM_1

Due to the food safety concern of AFM_1 in milk and dairy products,
methods for detecting AFM_1 in milk have recently been under vigorous development. We were first concerned with the production and
purification of AFM_1 for analytical standards from the rice cultures of A. flavus (13). The most significant interference comes
from AFB_2a, which is concurrently produced with AFM_1 in these cultures. In addition to the aforementioned BE and CM systems that
are capable of separating these two compounds, two other systems
recommended by the Association of Official Analytical Chemists
(AOAC) for analysis of AFM_1 in milk, i.e., chloroform-acetone-

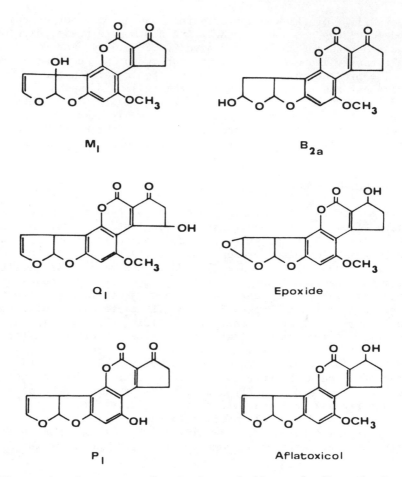

Figure 4.   Structure of animal metabolites of aflatoxin B_1.

2-propanol (CAI, 85:15:2.5) and ethyl ether-methanol-water (EMW,
95:4:1) according to Stoloff (14) were compared (Figure 6).   Separation of AFM_1 and AFB_{2a} was achieved only in the EMW system.
Using the recently available reverse-phase TLC plates, which are
prepared from silica gel coated with a chemically bonded organic
phase, and a mixture of acetonitrile and 0.5 M NaCl solution
(35:65) as the developing system, AFB_{2a} was found to be more
mobile than AFM_1, but separation of the two compounds was incomplete (Figure 6).   Therefore, the current TLC system of choice for

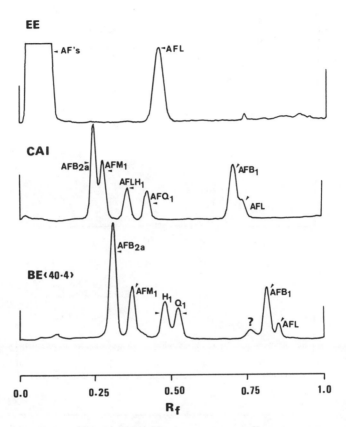

Figure 5. Mobility and spectrodensitometric response of some animal metabolites of aflatoxin B$_1$ on a silica gel plate developed in a sequence in ethyl ether (EE), chloroform-acetone-2-propanol (85 + 10 + 5) (CAI), and benzene-ethanol (40 + 4) (BE).

for analysis of AFM$_1$ in milk appears to be EMW. Using this system, the result of an analysis of a sample of California milk containing different amounts of AFM$_1$ is shown in Figure 7. The minimum detectability in this system is 0.5 ng per spot, representing 0.05 ppb of AFM$_1$ in milk according to the AOAC procedure.

The recently available high-performance TLC plates have not produced more desirable analytical results in our hands, but their superior resolution power is currently being explored.

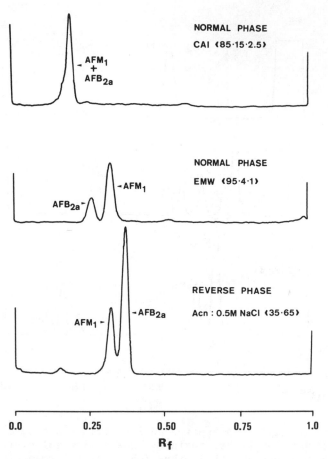

Figure 6. Mobility and spectrodensitometric response of aflatoxin M_1 and aflatoxin B_{2a} on a silica gel plate developed in chloroform-acetone-2-propanol (85 + 15 + 2.5) (CAI), ethyl ether-methanol-water (95 + 4 + 1) (EMW), and on a reversed-phase thin layer plate developed in acetonitrile - 0.5 N NaCl solution (35 + 65).

ACKNOWLEDGMENT

This research was supported by NIEHS Grant ES 00612, Western Regional Research Project, W-122, and Dairy Council of California.

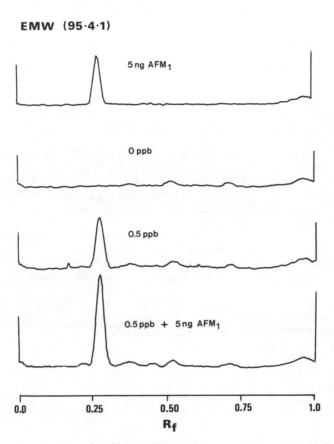

Figure 7. Analysis of aflatoxin M_1 in a sample of California milk on a silica gel plate developed in ethyl ether-methanol-water (95 + 4 + 1).

REFERENCES

1. L. S. Lee, J. W. Bennett, A. F. Cucullu, and J. B. Stanley, J. Agric. Food Chem. 23, 1132 (1975).
2. J. J. Wong and D. P. H. Hsieh, Mut. Res. 44, 447 (1977).
3. Z. A. Wong and D. P. H. Hsieh, Science 200, 325 (1978).
4. J. J. Wong and D. P. H. Hsieh, Proc. Nat. Acad. Sci. USA 73, 2241 (1976).

5. D. P. H. Hsieh, Z. A. Wong, J. J. Wong, C. Michas, and B. H. Ruebner, in Mycotoxins in Human and Animal Health, edited by J. V. Rodricks, C. W. Hesseltine, and M. A. Mehlman (Pathotox, Park Forest South, Ill., 1977), p. 37.

6. T. L. Campbell and J. R. Hayes, Toxicol. Appl. Pharmacol. 35, 119 (1976).

7. G. N. Wogan and S. Paglialunga, Food Cosmet. Toxicol. 12, 381 (1974).

8. J. V. Rodricks and L. Stoloff, in Ref. 5, p. 67.

9. Association of Official Analytical Chemists, in Official Methods of Analysis of the AOAC (AOAC, Washington, D. C., 1975), chap. 26.

10. J. I. Teng and P. C. Hanzas, J. Assoc. Off. Anal. Chem. 52, 83 (1969).

11. D. P. H. Hsieh, J. I. Dalezios, R. I. Krieger, M. S. Masri, and W. F. Haddon, J. Agr. Food Chem. 22, 515 (1974).

12. W. A. Pons, J. A. Robertson, and L. A. Goldblatt, J. Am. Oil Chem. Soc. 43, 665 (1966).

13. R. D. Stubblefield, G. M. Shannon, and O. L. Shotwell, J. Am. Oil Chem. Soc. 47, 389 (1970).

14. L. Stoloff, J. Assoc. Off. Anal. Chem. 61, 340 (1978).

15. W. A. Pons and L. A. Goldblatt, J. Am. Oil Chem. Soc. 42, 471 (1965).

16. R. M. Eppley, J. Assoc. Off. Anal. Chem. 49, 473 (1966).

17. R. D. Stubblefield, G. M. Shannon, and O. L. Shotwell, J. Assoc. Off. Anal. Chem. 52, 669 (1969).

18. P. M. Scott and T. B. Hand, J. Assoc. Off. Anal. Chem. 50, 366 (1967).

19. R. E. Peterson and A. Ciegler, J. Chromatog. 31, 250 (1967).

20. L. Stoloff, S. Nesheim, L. Yin, J. V. Rodricks, M. Stack, and A. D. Campbell, J. Assoc. Off. Anal. Chem. 54, 91 (1971).

CHAPTER 17

Analysis of Mycotoxins
Other than Aflatoxins
in Foodstuffs

Peter M. Scott

Thin layer chromatography (TLC) is still the most commonly used
technique for analytical chromatography of mycotoxins. Like the
aflatoxins, some of the other mycotoxins are fluorescent, whereas
others require spray reagents to make them detectable in ultraviolet (uv) or visible light. TLC is especially useful for
screening purposes--"multimycotoxin analysis"--but it is also
employed as the determinative step in the five AOAC official
methods developed specifically for other mycotoxins--for ochratoxins, sterigmatocystin, zearalenone, and patulin. This paper
aims to review selectively the best of the available TLC detection methods for mycotoxins other than aflatoxins. Details of
extraction and cleanup will not be covered.

Table I lists those of the more important mycotoxins that
can be detected by fluorescence under uv light. The detection
limits given are approximate only and will of course vary with
the lamp used, chromatographic conditions, and the observer.
Generally, sensitivity does not approach that of the aflatoxins.
In some cases, for example sterigmatocystin and zearalenone, sensitivity is improved by use of spray reagents (Table II). Tables
III-XXI give further details on TLC of mycotoxins other than aflatoxins for which there has been at least some attempt at development of a method for their analysis in foods or grains. References are listed in the Bibliography under the relevant mycotoxin
or, for Tables XV-XX, under "Other Mycotoxins." Application of
these methods has shown many instances of these mycotoxins
occurring naturally, usually at low levels and in low incidence
but, it should be emphasized, in human foods or foodstuffs des-
tined for human consumption, as well as in feed grains (1-4).

Excluding aflatoxin occurrence in foods, the most important
mycotoxin contamination as far as the human diet is concerned
appears to be that of apple juice by patulin and of corn and corn

251

TABLE I. FLUORESCENT MYCOTOXINS: TLC DETECTION LIMITS (VISUAL)

Mycotoxin	Fluorescence Color under uv Light (L = longwave, S = shortwave)	Detection Limit (ng)
Aflatoxins B_1, G_1	Blue, green (L)	0.1
Aflatoxins B_2, G_2	Blue, green (L)	0.04
Ochratoxin A	Blue-green (L+S)	2
Sterigmatocystin	Brick-red (L)	40
Zearalenone	Greenish-blue (S)	20
Alternariol, alternariol methyl ether	Blue (L+S)	20
Citrinin	Yellow (L)	10
Citreoviridin	Yellow (L)	
PR toxin	Green (L after S)	25
Rubratoxin B	? (L+S, after heating)	500

TABLE II. DETECTION OF MYCOTOXINS USING SPRAY REAGENTS

Mycotoxin	Reagent	Detection Limit (ng)
Sterigmatocystin	$AlCl_3$, heat, uv	5
Zearalenone	Fast Violet B Salt	5
Patulin	MBTH, heat, U	10
Penicillic acid	Acidic anisaldehyde, heat, uv	10
Tenuazonic acid	$FeCl_3$	
T-2 toxin	Acid, heat, uv	54
Diacetoxyscirpenol	Acid, heat, uv	50[a]
Penitrem A	$FeCl_3$	1000
Cyclopiazonic acid	Oxalic acid (in layer)	100
	Ehrlich reagent	50
Cyclochlorotine	Cl_2, o-tolidine	
Cytochalasins	65% H_2SO_4, heat, uv	

[a] By fluorodensitometry.

products by zearalenone. The possible significance of this to health has not been assessed. Patulin is the only mycotoxin other than aflatoxin for which any country has set an action limit--50 µg/l in fruit juices in Sweden (5).

SPECIFIC MYCOTOXINS

Table III summarizes TLC parameters for the ochratoxins. Note that the solvent systems contain acid; ochratoxins A and B are carboxylic acids and scarcely migrate in neutral solvent systems. Because they are highly fluorescent, fluorodensitometry can be used successfully to measure the ochratoxins and sensitivity if ochratoxin A is doubled, from 0.5 to 0.25 ng, after exposing the TLC plate to ammonia fumes (6).

TABLE III.   TLC OF OCHRATOXINS

Sorbent	Silica gel
Solvents	C_6H_6-AcOH-MeOH (18:1:1), not equilibrated toluene-EtAc-HCOOH (5:4:1) hexane-acetone-AcOH (18:2:1) for esters
Standards	1-5 µg/ml C_6H_6-AcOH (99:1)
Detection	Longwave and shortwave uv Blue-green fluorescence (blue with NH_3 or $NaHCO_3$)
Overall detection limit of AOAC methods	12 µg/kg barley; 20 µg/kg green coffee
Fluorodensitometry	Ochratoxin A, 340 nm excitation, 475 nm emission
Natural occurrence	Grains, white beans, green coffee beans

As previously indicated, sterigmatocystin is not very fluorescent by itself and spraying and heating with aluminum chloride solution is necessary for good visualization (Table IV). The

TABLE IV.   TLC OF STERIGMATOCYSTIN

Sorbent	Silica gel
Solvents	C_6H_6-AcOH-MeOH (18:1:1) C_6H_6-AcOH (9:1)
Standard	5 µg/ml C_6H_6
Detection	Spray 20% $AlCl_3$ · 6 H_2O in EtOH, heat 80° for 10 min shortwave uv : yellow fluorescence
Overall detection limit of methods	30-60 µg/kg grains, soybeans, green coffee
Fluorodensitometry	$AlCl_3$ derivative, 360 nm excitation, 500 nm emission
Natural occurrence	Feed grain, rice

official AOAC method for sterigmatocystin (7,8) is not applicable to corn and oats because of an oily interference. However, alternative methods for these and other agricultural commodities have been published (9,10).

The AOAC method for zearalenone in corn (11) is a modification of part of the multimycotoxin method of Eppley (12). Detection of zearalenone is by fluorescence under shortwave uv light (Table V) and the change in fluorescence intensity on switching between longwave and shortwave uv light is a useful test. Recently, diazonium salt spray reagents, in particular Fast Violet B Salt, have been shown to offer more sensitivity and specificity (13).

MBTH (Table VI) is the TLC spray reagent used for patulin in the official AOAC method for its analysis in apple juice (14,15). The yellow-brown fluorescence is less intense after TLC development in neutral solvent systems and it is useful to add 5% of formic acid to the spray reagent. In some cases two-dimensional TLC is necessary to effect separation of patulin from interfering spots (16). The MBTH spray has also been used for detection of penicillic acid in corn, beans, and peanuts (17) (Table VII).

Thin layer chromatograms and fluorescence of alternariol methyl ether (Table VIII) and zearalenone are similar but the compounds can be separated in benzene-ethanol-0.4 N NaOH (96:4;

TABLE V.   TLC OF ZEARALENONE

Sorbent	Silica gel
Solvents	$CHCl_3$-EtOH (95:5) $CHCl_3$-MeOH (93:7) C_6H_6-AcOH (95:5)
Standard	50 µg/ml C_6H_6-CH_3CN (98:2) 5 µg/ml toluene
Detection	Shortwave uv:greenish-blue fluorescence (blue under longwave uv after spraying 20% ethanolic $AlCl_3$, 130°/5 min) Spray 0.7% Fast Violet B Salt, then pH 9 buffer, then 50% H_2SO_4, 120°/5 min: mauve spot
Overall detection limits of methods	100 µg/kg corn (AOAC), 20 µg/kg cornflakes
Fluorodensitometry	313 nm excitation, 443 nm emission
Natural occurrence	Grains, especially corn and corn products; pecans

5 drops) (18) and differentiated by the former metabolite's failure to form a colored spot with Fast Violet B Salt (13).

TLC procedures for tenuazonic acid are not sensitive (Table IX) and the toxin forms an elongated spot, as does citrinin (Table X), in acidic solvent systems such as toluene-ethyl acetate-formic acid (5:4:1). Fortunately, citrinin forms an almost round spot at an R_f of 0.5 in ethyl acetate-acetone-water (19).

Although gas-liquid chromatography is preferred for the determination of the trichothecene mycotoxins in grains, TLC is most frequently used for screening purposes. There are about 40 known fungal trichothecenes and usually a spray reagent is required for their detection (Table XI). Trichothecenes such as T-2 toxin, HT-2 toxin, neosolaniol, and diacetoxyscirpenol that lack the α,β-enone system give a blue fluorescence with sulfuric acid and heat, whereas those possessing the α,β-enone system, for example deoxynivalenol (vomitoxin), nivalenol, and fusarenone-X, give a nonfluorescent spot (20). However, the latter compound

TABLE VI.   TLC OF PATULIN

Sorbent	Silica gel
Solvents	Toluene-EtAC-HCOOH (5:4:1) hexane-ether (1:3) $CHCl_3$-MeOH (95:5) $CHCl_3$-acetone (9:1)
2D-TLC	(1) C_6H_6-MeOH-AcOH (19:2:1) (2) toluene-EtAc-AcOH (5:4:1)
Standard	10 µg/ml $CHCl_3$
Detection	Spray 0.5% 3-methyl-2-benzothiazolinone hydrazone (MBTH) hydrochloride, heat 130° C for 15 min, longwave uv: yellow-brown fluorescence
Overall detection limit of methods	20 µg/l apple juice (AOAC), 40 µg/kg corn
Densitometry	SIL G-25 uv_{254}, reflectance, 273 nm, 5 ng detection limit
Natural occurrence	Apple juice, bread, silage

---

has been reported to fluoresce on heating with aluminum chloride.
Problems of identification arise when extracts contain interfering spots. For full information on separation of trichothecenes,
the papers by Pathre and Mirocha (21) and Ueno et al. (20) should
be consulted.

There are many other mycotoxins, including ergot alkaloids,
for which TLC data have been recorded. Some of these are covered
in Tables XII-XXI. Apart from cyclopiazonic acid (Table XVII),
sensitive TLC methods for their detection in foodstuffs have not
been developed.

MULTIMYCOTOXIN TLC AND METHODS

The foregoing tables are necessarily simplified and not all possible mobile phases or means of detection are given. It is apparent,
however, that TLC of these mycotoxins have many features in common.
Not surprisingly therefore, schemes for TLC of several mycotoxins

TABLE VII. TLC OF PENICILLIC ACID

Sorbent	Silica gel
Solvents	Toluene-EtAc-HCOOH (6:3:1) EtAc-CHCl$_3$-HCOOH (60:40:1)
Standard	100-200 µg/ml CHCl$_3$
Detection	Spray MBTH, heat at 130° C for 10 min, longwave uv : yellow fluorescence (50 ng limit)
	Spray 0.5% p-anisaldehyde in MeOH-AcOH-c·H$_2$SO$_4$ (85:10:5), heat at 130° C for 8 min, longwave uv : blue fluorescence
Overall detection limit of method	300 - 400 µg/kg corn
Fluorodensitometry	NH$_3$ derivative, 350 nm excitation 440 nm emission, 1 µg limit
Natural occurrence	Corn and beans (by GLC)

TABLE VIII. TLC OF ALTERNARIOL AND ITS MONOMETHYL ETHER

Sorbent	Silica gel G-HR or G
Solvents	CHCl$_3$-acetone (88:12) C$_6$H$_6$-EtOH-0.4 N NaOH (96:4, 5 drops)
Standards	16-17 µg/ml C$_6$H$_6$-CH$_3$CN (98:2)
Detection	Longwave + shortwave uv : blue fluorescence
Overall detection limit of method	100 µg/kg sorghum
Fluorodensitometry	Fading noted with alternariol methyl ether
Natural occurrence	Grain sorghum, pecans

TABLE IX.   TLC OF TENUAZONIC ACID

Sorbent	Silica gel F-254
Solvents	Hexane-EtAc-AcOH (70:30:4) toluene-EtAc-HCOOH (5:4:1)
Standard	2000 µg/ml toluene
Detection	Quenching of shortwave uv (0.5 µg limit) Spray 2% $FeCl_3$
Natural occurrence	Blast-diseased rice plants

TABLE X.   TLC OF CITRININ

Sorbent	Silica gel
Solvents	EtAc-acetone-H_2O (5:5:2), not equilibrated EtAc-acetone-1 N HCl (5:5:2)
Standard	100 µg/ml $CHCl_3$ 10 µg/ml toluene
Detection	Longwave uv, esp. after spraying 20% H_2SO_4 : yellow fluorescence
Overall detection limit of methods	200 µg/kg barley, corn
Fluorodensitometry	Excitation longwave uv, emission 508 nm
Natural occurrence	Feed grains, groundnuts

have been devised (Table XXII). The mobile phase toluene-ethyl acetate-formic acid, for example, has proved widely useful and requires little maintenance, other than occasionally topping up the amount in the solvent trough, in order to obtain consistent separations. Examples of R_f values of several mycotoxins in this system (proportions 6:3:1) and another commonly used mobile phase,

TABLE XI. TLC OF TRICHOTHECENES

Sorbent	Silica gel
Solvents	Toluene-EtAc-HCOOH (6:3:1) EtAc-toluene (3:1) $CHCl_3$-MeOH (95:5)
Standards	100 µg/ml $CHCl_3$ (T-2 toxin) 1000 µg/ml ethanol (nivalenol)
Detection	Spray 20-30% H_2SO_4, heat at 120° C for 15 min, longwave uv : blue fluorescence (T-2 toxin, diacetoxyscirpenol, neosolaniol)
	Spray 0.5% p-anisaldehyde in MeOH-AcOH-c·H_2SO_4 (85:10:5), heat at 130° C for 8 min, longwave uv : blue fluorescence (T-2 toxin)
	Spray 50% $AlCl_3$ heat, uv : fluorescence (fusarenone X)
Overall detection limit of method	2500 µg/kg barley (T-2 toxin)
Fluorodensitometry	Diacetoxyscirpenol + H_2SO_4 (down to 50 ng)
	T-2 toxin + H_2SO_4, 365 nm excitation
	Fusarenone X + $AlCl_3$, excitation 335 nm, emission 455 nm
Natural occurrence	Feed grains

benzene-methanol-acetic acid (24:2:1), are shown in Table XXIII
(22). The systematic analysis of Duracková et al. (23) includes
"chromatographic spectra" of aflatoxin B_1, alternariol methyl
ether, kojic acid, and trichothecin in eight solvent systems. R_f
values of 39 other mycotoxins and fungal metabolites in these
solvent systems are also presented by these authors. Their
results illustrate the dependency of R_f values on the type of
silica gel used. As this is but one factor affecting reproducibility of R_f values, those found in the literature should obviously be used only as a guide, for solvent selection and to judge
relative polarities of different mycotoxins. A universal spray
reagent for mycotoxins that has reasonable sensitivity has not
been found, but various sulfuric acid sprays, followed by heating
and examination under uv light, have been used for detection of

TABLE XII. TLC OF TREMORTINS (PENITREMS) AND VERRUCULOGEN

Sorbent	Silica gel
Solvents	$CHCl_3$-acetone (93:7) ether-cyclohexane (3:1)
Standard	100 µg/ml (penitrem A)
Detection	Spray 1% $FeCl_3$ in butanol, warm : greenish-blue spots (tremortins A, B, C)
	Spray 50% ethanolic H_2SO_4, longwave uv light, fluorescence : mustard (verruculogen)
Natural occurrence	Not yet demonstrated

TABLE XIII. TLC OF RUBRATOXIN B

Sorbent	Silica gel, predeveloped $CHCl_3$-MeOH-AcOH (80:20:1)
Solvents	$CHCl_3$-MeOH-AcOH-H_2O (80:20:1:1)
Standard	105 µg/ml EtAc, spotted under N_2
Detection	Heat at 200° C for 10 min, longwave and shortwave uv : fluorescence
Overall detection limit of method	70,000 µg/kg corn
Natural occurrence	Not yet demonstrated

TABLE XIV. TLC OF CYCLOCHLOROTINE

Sorbent	Silica gel F-254
Solvents	n-butanol-AcOH-H$_2$O (4:1:2)
Detection	Cl$_2$ for 5 min; then spray with 0.032% o-tolidine in solution of KI (0.2%) in AcOH-H$_2$O (3:47) : yellow spot
Overall detection limit of method	1000 μg/kg grains
Natural occurrence	Not yet demonstrated

TABLE XV. TLC OF CITREOVIRIDIN

Sorbent	Silica gel G-1500
Solvent	EtAc-toluene (3:1)
Standard	5-1000 μg/ml CHCl$_3$
Detection	Longwave uv : yellow fluorescence
Densitrometry	uv absorption 360 nm Fluorescence : 362 nm excitation, 525 nm emission, 50 ng detectable
Natural occurrence	Not yet demonstrated

many mycotoxins; in the case of the trichothecenes this is one of the reagents of choice.

Detection of standards alone on the TLC plate is one thing; detection and measurement in the presence of good extracts is quite another. Part of the problem is that visual estimation contributes to quite large coefficients of variation, as observed in collaborative studies that have been carried out on methods for the ochratoxins (24), sterigmatocystin (8), zearalenone (11), and patulin (15) (Table XXIV). It is of interest to note, however, that use of a spray reagent--for sterigmatocystin and patulin--does not appear to give any worse variation than with the

TABLE XVI.   TLC OF LUTEOSKYRIN

Sorbent	Oxalic acid - impregnated silica gel
Solvents	Acetone-hexane-H_2O (6:3:1.5)
Standard	1000 µg/ml $CHCl_3$
Detection	Visible light : yellow
Overall detection limit of method	< 1000 µg/kg oats
Natural occurrence	Not yet demonstrated

TABLE XVII.   TLC OF CYCLOPIAZONIC ACID

Sorbent	Oxalic acid - impregnated silica gel
Solvents	Chloroform-methylisobutylketone (4:1)
Detection	Heat at 110° C for 1-2 min ; purple spot Ehrlich reagent : purple-blue spot
Overall detection limit of method	100 µg/kg cheese
Natural occurrence	Overripe Camambert cheese

fluorescent mycotoxins. Some indication of detection limits of methods for the analysis of one mycotoxin or a small group of similar mycotoxins in foodstuffs has been given in Tables III-XXI where available. In order to extend the scope of a method to other mycotoxins some of this sensitivity is usually sacrificed. Detection limits of such "multimycotoxin methods" are summarized in Table XXV. Some of these methods feature a separation of acidic from neutral toxins (25-27), whereas others include an adsorption column chromatographic separation (12,28,29). The low levels of aflatoxins, ochratoxin A, zearalenone, sterigmatocystin, and patulin detectable by the method of Josefsson and Müller (30) are noteworthy. This method involves cleanup by gel filtration chromatography.

TABLE XVIII.   TLC OF PR TOXIN

Sorbent	Silica gel
Solvents	Toluene-EtAc-HCOOH (5:4:1)
Standard	25 µg/ml $CHCl_3$
Detection	(1) Shortwave uv 0.5 min; (2) Longwave uv : green fluorescence
Overall detection limit of method	250 µg/kg blue cheese (unstable)
Natural occurrence	Silage

TABLE XIX.   TLC OF MONOLIFORMIN

Sorbent	Silica gel F-254
Solvents	$CHCl_3$-MeOH (3:2)
Detection	Shortwave uv : detection limit 150 ng
Natural occurrence	Not yet demonstrated

HIGH-PERFORMANCE LIQUID CHROMATOGRAPHY

Finally, the increasing application of high-performance liquid chromatography (HPLC) to the determination of mycotoxins must be mentioned. HPLC methodology for the aflatoxins is already well developed. Table XXVI indicates that generally detection limits for standards are lower than those observed using TLC (Tables I and II). Several HPLC methods for use with foodstuffs have been proposed, although at present there is no official AOAC method for mycotoxins that utilizes this detection technique. While Engstrom et al. (31) have resolved a mixture of seven mycotoxin

TABLE XX.   TLC OF CYTOCHALASINS

Sorbent	Silica gel G or DB
Solvents	$CHCl_3$-MeOH (95:5) Toluene-EtAc-HCOOH (5:4:1) $CHCl_3$-acetone (9:1)
Standards	100 µg/ml $CHCl_3$ (cytochalasins A, B, C, D) 200 µg/ml $CHCl_3$ (cytochalasin E)
Detection	Spray 65% H_2SO_4, heat at 100° C for 3-4 min, longwave uv : blue fluorescence (cytochalasins A, B), yellow fluorescence (cytochalasins C, D), red-brown fluorescence (cytochalasin E)
Natural occurrence	Tomato paste (cytochalasin B)

TABLE XXI.   TLC OF ERGOT ALKALOIDS

Sorbents	(1) Silica gel (2) Alumina (3) $HCONH_2$-impregnated silica gel
Solvents	(1,2) $CHCl_3$-MeOH (4:1), acetone, etc. (1) $CHCl_3$-MeOH-conc. NH_4OH (80:20:0.2) (3) isopropylether-THF-toluene-$NHEt_2$ (70:15:15:0.1)
Standards	(3) 200 µg/ml $CHCl_3$-C_6H_6 (1:1) + HCONH
Detection	Longwave uv (ergotamine, ergocristine, etc. 5% p-dimethylaminobenzaldehyde in conc. HCl Various Π-acceptors
Photodensitometry	(Van Urck spray), 580 nm
Fluorodensitometry	313 or 325 nm excitation, 445 nm emission
Natural occurrence	Ergot-infected rye, wheat, millet

TABLE XXII. GENERAL TLC SYSTEMS FOR MYCOTOXINS

Sorbent	Mobile Phases	Means of Detection	Mycotoxins	Reference
Oxalic acid-impregnated silica gel G	$CHCl_3$-methyl-isobutyl ketone (4:1)	(i) uv, (ii) H_2SO_4, (iii) $FeCl_3$	cyclopiazonic acid secalonic acid D, aspertoxin + 8 others	34
Adsorbosil 5 silica gel	toluene-EtAc-HCOOH (6:3:1) benzene-MeOH-AcOH (24:2:1)	(i) visible light (ii) uv (iii) acidic anisaldehyde	Luteoskyrin, penicillic acid + 17 others	22
Silica gel G, "Silufol" sheets	8 mobile phases of low to high polarity	(i) uv (ii) acidic anisaldehyde (iii) $FeCl_3$	37 mycotoxins + 6 other fungal metabolites	23

TABLE XXIII.   TYPICAL R_f VALUES OF SOME MYCOTOXINS ON SILICA GEL
               (ADSORBOSIL 5)

Mycotoxin	Tol.-EtAC-90% HCOOH (6:3:1)	C_6H_6-MeOH-AcOH (24:2:1)
Citrinin	0.16-0.48	0-0.20
Luteoskyrin	0-0.47	0-0.23
Nivalenol	0-0.02	0-0.01
Butenolide	0.10	0.03
Kojic acid	0.16	0.03
Aflatoxin G_2	0.17	0.13
Nivalenol acetate (fusarenone X)	0.19	0.09
Aflatoxin G_1	0.23	0.14
Aflatoxin B_2	0.26	0.20
Aflatoxin B_1	0.31	0.23
Diacetoxyscirpenol	0.33	0.24
Aspertoxin	0.35	0.13
T-2 toxin	0.36	0.28
Patulin	0.41	0.21
Penicillic acid	0.47	0.22
Gliotoxin	0.53	0.39
Ochratoxin A	0.55	0.35
Zearalenone	0.78	0.42
Sterigmatocystin	0.85	0.75

Reference: Scott et al. (22)

standards by using an UV detector, reverse phase C_{18} column and a
solvent system of acetonitrile-water-acetic acid (55:45:2), there
have been only two attempts at multimycotoxin HPLC method development for foods (32,33). Detection limits of these methods for
aflatoxins and ochratoxin A, or zearalenone, patulin, and penicillic acid in foods were comparable to or better than the best of
the TLC multimycotoxin methods (Table XXV). However, except for
the penicillic acid, a preparative TLC or HPLC cleanup step was
required.

TABLE XXIV. COLLABORATIVE STUDIES--SPIKED SAMPLES

Mycotoxin	Foodstuff	Added Level μg/kg	C.V. %	
			Visual	Visual or Densitometric
Ochratoxin A	Barley	44.9		30.5
		89.7	36.3	23.3
Ochratoxin B	Barley	48.8	37.7	34.4
		97.6		54.4
Et Ochratoxin A	Barley	60		50.2
		120		33.2
Et Ochratoxin B	Barley	61		76.0
		122		36.8
Ochratoxin A	Green coffee	57	49.6	
		154	37.0	
		230	39.1	
Sterigmatocystin	Barley	100	16.2	
		200	8.2	
	Wheat	100	20.3	
		200	22.5	
Zearalenone	Corn	300	53.0	
		1000	38.2	
		2000	27.0	
Patulin	Apple juice	50	44.0	
		120	38.3	
		340	48.7	

TABLE XXV.   MULTIMYCOTOXIN METHODS--

Method	Commodity	Aflatoxin	Ochra-toxin A	Ochra-toxin C	Zeara-lenone
Eppley (12)	Various	<32 (B$_1$)	<55		<500
	Corn	1-3 (B$_1$ or G$_1$)	50		200
Vorster (25)	Grains, Peanuts	4 (B$_1$)	20	20	
Stoloff et al. (35)	Grains	20 (B$_1$ or G$_1$)	45-90	50-100	200-500
+ acid Scott et al. (36)	Grains	+	+		
BF (Scott et al. [36])	Grains	+	+		+
BF + CuCO$_3$ (Thomas et al. [37])	Corn	2 (B$_1$)			100
Hagan and Tietjen (38)	Oils	+	+	+	+
Roberts and Patterson (25)	Feed	3 (B$_1$)	80		1000
Wilson et al. (17)	Corn	2 (B$_1$ or G$_1$)	20 / 20		
	Beans	"	20		200-300
	Peanuts	"	40		
Takeda et al. (29)	Grains	10 (each)	40		
Balzer et al. (27)	Corn	2 (B$_1$)	40		200
Joseffson and Möller (30)	Cereals	5 (each)	10		35
Yamamoto (35)	Flours	5 (B$_1$)			

+ Toxin detectable but limit not determined

DETECTION LIMITS (µg/kg)

Sterigma-tocystin	Patulin	Penici-lic Acid	Citrinin	T-2 Toxin	Diacetoxyl-scirpenol	Peni-trem A
100						
60	400-1000	+				
		+	100		+	
+						
+						
330	600		+	+	4000	+
		300-400	100-200			
		300-500	400-500			
		1000	ND			
40			80			
10	50					
40						

TABLE XXVI. HPLC OF MYCOTOXINS OTHER THAN AFLATOXINS

Mycotoxin	Means of Detection	Detection Limit Standard (ng)	Overall Method (μg/kg)
Ochratoxin A (6, 24, 40-43)	Fluorescence uv (254 nm)	0.04	12.5 (beans, etc.)
Sterigmatocystin (7-10, 44-47)	uv (365 nm) uv (254 nm) uv (320 nm)	3	6 (pistachio) <25 (corn, oats)
Zearalenone 11-13, 37, 48-52)	uv (254 nm) uv (236 nm) Fluorescence (310 nm excitation) Fluorescence (236 nm excitation)	1 0.6 <1 <5	10 (animal chow) 2 (cereals) 5-15 (corn foods) 10 (corn)
Patulin (14-16, 53-55)	uv (254 nm) uv (230 nm)	1-4 102	11 (apple juice) 10 (rice)
Penicillic acid (17, 22, 56-58)	uv (230 nm) uv (254 nm)	50 10	5 (rice)
Alternariol, alternariol methyl ether (18, 59, 60)	uv (350 rm) Fluorescence	10 1-2	100 (grain sorghum)
Tenuzonic acid (61,62)	Underdevelopment - uv (350 nm)	5-10	
Citrinin (19, 63)	Fluorescence	0.3	

270

Tricothecenes (20, 21, 64-68)	Underdevelopment	
Rubratoxins A and B (69-73)	uv (254 nm)	3-5 (B) 15-20 (A)
Ergot alkaloids	uv (225, 240, 254, 282, 310 nm)	
Xanthomegnin	Absorption (405 nm)	750 (corn)

References are listed in parentheses under Mycotoxin.

REFERENCES

1. L. Stoloff, Mycotoxins and Other Fungal Related Food Problems, edited by J. V. Rodricks, Adv. Chem. Ser. 149 (Amer.
   Chem. Soc., Washington, D. C., 1976), pp. 23-50.
2. J. V. Rodricks, C. W. Hesseltine, and M. A. Mehlman, Eds.,
   Mycotoxins in Human and Animal Health (Pathotox Publishers,
   Inc., Park Forest South, Ill., 1971), p. 807.
3. T. D. Wyllie and L. G. Morehouse, Eds., Mycotoxic Fungi,
   Mycotoxins, Mycotoxicoses (Marcel Dekker, Inc., New York,
   1971), vols. I-III.
4. P. M. Scott, J. Food Prot. 41, 385 (1978).
5. L. Stoloff, J. Assoc. Off. Anal. Chem. 61, 340 (1978).
6. H. L. Trenk and F. S. Chu, J. Assoc. Off. Anal. Chem. 54,
   1307 (1971).
7. M. Stack and J. V. Rodricks, J. Assoc. Off. Anal. Chem. 54,
   86 (1971).
8. M. E. Stack and J. V. Rodricks, J. Assoc. Off. Anal. Chem.
   56, 1123 (1973).
9. A. K. Athnasios and G. O. Kuhn, J. Assoc. Off. Anal. Chem.
   60, 104 (1977).
10. G. M. Shannon and O. L. Shotwell, J. Assoc. Off. Anal. Chem.
    59, 963 (1976).
11. O. L. Shotwell, M. L. Goulden, and G. A. Bennett, J. Assoc.
    Off. Anal. Chem. 59, 666 (1976).
12. R. M. Eppley, J. Assoc. Off. Anal. Chem. 51, 74 (1968).
13. P. M. Scott, T. Panalaks, S. Kanhere, and W. F. Miles, J.
    Assoc. Off. Anal. Chem. 61, 593 (1978).
14. P. M. Scott and B. P. C. Kennedy, J. Assoc. Off. Anal. Chem.
    56, 813 (1973).
15. P. M. Scott, J. Assoc. Off. Anal. Chem. 57, 621 (1974).
16. S. Lindroth and A. Niskanen, J. Food Sci. 43, 446 (1978).
17. D. M. Wilson, W. H. Tabor, and M. W. Trucksess, J. Assoc.
    Off. Anal. Chem. 59, 125 (1976).
18. L. M. Seitz, D. B. Sauer, H. E. Mohr, and R. Burroughs,
    Phytopathology 65, 1259 (1975).
19. B. Hald and P. Krogh, J. Assoc. Off. Anal. Chem. 56, 1440
    (1973).
20. Y. Ueno, N. Sato, K. Ishii, K. Sakai, H. Tsunoda, and M.
    Enomoto, Appl. Microbiol. 25, 699 (1973).
21. S. V. Pathre and C. J. Mirocha, in Ref. 2, pp. 229-253.
22. P. M. Scott, J. W. Lawrence, and W. van Walbeek, Appl. Micro-
    biol. 20, 839 (1970).
23. Z. Duracková, V. Betina, and P. Nemec, J. Chromatogr. 116,
    141 (1976).
24. S. Nesheim, J. Assoc. Off. Anal. Chem. 56, 822 (1973).

25. B. A. Roberts and D. S. P. Patterson, J. Assoc. Off. Anal.
    Chem. 58, 1178 (1975).
26. D. M. Wilson, W. H. Tabor, and M. W. Trucksess, J. Assoc.
    Off. Anal. Chem. 59, 125 (1976).
27. I. Balzer, C. Bogdanic, and S. Pepeljnjak, J. Assoc. Off.
    Anal. Chem. 61, 584 (1978).
28. L. J. Vorster, Analyst 94, 136 (1969).
29. Y. Takeda, E. Isohata, R. Amano, M. Uchiyama, Y. Naoi, and
    M. Nakao, J. Food Hyg. Soc. Japan 17, 193 (1976).
30. B. G. E. Josefsson and T. E. Müller, J. Assoc. Off. Anal.
    Chem. 60, 1369 (1977).
31. G. W. Engstrom, J. L. Richard, and S. J. Cysewski, J. Agr.
    Food Chem. 25, 833 (1977).
32. D. C. Hunt, A. T. Bourdon, and N. T. Crosby, J. Sci. Food
    Agr. 29, 239 (1978).
33. D. C. Hunt, A. T. Bourdon, P. J. Wild, and N. T. Crosby, J.
    Sci. Food Agr. 29, 234 (1978).
34. P. S. Steyn, J. Chromatogr. 45, 473 (1969).
35. L. Stoloff, S. Nesheim, L. Yin, J. V. Rodricks, M. Stack,
    and A. D. Campbell, J. Assoc. Off. Anal. Chem. 54, 91 (1971).
36. P. M. Scott, W. van Walbeek, B. Kennedy, and D. Anyeti, J.
    Agr. Food Chem. 20, 1103 (1972).
37. F. Thomas, R. M. Eppley, and M. W. Trucksess, J. Assoc. Off.
    Anal. Chem. 58, 114 (1975).
38. S. N. Hagan and W. H. Tietjen, J. Assoc. Off. Anal. Chem. 58,
    620 (1975).
39. K. Yamamoto, J. Nara Med. Assoc. 26, 245 (1975).
40. Official Methods of Analysis, 12th ed., (AOAC, Washington,
    D. C., 1975), Secs. 26.091-26.098, 26.A18-26.A24.
41. S. Nesheim, F. Hardin, O. J. Francis, Jr., and W. S. Langham,
    J. Assoc. Off. Anal. Chem. 56, 817 (1973).
42. C. P. Levi, J. Assoc. Off. Anal. Chem. 58, 258 (1975).
43. F. S. Chu, J. Assoc. Off. Anal. Chem. 53, 696 (1970).
44. Ref. 40, Secs. 26.105-26.111.
45. G. Sullivan, D. D. Maness, G. J. Yakatan, and J. Scholler,
    J. Chromatogr. 116, 490 (1976).
46. J. Reiss, Z. Anal. Chem. 275, 30 (1975).
47. G. Engel, J. Chromatogr. 136, 182 (1977).
48. O. L. Shotwell, in Ref. 2, pp. 403-413.
49. L. Stoloff, S. Henry, and O. J. Francis, Jr., J. Assoc. Off.
    Anal. Chem. 59, 118 (1976).
50. C. J. Mirocha, B. Schauerhamer, and S. V. Pathre, J. Assoc.
    Off. Anal. Chem. 57, 1104 (1974).
51. L. M. Seitz and H. E. Mohr, J. Assoc. Off. Anal. Chem. 59,
    106 (1976).
52. P. M. D. Martin and P. Keen, Sabouraudia 16, 15 (1978).

53. Ref. 40, Secs. 26.099-26.104.
54. A. E. Pohland and R. Allen, J. Assoc. Off. Anal. Chem. 53, 686 (1970).
55. K. Polzhofer, Z. Lebensm. Unters.-Forsch. 163, 183 (1977).
56. A. Ciegler, H.-J. Mintzlaff, D. Weisleder, and L. Leistner, Appl. Microbiol. 24, 114 (1972).
57. P. M. Scott and E. Somers, J. Agr. Food Chem. 16, 483 (1968).
58. A. Ciegler and C. P. Kurtzman, J. Chromatogr. 51, 511 (1970).
59. L. M. Seitz, D. B. Sauer, H. E. Mohr, R. Burroughs, and J. V. Paukstelis, J. Agr. Food Chem. 23, 1 (1975).
60. R. W. Pero, R. G. Owens, and D. Harvan, Anal. Biochem. 43, 80 (1971)
61. N. Umetsu, J. Kaji, K. Aoyama, and K. Tamari, Agr. Biol. Chem. 38, 1867 (1974).
62. R. A. Meronuck, J. A. Steele, C. J. Mirocha, and C. M. Christensen, Appl. Microbiol. 23, 613 (1972).
63. P. M. Scott, W. van Walbeek, B. Kennedy, and D. Anyeti, J. Agr. Food Chem. 20, 1103 (1972).
64. N. Nakano, T. Kunimoto, and K. Aibara, J. Food Hyg. Soc. Japan 14, 56 (1973).
65. R. Puls and J. A. Greenway, Can. J. Comp. Med. 40, 16 (1976).
66. F. N. Kotsonis and R. A. Ellison, App. Microbiol. 30, 33 (1975).
67. Y. Naoi, K. Saito, E. Kazama, H. Ogawa, and Y. Kimura, Tokyo Toristu Eisei Kenkyusho Kenkyu Nempo 23, 175 (1972).
68. R. M. Eppley, in Ref. 2, pp. 285-293.
69. A. W. Hayes and H. W. McCain, Food Cosmet. Toxicol. 13, 221 (1975).
70. G. Rücker and A. Taha, J. Chromatogr. 132, 165 (1977).
71. R. V. Bhat, D. N. Roy, and P. G. Tulpule, Toxicol. Appl. Pharmacol. 36, 11 (1976).
72. M. Prosek, E. Kucan, M. Katić, and M. Bano, Chromatographia 9, 273 (1976).
73. E. Eich and W. Schunack, Planta Med. 27, 58 (1975).

CHAPTER 18

Analysis of Nitrosamines in Foods and Body Fluids by TLC

N. P. Sen

The detection and quantitative estimation of N-nitrosamines*
(simply called nitrosamines) in food and biological materials
poses a number of problems because many interfering substances
are inherently present in such items. This is particularly
true for measurement of these compounds at low ppb (1 in 10^{-9})
levels. The reliability and choice of the detection technique
will also depend, to a great extent, on the purity of the
extracts. To be applicable for trace analysis, the detection
technique used must be highly specific (selective) and mode-
rately accurate and precise. Although there are many methods
available for this purpose, only the TLC techniques will be
discussed in the present paper. A comprehensive review of the
subject "Analysis of Volatile Nitrosamines in Food" and a list
of recommended methods have been recently published (1). As it
is beyond the scope of the paper to discuss all the details of
cleanup and extraction procedures, only the final TLC determinative steps and the underlying principles behind these techniques
will be discussed. The reader is advised to consult the above-
mentioned monograph for information on the type of cleanup procedures to be used for the analysis of a particular type of
material. A great deal of information on the various chromatographic techniques, including TLC, for the analysis of nitrosamines can also be obtained from an earlier review by Fishbein
and Falk (2).

TLC has been used for both qualitative and quantitative
determination of nitrosamines. The main advantages of TLC
over other methods are its simplicity and low cost. Preussmann
et al. (3) were the first to report a sensitive TLC technique

---

* For abbreviations on the names of the various nitrosamines see
  Table I.

TABLE I. ABBREVIATIONS USED FOR N-NITROSO COMPOUNDS

Name	Abbreviations
Nitrosomethylallylamine	NMA1A
Nitrosodimethylamine	NDMA
Nitrosodiethylamine	NDEA
Nitrosodi-n-butylamine	NDBA
Nitrosodi-n-propylamine	NDPA
Nitrosodi-iso-propylamine	NDiPA
Nitrosodi-iso-butylamine	NDiBA
Nitrosodi-n-pentylamine	NDPenA
Nitrosodi-n-hexylamine	NDHexA
Nitrosodi-n-heptylamine	NDHepA
Nitrosodi-n-octylamine	NDOctA
Nitrosodiallylamine	NDA1A
Nitrosodicyclohexylamine	NDCHexA
Nitrosodibenzylamine	NDBenA
Nitrosomethylbutylamine	NMBA
Nitrosoethylbutylamine	NEBA
Nitrosomethylphenylamine	NMPHA
Nitrosopropylphenylamine	NPPhA
Nitrosoethylbenzylamine	NEBenA
Nitrosophenylbenzylamine	NPhBenA
Nitrosopyrrolidine	NPYR
Nitrosopiperidine	NPIP
Nitrosoproline	NPRO
Nitrososarcosine	NSAR
Nitrosohydroxyproline	NHPRO
Nitrosohydroxypyrrolidine	NHPYR
Nitrosohydroxybutylbutylamine	NHBBA
Nitrosonornicotine	NNN
Nitrosoanatabine	NAT
Nitrosoanabasine	NAB
Mononitrosopiperazine	MNPZ
Dinitrosopiperazine	DNPZ
Carboxymethylnitrosourea	CMNU
Nitrosomethylpiperazinylamine	NMPz
Nitrosocarbazolylamine	NCARB

for detecting nitrosamines in microgram quantities. A multitude of methods have been published since then. According to Fiddler (4) these methods can be roughly divided into two main categories: (1) Cleavage methods--those based on the cleavage of the N-N=O bond and (2) Oxidation-Reduction methods--those based on TLC of oxidized or reduced derivatives of nitrosamines. These can be further subdivided as shown in Figures 1 and 2.

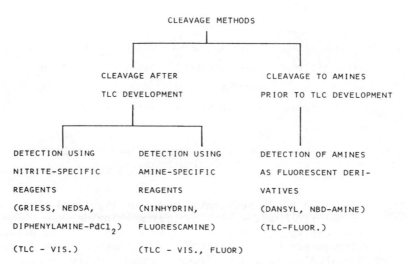

Figure 1. Various cleavage methods for the TLC analysis of nitrosamines.

TLC TECHNIQUES

Cleavage Methods

Cleavage After TLC Development. As mentioned above, Preusmann et al. (3) first described a TLC technique for the semiquantitative estimation of volatile nitrosamines. In this method, the developed plate containing the nitrosamine spots is first sprayed with a solution containing diphenylamine and palladium chloride and then irradiated under uv light (λ_{max} = 240 mμ) for a few minutes. Most nitrosamines appear as blue-to-violet spots. The detection limit is about 0.5-2 μg per spot depending on the compound tested. The principle of the method is believed to be photochemical

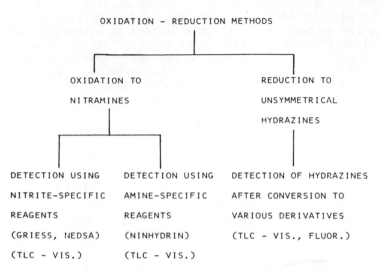

Figure 2.   TLC analysis of nitrosamines based on oxidation-
reduction products and their derivatives.

<u>trans</u>-nitrosation of diphenylamine under uv light and the forma-
tion of a colored complex with palladium chloride.
        Various dialkyl-, diaryl-, alkylaryl-, cyclic, and hetero-
cyclic nitrosamines were subjected to the above-mentioned test.
All but one (NDA1A) gave a positive reaction to the spray
reagent, and NMA1A gave only a weak positive spot (5). Alkyl-
arylnitrosamines can also be detected by this technique. The
major drawback of the technique is that it is not quite specific
for nitrosamines. An extensive examination for specificity
showed that amines, amino acids, lower carbonyl compounds, and
phenols do not give the color reaction, but quinonoid and nitro
compounds, especially aromatic ones, give a nonspecific brown
coloration. Acids, especially dicarboxylic acids, give interfering blue coloration. A number of terpenes, unsaturated hydrocarbons, and compounds with α,β-unsaturated carbonyl functions also
interfere. Therefore, if these compounds are not removed during
the cleanup process falsely positive results can be obtained.
        Using this detection technique, Marquardt and Hedler (6) in
Germany reported the finding of rather high levels (up to 50 ppm)
of NDEA in wheat flour, especially in the heated samples. Subsequent work by other investigators (7-9), however, failed to confirm these findings. Thewlis (7,8) suggested that the positive

TLC result obtained by Marquardt and Hedler (6) could have
resulted from interfering phenolic compounds present in flour.
Kroeller (10) observed that linoleic acid, tocopherol, xantho-
phyll, and carotene can also give falsely positive spots by this
method. Van Ginkel (11) applied the technique to the determination of nitrosamines in cheese. Some samples gave a positive
indication of the presence of NDMA by this method but gave negative results with Griess reagent (see below). The compounds in
question were, therefore, not nitrosamines. Because of these
problems, the diphenylamine-$PdCl_2$ reagent has not found much use
for the TLC detection of nitrosamines in recent years.

Preussmann et al. (5) later reported another TLC technique
in which the nitrosamines were detected by spraying the developed
plate with Griess reagent (sulfanilic acid + α-naphthylamine)
followed by exposure to uv light. The nitrosamines appear as
pink spots. Under uv irradiation at an acidic pH, nitrosamines
are cleaved to secondary amines and nitrous acid, which is then
detected by the well-known diazotization and coupling reaction.
The reagent is more specific than the diphenylamine-$PdCl_2$ reagent,
but positive reactions can also be obtained if nitrites, nitrates,
and esters of nitrous acid are present in the final extract. Of
about 60 nitrosamines tested only N-nitroso-N,O-dimethylhydroxyl-
amine failed to give a positive test to this reagent. The detection limit is about 0.5-5 µg per spot. Aromatic aldehydes have
been reported to produce yellow spots with Griess reagent (12).

The detection of secondary amines, produced by the photolytic
cleavage of nitrosamines, may also serve as the basis of a TLC
method for these compounds. Kroeller and Sen et al. (9) have
used the ninhydrin reagent for detecting the liberated amines.
In this method, the chromatogram is first sprayed with a dilute
acid (HCl or acetic acid) and then irradiated under uv light.
The plate is then briefly dried at room temperature to remove the
excess acid, sprayed with an ethanolic solution of ninhydrin, and
then heated in an oven (80°-100° C). Various amines show up as
colored spots on the plate. Because the colors of the reaction
products are different for different amines, these can be used
for identification purposes. For example, the colors produced
by dimethylamine (from NDMA), diallylamine (from NDAlA), and
diphenylamine (from NDPhA) are, respectively, light pink, yellow-
ish pink, and bluish-green. The technique is quite sensitive; as
little as 0.2 µg of a nitrosamine per spot can be detected. As
free amines also give the same color reactions, they must be completely removed by an efficient cleanup process before applying
the technique.

Griess reagent and ninhydrin are specific for nitrite and
amines, respectively. Therefore, if a compound on the TLC plate

(at a particular R_f zone) gives a positive reaction with both reagents, it is highly likely that the compound is a nitrosamine. Although organic nitrites and free amines interfere with the detection, these compounds are rarely present in the final extracts if the sample has undergone an efficient cleanup procedure. Moreover, nitrosamines give positive spots only after uv irradiation, whereas the organic nitrites and free amines do not require such a treatment (9). Therefore, these interfering compounds can be differentiated from nitrosamines by carrying out a duplicate TLC run in which the uv irradiation step is omitted.

While using the ninhydrin technique Sen et al. (13) observed that the ninhydrin reaction product of pyrrolidine (liberated from NPYR) on the TLC plate is highly fluorescent (orange) when observed under uv light in a dark viewing chamber. Other volatile nitrosamines (NDMA, NDEA, NDPA, NDBA, NPIP) did not fluoresce under the conditions employed. The formation of this fluorescent product was found to be dependent on the type of silica gel used for preparing the TLC plates. Of all the silica gel tested, only the MN-Silica gel GHR brand (Macherey, Nagel & Co., Germany) produced the orange fluorescent product. The technique is highly sensitive; as little as 10 ng NPYR can be detected by this method compared to 200 ng detection limit by the visualization method under normal daylight. The disadvantage of the technique is that it is only applicable to the determination of NPYR and not the other nitrosamines.

Both the Griess and ninhydrin reagents have been used by many researchers for the semiquantitative estimation of nitrosamines in foods and biological materials. These will be discussed later in the text. Since one of the components of Griess reagent, namely α-naphthylamine, contains a carcinogen (β-naphthylamine) as an impurity, its use in daily routine work should be discouraged. As a replacement, a combination of N-(1-naphthyl)ethylenediamine and sulfanilic acid (NEDSA reagent) can be used (14).

Although the main photolytic breakdown products of nitrosamines are nitrous acid and secondary amines, Young (15) observed that primary amines are also produced as minor products that can be detected with fluorescamine--a reagent that is specific for primary amines. As the reaction products of primary amines and fluorescamines are highly fluorescent, the technique allowed the detection of minute amounts of nitrosamines. For most nitrosamines, except for the highly volatile NDMA, the detection limit was in the range of 10-40 ng per spot. About 500 ng of NDMA per spot was necessary for the detection of NDMA. Quantitation was achieved with a fluorescence spectrophotometer equipped with a TLC accessory and strip chart recorder. The R_f values of 24 nitrosamines on two different adsorbents (silica gel and aluminum

oxide) and the respective detection limits are given in Table II.
The technique, however, has not been applied to foods. Therefore,
it is difficult to comment on the reliability and suitability of
the method for food analysis.

Klus and Kuhn (16) developed a technique in which the nitrosamines, after TLC development, were denitrosated in situ by
treatment with titanium chloride in hydrochloric acid. The
amines were then separated by carrying out a second TLC development in the perpendicular direction and detected by spraying the
plate with a mixture of p-aminobenzoic acid (in ethanol) and
phosphate buffer (pH 7) and exposing it to cyanogen bromide vapor.
The technique was used for the analysis of NNN, NAB and NAT; the
detection limit is about 0.5 µg/spot. The denitrosation technique, however, gave poor results (≈ 30% yield) with volatile
nitrosamines including NPYR and NPIP. The technique was tried
only with pure standards.

Cleavage Prior to TLC Development. In this method, nitrosamines
are first hydrolyzed to secondary amines, which are then converted
to suitable derivatives prior to final analysis by TLC. Eisenbrand
(17) first used this technique for the analysis of nitrosamines
in wheat flour and cheese. After denitrosation of the nitrosamines by treatment with hydrogen bromide in glacial acetic acid,
liberated amines were subjected to various cleanup steps and
reacted with excess dansyl chloride (1-dimethyl-aminonaphthalene-
5-sulfonyl chloride) in an acetone-water system at pH 8.5. The
resulting dansyl derivatives, which are highly fluorescent, were
analyzed by TLC using the solvent system cyclohexane-n-pentane-
diethyl ether (4:4:3). The developed plates were viewed under uv
light (350 nm) inside a dark viewer. The method is fairly sensitive; as little as 1 nmole (74 ng NDMA) of a nitrosamine can be
easily detected. The above-mentioned workers were able to detect
five volatile nitrosamines (NDMA, NDEA, NDPA, NDBA, AND NDPenA)
in samples of wheat flour and cheese spiked with 10 ppb levels of
each of these nitrosamines. It should be emphasized that all
secondary amines must be removed before applying this technique to
any environmental sample extract, and that extreme care should be
taken to avoid contamination by amines from contaminated glassware
and reagents. Interference can also be obtained from compounds
that liberate amines upon treatment with an HBr-glacial acetic
acid mixture.

More recently, Klimisch and Stadler (18) and Wolfram et al.
(19) have used similar techniques for TLC determination of nitrosamines or a nitrosamino acid (NPRO). The nitroso compounds were
denisotrated by the method of Eisenbrand and Preussmann (20) as
described above, and the resulting amines or amino acid (proline)

TABLE II. R_f VALUES AND VISUAL FLUORESCENCE DETECTION LIMITS OF NITROSAMINES ON SILICA GEL AND ALUMINUM OXIDE PLATES[a]

Nitrosamine[b]	R_f Value[c]				Detection Limit
	Silica Gel		Aluminum Oxide		ng
	A	B	C	D	
NDMA	0.35	0.44	0.24	0.44	500
NDEA	0.44	0.60	0.39	0.52	10-40
NDPA	0.58	0.65	0.49	0.58	7-10
NDBA	0.67	0.67	0.53	0.58	9-12
NDPenA	0.73	0.68	0.56	0.66	4- 6
NDHexA	0.76	0.76	0.57	0.68	9-12
NDHepA	0.77	0.78	0.58	0.69	9-12
NDOctA	0.79	0.79	0.59	0.70	8-11
NDA1A	0.62	0.68	0.50	0.55	10-12
NDiBA	0.69	0.69	0.56	0.58	8-11
NDCHexA	0.66	0.76	0.52	0.59	12-14
NDBenA	0.69	0.77	0.53	0.61	15-18
NMBA	0.45	0.58	0.38	0.54	7-10
NEBA	0.55	0.62	0.47	0.55	8-11
NMPhA	0.63	0.67	0.49	0.59	17-20
NPPhA	0.70	0.73	0.56	0.63	10-12
NEBenA	0.56	0.65	0.58	0.58	16-20
NPhBenA	0.70	0.75	0.55	0.60	7-10
NPYR	0.41	0.50	0.36	0.51	20-40
NPIP	0.22	0.34	0.20	0.39	17-20
NDPhA	0.71	0.75	0.57	0.63	nf
NMOR	0.21	0.33	0.23	0.39	nf
NMPZ	0.03	0.03	0.06	0.28	nf
NCARB	0.76	0.79	0.60	0.60	nf

[a] Source: Young (34).

[b] For abbreviations see Table I.

[c] Solvent systems: A, B, and C--hexane-ether-methylene chloride (3:4:2), (5:7:10), and (10:3:2) respectively; system D--ethyl acetate-hexane (4:1). Detection limit is the minimum amount of an N-nitrosamine that ultimately gives a detectable visual

were derivatized with 7-chloro-4-nitro-benzo-2-oxa-1,3-diazole
(NBD-Cl). Like the dansyl derivatives, the NBD-derivatives of
the amines are also highly fluorescent and this permitted the
detection of very small amounts of the compounds on TLC plates.
Wolfram et al. (19) used silica gel G as adsorbent and hexane-n-
butanol-ethyl acetate (80:10:10) as the solvent system for TLC
analysis of NBD-proline derivative. Based on the same principle,
Cross et al. (21) have developed a TLC-fluorometric technique
for the determination of volatile nitrosamines, mainly NDMA and
NPYR, in fried bacon and cooked-out bacon fat. The details of
this work will be presented by C. Cross in another paper.

   Oxidation-Reduction Methods

Oxidation to Nitramines. Nitrosamines can be oxidized to nitramines by treatment with peroxytrifluoroacetic acid, and the nitramines can then be analyzed by TLC using either Griess, NEDSA, or
ninhydrin reagent (14). The technique has three major advantages:
firstly it reduces the losses due to volatilization during TLC
analysis (nitramines are much less volatile than nitrosamines);
secondly the oxidation products are much more firmly adsorbed on
basic alumina than the nitrosamines, and this property can be
used for cleanup purposes; and thirdly during oxidation of food
extracts many interfering pigments are oxidized to colorless
products, thus reducing the chance of interference during the
final TLC analytical step. The technique has been used for the
detection of volatile nitrosamines in alcoholic beverages with a
sensitivity of 25 ppb in terms of the original sample.

Reduction to Unsymmetrical Hydrazines. The reduction of nitrosamines to unsymmetrical hydrazines and the subsequent formation of
colored or fluorescent derivatives form the basis of many of the
earlier TLC methods for nitrosamines. Neurath et al. (22)
reduced the nitrosamines to hydrazines with LiAlH$_4$ in anhydrous
ether, and then analyzed the hydrazines by TLC after formation of
red 5-nitro-2-hydroxybenzaldehyde derivatives. The overall yield
of nitrosamines from spiked samples of cigarette smoke condensate

---

Table II (continued)

fluorescence. Determined on activated silica gel by spotting
nitrosamine, developing, irradiating with uv light, spraying with
fluorescamine reagent, and viewing under long-wave uv light.
nf = nonfluorescent products.

was, however, very low ($\leq$ 20%). Serfontein and Hurter (23) converted the hydrazines (obtained from nitrosamines) to colored
derivatives by reacting them with 4-nitroazobenzene-4'-carboxylic
acid chloride, and separated the derivatives by two-dimensional
TLC. The method was used to detect nitrosamines in cigarette
smoke. The lowest detectable amount of a nitrosamine by this
method was about 1-2.5 µg per spot.

More recently, Yang and Brown (24) have described a technique in which the unsymmetrical hydrazines, resulting from the
reduction of nitrosamines, were converted to highly fluorescent
hydrazones before analysis by TLC. Tetrahydrofuran was reported
to be a better solvent than diethylether for reduction of nitrosamines with LiAlH$_4$. Two different aromatic aldehydes, namely,
9-anthraldehyde and 9-phenanthraldehyde, were used to make the
hydrazone derivatives. The overall yield was between 75-90%
except for NDA1A, which gave a very poor yield. The fluorescent
spots could be scraped off the TLC plate and their identity
further confirmed by their characteristic uv and mass spectra.
The technique was applied to the analysis of 21 nitrosamines.
Ultraviolet and mass spectral data and R_f values of the hydrazones in seven solvent systems were described. The technique,
however, has not been applied to the analysis of any food or biological materials. Hoffmann et al. (25) used diborane for the
reduction of nitrosamines to unsymmetrical hydrazines.

Miscellaneous Detection Techniques. Various other TLC techniques
have been used to detect nitrosamines. Most of them, however,
can be considered as general indicators of organic compounds and,
therefore, not specific for nitrosamines. These are the following: (a) detection of uv light as blue fluorescent spots (26,27);
(b) detection under uv light as dark spots on TLC plates containing fluorescent indicators (3,28); (c) detection by radioactive
tracer technique during TLC analysis of ^{14}C-labeled nitrosamines
(29,30); (d) detection as brown spots after treatment with iodine
vapors (27), iodoplatinate spray reagent (31), sulfuric acid (27),
or a mixture of sulfuric acid and ethanol (31); and (e) detection
of acidic nitrosamines with bromocresol green indicator (32).
In addition, two other spray reagents have been described that are
somewhat specific for aromatic nitrosamines. In one of them, the
nitrosamines are detected by heating with an acidic solution of
β-naphthol (33). This reagent, however, gives a positive test to
both C-nitroso and N-nitroso compounds containing an aromatic
nucleus, being more sensitive to C-nitroso compounds (e.g., p-
nitrosodimethylaniline) than N-nitroso compounds (e.g., N-nitroso-
dimethylaniline). The second technique consists of a sequential
spray of 2,4-dinitrophenylhydrazine and phosphomolybdic acid (34).

Of 24 nitrosamines tested, only the N-phenyl and carbazolyl compounds gave strong color reactions that varied from compound to compound (e.g., brown, blue-grey, or mauve). The detection limits ranged between 10 and 500 ng per spot. Carbonyl compounds, if present in the test solution, may interfere with the analysis. The method has not yet been tested on food extracts or extracts of body fluids.

SORBENTS AND MOBILE PHASE

Silica gel is probably the most commonly used sorbent employed for the TLC analysis of nitrosamines. Both the commercially available precoated silica gel plates, with or without fluorescent indicators, and the laboratory-made plates have been used. The stability of NBD-amine derivatives has been reported to be much better on the commercially available precoated plates than on the plates freshly made in the laboratory (21). Klimisch and Stadler (35), on the other hand, preferred polyamide sheets for TLC analysis of NBD-amines because of the breakdown problem on silica gel. Young (15,34) has used both silica gel and aluminum oxide (Table II) as sorbents for the TLC analysis of nitrosamines. In the majority of cases, the thickness of the sorbent layer varied between 0.25 and 0.6 mm, but a much thinner ($\approx$ 0.1 mm) coating was used with the precoated plates (15). Different degrees of activation of the coated sorbents have been used, depending on the reaction conditions needed for the spray reagent and the nature of the compounds studied. For example, Young (15) observed that it was absolutely necessary to activate the plate at 110° C for 1 hour just prior to spotting the nitrosamine standards and development of the chromatogram in order to obtain the specific color reaction with fluorescamine reagent. The course of the photolysis seemed to be affected by the water content of the sorbent; only the primary amines are formed from the nitrosamines if the water content is low.

For semiquantitative estimation both visual and TLC-densitometric (or fluorometric) determinations can be carried out and reasonably accurate results obtained (15,21,28,36,37). In the visual technique, varying amounts of the standard nitrosamines are spotted alongside the spot of the test solution, and the concentration of the nitrosamine in the test solution is calculated by comparing the relative color intensities of the standard and test spots. Although visual comparisons are not as accurate as the instrumental methods and are prone to subjective judgment, an experienced analyst is able to obtain results within $\pm$ 20% of the true value. As the most important criterion is the specificity

and not the precise accuracy of the method, such errors (+ 20%)
in measurements are acceptable. However, if proper instrumentation facilities, as mentioned above, are available they should be
utilized. Alternatively, the spots can be scraped off the plate,
the compound eluted with a suitable solvent, and its concentration measured quantitatively with a spectrophotometer or a fluorometer (19,38). If the compound in question is radioactively
labeled, the quantity can be measured by scintillation counting
(29,30).

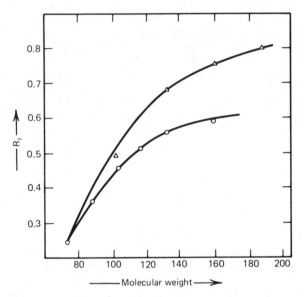

Figure 3. R_f values of some nitrosamines versus molecular weight.
o, alkylmethylnitrosamines; Δ, symmetrical dialkyl nitrosamines.
Sorbent: silica gel G (0.25 mm). Developing solvent: n-hexane-
diethylether-dichloromethane (4:3:2).

In an attempt to correlate chemical structure with adsorption properties of homologous nitrosamines, Gunatilaka (31)
plotted the R_f values of nitrosamines against the molecular
weights. As can be seen from Figure 3 the curves are smooth, the
R_f values steadily increasing with increases in molecular weights
within the particular mologous series. A mathematical

description of the effect of side chains on chromatographic
adsorption can be represented by the following equation:

$$f = (1-R_f)/R_f = K(M_{sc})^{-1}$$

where f is the adsorption affinity, R_f the ratio of distance
moved by sorbate and by the solvent, M_{sc} the molecular weight of
the side-chain attached to the functional group, and K the sum of
the interaction tendencies between the adsorbent and the adsorbed
compound. With the help of such an equation one would be able to
predict the R_f value of a new nitrosamine in the same homologous
series. It was further observed that within an isomeric pair,
the alkyl-methyl nitrosamine (e.g., methyl-n-amylnitrosamine,
R_f = 0.57) is always more strongly bound than the symmetrical
counterpart (i.e., di-n-propylnitrosamine, R_f = 0.69) to the
adsorbent. The above author suggested that this property could
be used to differentiate between isomeric pairs of nitrosamines.

Preussmann et al. (3,5) reported three mobile phases for TLC
analysis of various nitrosamines on silica gel plates. For symmetrical dialkylnitrosamines and methyl-alkylnitrosamines they
used a mixture of n-hexane-diethyl-ether-methylene chloride
(4:3:2). The same solvents but in different proportions (as indicated in parentheses) were used for cyclic (5:7:10) and diaryl-
or alkylaryl-nitrosamines (10:3:2). For polar nitrosamines such
as hydroxyalkylnitrosamines, nitrosaminoacids, or their derivatives, however, a much more polar mobile phase is needed. A mixture of chloroform-methanol-formic acid (50:50:1) (27), n-butanol-
acetic acid-water (4:1:1), or n-hexane-n-butanol-ethyl acetate
(80:10:10) can be used for these compounds (19,27,39,40).

Liberek et al. (32) reported a number of mobile phases for
TLC analysis of α-N-nitroso-N-alkylamino acids. These workers
were also able to separate the Z and E isomers of the various
nitrosamino acids investigated (Table III). Comparative studies
revealed that TLC offers a more sensitive and rapid technique for
checking the conformational purity of isomers than nmr spectroscopy. The TLC analysis was carried out at 0-2° C on silica gel G
layers. The NBD-amine derivatives and the anthraldehyde hydrazones can be chromatographed using the system cyclohexane-ethyl
acetate (1:1) and cyclohexane-chloroform (3:2), respectively (21,
24). Since nitrosamines are sensitive to light, especially under
acidic conditions, the TLC analysis should preferably be carried
out in the dark, and if possible at 4° C (9,41).

TABLE III. R_f VALUES OF Z AND E ISOMERS OF N-NITROSAMINO-N-ALKYLAMINO ACIDS ON SILICA GEL G PLATES[a]

Compound[b]	S1[c]		S2	
	Z	E	Z	E
NSAR	-	-	0.19	0.24
N-Nitroso-N-ethylglycine	-	-	0.19	0.24
N-Nitroso-N-propylglycine	0.21	0.31	-	-
N-Nitroso-N-isopropylglycine	0.20	-	-	-
N-Nitroso-N-methyl-DL-alanine	0.20	0.25	-	-
N-Nitroso-N-ethyl-DL-alanine	0.22	0.32	-	-
N-Nitroso-N-propyl-DL-alanine	0.24	0.32	-	-
N-Nitroso-N-isopropyl-DL-alanine	0.21	-	-	-
N-Nitroso-N-methyl-DL-phenylalanine	-	-	0.29	0.35
N-Nitroso-N-methyl-L-valine	-	-	0.25	0.31
NPRO	-	-	0.16	0.20

[a] Source: Liberek et al. (32).

[b] Detection reagent: 0.1% Bromocresol green in methanol or iodine vapor was used to locate the compounds on the TLC plate.

[c] Solvent systems: S1--n-butanol-25% ammonia-water (75:6:75); S2--n-butanol-25% ammonia (6:1). Temp. 0-2° C.

APPLICATION OF TLC FOR NITROSAMINE ANALYSIS

Model System Studies

TLC offers a simple and sensitive technique for studying the formation of nitrosamines in model systems. In such systems, the identity of all the reacting ingredients is known and the system is well defined. Therefore, the chance of obtaining a falsely positive test from interfering materials is minimized. The technique can also be used to monitor the course of a derivatization reaction of a nitrosamine standard (42). Yamamoto et al. (28) have used the TLC technique for studying the reaction products of nitrite and glycocyamine, a naturally occurring guanidine compound. The main product was identified as CMNU. Hotchkiss et al. (39) have used TLC for identifying the reaction products of nitrite and spermidine; a total of nine nitrosamines were

identified in the reaction mixture. Earlier, Preussmann (43)
used the TLC technique for studying the oxidative degradation of
nitrosamines by enzyme-free model systems. The use of the hydroxylase system of Udenfriend et al. (44) led to the formation of
several oxidative degradation products that were identified and
characterized by TLC. Ender and Ceh (45) and Sen et al. (46,47)
used TLC for studying the formation of nitrosamines as a result
of the interaction of nitrite and various amino compounds such as
dimethylamine, diethylamine, piperazine (a drug), and several
pesticides. The TLC technique was found to be superior to the
spectrophotometric techniques (used by other workers) for studying the formation of nitrosopiperazines because it allowed the
simultaneous detection and measurement of both MNPZ and DNPZ
whereas the latter technique only measured the total of the two
nitrosamines. Similarly, Yamada et al. (48) and Boyland et al.
(49) used the TLC technique for studying the catalytic effect of
thiocyanate in the formation of nitrosamines. The heat-induced
formation of NHPYR from NHPRO in model systems was demonstrated
by TLC (50).

Food Analysis

In earlier work most of the information about the occurrence of
nitrosamines in foods originated from TLC analysis. In fact, the
first evidence of the occurrence of NDMA in nitrite-treated fish
meals, which were implicated in the death of fur-bearing animals
in Norway, was obtained by TLC and colorimetric analysis (51).
Some of the other applications (e.g., wheat flour, cheese, etc.)
have already been mentioned. A list of the major applications of
TLC to food analysis is presented in Table IV. The subject was
also reviewed earlier by Sen (52), Scanlan (53), and Magee et al.
(54).
    It should be emphasized, however, that many of the earlier
data obtained by TLC techniques have not been confirmed in later
studies (using more sophisticated techniques), and therefore their
validity is questionable. Moreover, the nature of the extensive
cleanup procedures, which are necessary to prepare suitable food
extracts for TLC analysis, makes the whole method very tedious
and time-consuming. The sensitivity is also not adequate. With
the exception of the fluorescent methods the rest are unsuitable
for analysis below 20 ppb levels. The fluorescent methods,
especially those based on the formation of amine derivatives
prior to TLC, also have many drawbacks, the most important being
the chance of getting falsely positive results due to contamination from the amines that are abundant in foods. Therefore, these
TLC methods should only be used for screening purposes, and all

TABLE IV. ANALYSIS OF FOODS FOR NITROSAMINES BY TLC TECHNIQUE

Type of Food or Feed Analyzed	Detection Method[a]	Nitrosamines Analyzed or Detected	Sensitivity of Method (ppb)	Reference(s)
Nitrite-treated herring meal	A	NDMA	10,000	51
Fish meal	B, C	NDMA	25	68
Smoked meats, smoked fish, mushrooms	D, E	NDMA		44
Salted dried fish		NDMA, NDEA		69
Fish and sea foods	A, B, C	NDMA, NDEA, NPYR, NPIP	2-20	70
Miscellaneous cured meat products	A, B, C	NDMA, NDEA, NPYR, NPIP	2-20	13, 37, 55, 71, 72
Fried bacon, cooked-out fat	F	NDMA, NPYR	5	21
Raw bacon	B, F	NPRO		73, 19
Raw bacon	G	NPRO		74
Raw bacon	C	NPRO		75

Cheese, wheat flour, spinach	A, B, C	NDMA, NDEA	50-150	9
Cheese	A, B	NDMA, NDEA	125	11
Havarti and Gouda cheese	A, B, C	NDMA, NDEA, NDPA, NDiPA	5	10
Tollenser cheese	B, C	NDMA, NDEA		76
Uncooked sauage, cured pork, gelatin-containing infusion liquid	B, C	NDMA, NDEA, NDPA		77
Wheat flour	A	NDEA		6, 7, 8
Tobacco smoke condensate, tobacco leaf	D	NDMA, NDEA, NPIP, NDBA, NMBA		23, 78
Nigerian palm wine	A	NDMA		79
Alcoholic beverages	B, C, H	NDMA, NDEA, NPYR, NPIP	25	14

a Abbreviations for the detection methods: A--Diphenylamine-PdCl$_2$ + uv; B--Griess or NEDSA + uv; C--acetic acid + uv + ninhydrin (heat); D - reduce to hydrazine, then derivatize with 4-nitroazobenzene-4'-carboxylic acid chloride; E--reduce to hydrazine, then derivatize with 5-nitro-2-hydroxybenzaldehyde; F--As NBD-amines; G--uv-fluorescamine; H--oxidation to nitramines, then use spray 'B' or 'C' as above. For details, see text.

291

positive results should be confirmed by some reliable techniques
such as GLC-TEA or GLC-MS (1).

Despite the various drawbacks and limitations of the TLC
methods, reliable results can be obtained if adequate care is
taken while carrying out the analyses. For example, Sen et al.
(36,37) applied the technique for the analysis of nitrosamines in
fried bacon and other cured meat products, and compared the TLC
data with those obtained by the more reliable GLC-MS method.
Good correlation was obtained in the majority of cases (Table V).
More recently, Cross et al. (21) made similar comparisons and
demonstrated that reliable results can also be obtained for the
analysis of NDMA and NPYR in cooked-out bacon fat by the fluorometric NBD-amine method mentioned earlier. The use of the TLC-
fluorometric method of Sen et al. (55) for NPYR gave valuable
data (Table VI) on the effect of nitrite concentration on the
NPYR levels in fried bacon. The TLC technique was also very useful in the research that led to the discovery of high levels of
nitrosamines in spice-nitrite premixes that until recently were
widely used for the preparation of a variety of cured meat products (56). The use of such premixes was later banned both in the
United States and Canada.

Analysis of Biological Materials and Body Fluids

Unfortunately, only limited information is available on the application of one TLC technique to the analysis of biological materials. The main obstacle is the lack of adequate sensitive and
specific methods. Moreover, previous studies have suggested that
nitrosamines are metabolized very rapidly, causing their concentration in body fluids or body tissues to be extremely low. However, if sufficiently high doses of nitrosamines are administered
to laboratory animals, traces of residues or of degradation products can be detected. For example, Mohr et al. (57) used the TLC
technique to study the placental transfer of NDEA in the golden
hamster. NDEA was detected on TLC plates with the diphenylamine-
$PdCl_2$ reagent or by its fluorescence under uv light. A dose of
10-30 mg NDEA was administered to each animal on the 15th day of
pregnancy. The animals were killed 2 minutes to 12 hours after
the dose, and liver, placenta, and fetuses were analyzed for
nitrosamines. Similarly, Okada and Uki (58) and Krueger and
Bertram (59) used TLC as one of the techniques for studying the
metabolism of NHBBA and NPYR, respectively.

The subject of bacterial synthesis and degradation of nitrosamines has received a great deal of attention in recent years.
TLC techniques were used in many of these studies. In these
cases also, the bacterial cultures were grown in the presence of
fairly high levels of nitrate, nitrite, and secondary amines

TABLE V.   CORRELATION OF TLC RESULTS WITH GLC-MS DATA (ppb)[a]

Sample	NPIP[b]		NPYR	
	TLC	GLC-MS	TLC	GLC-MS
Chinese sausage	Neg	Neg	30	33
Chinese sausage	Neg	Neg	20	29
Bologna	10	14	20	19
Bologna	20	14	10	7
Wiener	50	Neg	Neg	Neg
Wiener	22	15	10	8
Mock chicken	Neg	Neg	10	12
Bologna	Neg	21	15	19
Meat loaf	Neg	Neg	10	18
Fried bacon	-	-	44	44
Fried bacon	-	-	4.4	4.4
Fried bacon	-	-	38.8	22.2
Fried bacon	-	-	5.5	6.6
Fried bacon	-	-	24.4	22.2
Fried bacon	-	-	30.0	11.1
Fried bacon	-	-	6.6	5.5
Fried bacon	-	-	2.8	Neg[d]

[a] Source: Sen et al. (36,37).

[b] Dashes signify not compared.

[c] Detection reagents: NEDSA and Ninhydrin.

[d] Detection limit 5.5 ppb for the sample size used.

which resulted in the formation of sufficient levels of nitrosamines for detection by TLC. In three studies (60-62) Griess,
diphenylamine-$PdCl_2$, and ninhydrin reagents were used to detect
the nitrosamines on the TLC plates. In one case (38) the concentration of the nitrosamine in question (NSAR) was quantitatively
measured by uv-spectrophotometry after elution of the spot from
the TLC plate. In two other studies (29,32) ^{14}C-labeled nitrosamines or precursors were used. This allowed a more sensitive
detection of the nitrosamines on TLC plates.

    There is very little information available on the applica-
tion of the TLC technique to the analysis of body fluids. The

TABLE VI.   TLC ANALYSIS OF NPYR IN FRIED BACON PREPARED WITH
              DIFFERENT LEVELS OF SODIUM NITRITE[a]

Sample No.	Initial Sodium Nitrite Level (ppm) Used to Prepare the Bacon	NPYR Levels (ppb) in Fried Bacon
1-5	0	all neg.
6-9	50	2
		2
		4
		4
10-14	100	8
		7
		8
		8
		8
15-19	150	10
		20
		10
		5
		5
20-24	20	20
		20
		20
		12
		12

[a] Source: Sen et al. (55).

[b] Detection reagent: acetic acid-uv-ninhydrin (fluorescence spots).

only studies that made significant use of TLC were those of
Juszkiewicz and Kowalski (63), who investigated the passage of
nitrosamines in the blood and milk of goats after oral administration of various nitrosamines. Similar studies were performed

TABLE VII.   MILK AND BLOOD CONCENTRATIONS[a] OF NITROSAMINES IN
MILKING GOATS[b]

Time	NDMA (mg/kg)		NDEA (mg/kg)		NDPA (mg/kg)	
(Hours)	Milk	Blood	Milk	Blood	Milk	Blood
½	9.696	-[d]	4.941	-	5.833	-
1	12.179	10.058	11.429	11.905	4.900	1.578
2	12.260	-	14.683	-	2.100	-
4	8.947	-	8.061	-	0.027	-
6	7.051	9.615	4.841	-	0.011	0.004
8	6.851	-	0.988	-	0.005	-
12	3.462	-	0.218	-	0.004	-
24	0.046	-	0.008	-	Tr[f]	-
36	Tr[e]	-	Tr	-	ND[f]	-

[a] A single oral dose of 30 mg/kg of each compound was administered.

[b] Source: Juszkiewicz and Kowalsky (63).

[c] Detection reagents: Griess and ninhydrin.

[d] Not analyzed.

[e] Traces.

[f] Not detected.

on rabbits but only blood levels were measured. TLC analysis was
carried out by the method of Sen et al. (9) and the results were
further confirmed by GLC using two different types of detectors.
These studies produced some interesting results; these are presented in Table VII. It should be noted, however, that the doses
of various nitrosamines administered were quite high (Table VII).
No nitrosamines were detected in the blood or milk of the goats
before they received the nitrosamine dose. Recently Juszkiewicz
and Kowalsky (64) used the TLC technique for the analysis of NDMA
in the blood, muscles, liver, brain, and eggs of hens after oral
administration of fairly large doses of NDMA.
    There appears to be no report of the application of TLC tech-
niques to the analysis of nitrosamines in human body fluids. The
only report of the finding of nitrosamines in human blood was

obtained by using a Thermal Energy Analyzer Technique (TEA), which
was developed by Fine et al. (65). The concentrations of NDMA
and NDEA in human blood after a fried bacon and raw spinach diet
were found to reach 0.77 and 0.46 ppb, respectively (66). The
normal levels of these two nitrosamines in the resting (before
the meal) blood were much lower (0.35-0.09 ppb). Preliminary
results from our laboratory (67) confirmed the finding that minute
traces (0.12-0.14 ppb) of NDMA may be present in the resting human
blood (2 out of 10 were positive). No NDEA were detected (detection limit, 0.05 ppb) in any of the samples analyzed. As in the
study of Fine et al. (64) the GLC-TEA technique was used. It is
doubtful whether the presently available TLC techniques would be
sensitive and specific enough to measure such low levels of
nitrosamines.

REFERENCES

1. R. Preussmann, M. Castegnaro, E. A. Walker, and A. E. Wasserman, Environmental Carcinogens: Selected Methods of Analysis,
   vol. 1--Analysis of Volatile Nitrosamines in Food (International Agency for Research on Cancer, Lyon, 1978).
2. L. Fishbein and H. L. Falk, Chromatog. Rev. 11, 365 (1969).
3. R. Preussmann, D. Daiber, and H. Hengy, Nature (London) 201,
   502 (1964).
4. W. Fiddler, Toxicol. Appl. Pharmacol. 31, 352 (1975).
5. R. Preussmann, G. Neurath, G. Wulf-Lorentzen, D. Daiber, and
   H. Hengy, Z. Anal. Chem. 202, 187 (1964).
6. P. Marquardt and L. Helder, Arzneim.-Forsch. 16, 778 (1966).
7. B. H. Thewlis, Food Cosmet. Toxicol. 5, 333 (1967).
8. B. H. Thewlis, Food Cosmet. Toxicol. 6, 822 (1968).
9. N. P. Sen, D. C. Smith, L. Schwinghamer, and J. J. Marleau,
   J. Assoc. Off. Anal. Chem. 52, 47 (1969).
10. E. Kroeller, Deut. Lebensm. Rudsch. 63, 303 (1967).
11. J. G. Van Ginkel, Ann. Bull. Int. Dairy Fed., Part 6 (1970).
12. R. Schoental and S. Gibbard, Nature (London) 216, 612 (1967).
13. N. P. Sen, B. Donaldson, J. R. Iyengar, and T. Panalaks,
    Nature (London) 241, 473 (1973).
14. N. P. Sen and C. Dalpé, Analyst 97, 216 (1972).
15. J. C. Young, J. Chromatog. 124, 17 (1976).
16. H. Klus and H. Kuhn, J. Chromatog. 109, 425 (1975).
17. G. Eisenbrand, in N-Nitroso Compounds--Analysis and Formation
    (International Agency for Research on Cancer, Lyon, 1972),
    IARC Scientific Publication No. 3, pp. 64-70.
18. H. J. Klimisch and L. Stadler, J. Chromatog. 90, 141 (1974).
19. J. H. Wolfram, J. I. Feinberg, R. C. Doerr, and W. Fiddler,
    J. Chromatog. 132, 37 (1977).

20. G. Eisenbrand and R. Preussmann, Arzneim.-Forsch. 20, 1513 (1970).
21. C. K. Cross, K. R. Barucha, and G. M. Telling, J. Agric. Food Chem. 26, 657 (1978).
22. G. Neurath, B. Pirman, and H. Wichern, Beitr. Tabakforsch. 2, 311 (1964).
23. W. J. Serfontein and P. Hurter, Nature 209, 1238 (1966).
24. K. W. Yang and E. V. Brown, Anal. Lett. 5, 293 (1972).
25. D. Hoffmann, G. Rathkamp, and Y. Y. Lin, in N-Nitroso Compounds in the Environment, edited by P. Bogorsei and E. A. Walker (International Agency for Research on Cancer, Lyon, 1974), Scientific Publication No. 9, pp. 159-165.
26. K. Moehler and O. L. Mayrhofer, Z. Lebensm. Untersuch.-Forsch. 135, 313 (1968).
27. G. S. Rào and E. A. Bejnarowicz, J. Chromatog. 123, 486 (1976).
28. M. Yamamoto, T. Yamada, and A. Tanimura, J. Food Hyg. Soc. 17, 176 (1976).
29. I. R. Rowland and P. Grasso, Appl. Microbiol. 29, 7 (1975).
30. P. Klubes and W. R. Jondorf, Res. Commun. Chem. Pathol. Pharmacol. 2, 24 (1971).
31. A. A. L. Gunatilaka, J. Chromatog. 120, 229 (1976).
32. B. Liberek, J. Augustyniak, J. Ciarkowski, K. Plucinska, and K. Stachowiak, J. Chromatog. 95, 223 (1974).
33. K. Yasuda and K. Nakashima, Buneski Kagaku 17, 732 (1968).
34. J. C. Young, J. Chromatog. 124, 115 (1976).
35. H. J. Klimisch and L. Stadler, J. Chromatog. 90, 223 (1974).
36. N. P. Sen, B. Donaldson, S. Seaman, J. R. Iyengar, and W. F. Miles, J. Agric. Food Chem. 24, 397 (1976).
37. N. P. Sen, J. R. Iyengar, W. F. Miles, and T. Panalaks, in Environmental N-Nitroso Compounds, edited by E. A. Walker, P. Bogovski, and L. Griciute (International Agency for Research on Cancer, Lyon, 1976), Scientific Publication No. 14, pp. 333-342.
38. M. A. Friedman, Bull. Environ. Cont. Toxicol. 8, 375 (1972).
39. J. H. Hotchkiss, R. A. Scanlan, and L. M. Libby, J. Agric. Food Chem. 26, 1183 (1977).
40. N. P. Sen and B. Donaldson, in Ref. 25, pp. 103-106.
41. G. Eisenbrand, K. Spaczynski, and R. Preussmann, J. Chromatog. 51, 503 (1970).
42. N. P. Sen, W. F. Miles, S. Seaman, and J. F. Lawrence, J. Chromatog. 128, 169 (1976).
43. R. Preussmann, Arzneim.-Forsch. 14, 769 (1964).
44. S. Udenfriend, C. T. Clark, J. Axelrod, and B. B. Brodie, J. Biol. Chem. 208, 731 (1954).
45. F. Ender and L. Ceh, Alkylierend Wirkende Verbindungen, 2nd Conference on Tobacco Research, Freiburg, 1967), pp. 83-91.

46. N. P. Sen, D. C. Smith, and L. Schwinghamer, Food Cosmet.
    Toxicol. 7, 301 (1969).
47. N. P. Sen, B. A. Donaldson, and C. Charbonneau, in Ref. 25,
    pp. 75-79.
48. T. Yamada, M. Yamamoto, and A. Tanimura, Shokuhin Eiseigaku
    Zasshi 15, 201 (1974).
49. E. Boyland, E. Nice, and K. Williams, Food Cosmet. Toxicol.
    9, 1 (1971).
50. J. S. Lee, D. D. Bills, R. A. Scanlan, and L. M. Libby, J.
    Agric. Food Chem. 25, 421 (1977).
51. J. Sakshaug, E. Soegnen, M. Aas Hansen, and N. Koppang,
    Nature (London) 206, 1261 (1965).
52. N. P. Sen, in Toxic Constituents of Animal Foodstuffs,
    edited by I. E. Liener (Academic Press, New York, 1974), pp.
    131-194.
53. R. A. Scanlan, in Critical Reviews in Food Technology,
    (Chemical Rubber Co., Cleveland, Ohio, 1975), vol. 5, pp.
    357-402.
54. P. N. Magee, R. Montesano, and R. Preussmann, in Chemical
    Carcinogens, edited by C. E. Searle (American Chemical Society,
    New York, 1976), pp. 491-625.
55. N. P. Sen, J. R. Iyengar, B. A. Donaldson, and T. Panalaks,
    J. Agric. Food Chem. 22, 540 (1974).
56. N. P. Sen, W. F. Miles, B. Donaldson, T. Panalaks, and J. R.
    Iyengar, Nature (London) 245, 104 (1973).
57. U. Mohr, J. Althoff, and A. Authaler, Cancer Res. 26, 2349
    (1966).
58. M. Okada and E. Uki, Gann 63, 391 (1972).
59. F. W. Krueger and B. Bertram, Z. Krebsforsch. 83, 255 (1975).
60. G. M. Hawksworth and M. J. Hill, Brit. J. Cancer 25, 520
    (1971).
61. J. Sander, Z. Physiol. Chem. 349, 429 (1968).
62. D. L. Collins-Thompson, N. P. Sen, B. Aris, and L. Schwing-
    hamer, Can. J. Microbiol. 18, 1968 (1972).
63. T. Juszkiewicz and B. Kowalski, in Ref. 25, pp. 173-176.
64. T. Juszkiewicz and B. Kowalski, in Environmental Aspects of
    the N-Nitroso Compounds, edited by E. A. Walker, M. Castegnaro,
    L. Griciute, and R. E. Lyle (International Agency for Research
    on Cancer, Lyon, 1978), Scientific Publication No. 19, pp.
    433-439.
65. D. H. Fine, F. Rufe, D. Lieb, and D. A. Rounbehler, Anal.
    Chem. 47, 1188 (1975).
66. D. H. Fine, R. Ross, D. P. Rounbehler, A. Silvergleid, and L.
    Song, Nature (London) 265, 753 (1977).
67. B. Kowalski, C. Miller, and N. P. Sen, unpublished results
    (1978).

68. N. P. Sen, L. A. Schwinghamer, B. A. Donaldson, and W. F. Miles, J. Agric. Food Chem. 20, 1280 (1972).
69. M. Ishidate, A. Tanimura, Y. Ito, A. Sakai, H. Sakuta, T. Kawamura, K. Sakai, F. Miyazawa, and H. Wada, in Topics in Chemical Carcinogenesis, Proceedings of the Second International Symposium, 1971, edited by Y. Nakahara and Y. Waro (University Park Press, Baltimore, Md., 1972), pp. 313-322.
70. J. R. Iyengar, T. Panalaks, W. F. Miles, and N. P. Sen, J. Sci. Food Agric. 27, 527 (1976).
71. T. Panalaks, J. R. Iyengar, and N. P. Sen, J. Assoc. Off. Anal. Chem. 57, 806 (1974).
72. N. P. Sen, Food Cosmet. Toxicol. 10, 219 (1972).
73. I. Kushir, J. I. Feinberg, J. W. Pensabene, E. G. Piotrowski, W. Fiddler, and A. E. Wasserman, J. Food Sci. 40, 427 (1975).
74. J. C. Young, J. Chromatog. 151, 215 (1978).
75. N. P. Sen, B. A. Donaldson, S. Seaman, J. R. Iyengar, and W. F. Miles, in Ref. 64, pp. 373-393.
76. U. Freimuth and E. Glaeser, Nahrung 14, 357 (1970).
77. K. Moehler and O. L. Mayrhofer, in Proceedings of the 15th European Meeting of Meat Research Workers, Helsinki, Finland, August 17-24, 1969), pp. 302-304.
78. G. Neurath, B. Pirman, W. Leuttich, and H. Wichern, Beitr. Tabakforsch. 3, 251 (1965).
79. O. Bassir and E. N. Maduagwu, J. Agric. Food Chem. 26, 200 (1978).

CHAPTER 19

Analysis of Volatile Nitrosamines in Bacon

Charles K. Cross

In recent years there have been many investigations of foodstuffs in which the presence of N-nitrosamines was demonstrated (1-4). For the most part, the nitrosamines have been found in cured meat products. The presence of nitrosamines in cured meats is a direct result of the use of sodium nitrite as one of the essential curing ingredients. The only cured meat in which nitrosamines have been consistently found in quantities larger than 1-2 ppb is fried side bacon. The two nitrosamines most commonly detected in fried bacon are dimethyl nitrosamine (NDMA) and nitrosopyrrolidine (NPYR), which occur at levels up to a total of about 50 ppb. Smaller amounts of diethylnitrosamine (NDEA) and nitrosopiperidine (NPIP) are occasionally found.

The official method of analysis, recognized by The Food and Drug Administration and The Department of Agriculture, is based on combined gas chromatography and high-resolution mass spectrometry. A semi-official or rapid screening method is based on gas chromatography with a Thermal Energy Analyzer as detector (5). Both of the methods involve expensive instrumentation and it was therefore desirable to develop a rapid, simple, and reliable method that could be used in many laboratories not equipped with a Mass Spectrometer or Thermal Energy Analyzer. In a recent publication (6) we have shown a good correlation between our method and the GC-MS method used at Unilever in Great Britain (7).

It should be emphasized that the method to be described, in common with any non-GS-MS method, can only be used as a screening technique. The analysis of the amine portion of the molecule produced by splitting of the nitrosamine with hydrogen bromide in acetic acid, as described by Eisenbrand and Preussmann in 1970 (8), was used. Figure 1 shows the reactions involved in this analytical method. The analysis of both the nitrite and amine portions produced in the splitting reaction was described in a previous

300

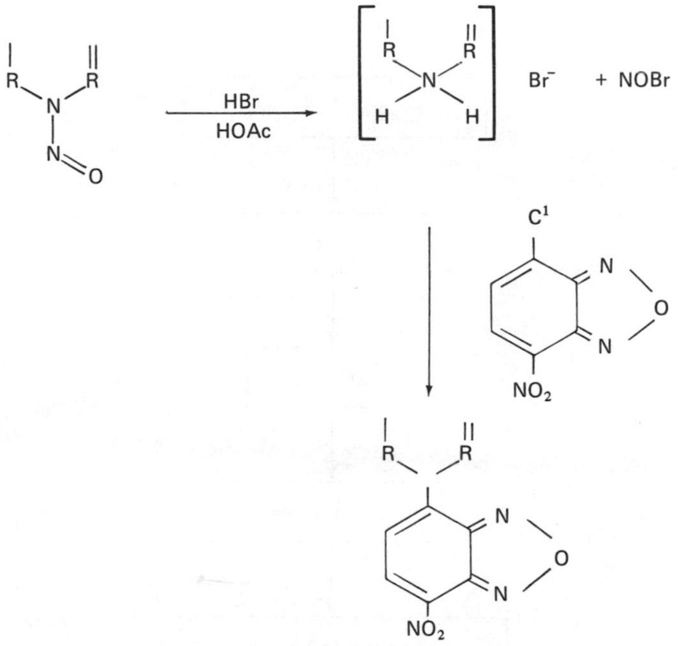

Figure 1.   Reactions involved in analysis of nitrosamines.

communication. One procedure could then substantiate the other.
The modified Griess method for determining nitrites did not lend
itself to the analysis of small samples and, in addition, was
found to be very sensitive to interferences, particularly when
bacon rasher was being analyzed and the amount of nitrosamine
present was less than 5 ppb. On the other hand, extremely sensitive and not subject to the interference was the determination of
the amine portion by fluorescence densitometry of its derivative
with 7-chloro-4-nitrobenzo-2-oxa-1,3-diazole (NBD-Cl) after thin
layer chromatographic separation. The analysis of the amines as a
measure of nitrosamines was described by Klimisch and Stadler (9).

Because the primary interest was the analysis of fried bacon, the analysis of the nitrite portion of the molecule was eliminated. This permitted considerable simplification of the procedure. The modified procedure can be applied to samples as small as 10-20 g of cooked bacon.

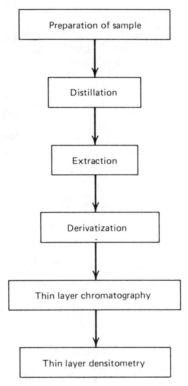

Figure 2.   Schema for preparation of samples.

Outlined in Figure 2 are the steps involved in the analysis. The only cleanup involved is a vacuum distillation and a liquid-liquid extraction. Preparation of the sample involves frying the bacon in a standardized way and sampling the product to obtain a uniform aliquot that is representative of the whole. This step

is critical in the analysis of such a small sample of material as
inhomogeneous as fried bacon. Cold 3N sodium hydroxide is added
to the sample and homogenization is performed. The nitrosamines
are separated from the matrix by distillation in a closed vacuum
system. The distillate is melted, acidified, and extracted with
dichloromethane. The dichloromethane is removed through a Snyder
column and the residue treated with hydrogen bromide in acetic
acid to give the amine hydrobromides that were shown in Figure 1.
The amine hydrobromides react quantitatively with NBD-Cl to give
the NBD amines, which separate well on thin layers of silica gel.
The derivatives fluoresce intensely under longwave ultraviolet
light and this property is used for the quantitative determination.

Preparation of the sample continues as follows: Ten slices
of bacon are fried in an electric frying pan with the pan starting at room temperature. The thermostat is set to give a final
temperature of 172° C, which is reached in about 5 minutes.
After frying for 4 minutes, the bacon is turned frequently and
removed in 12 minutes. These frying conditions result in maximum
nitrosamine formation. The fried rasher is patted dry on paper
towels to remove excess cook-out fat and then placed in a poly-
ethylene bag. The rasher is then frozen by covering the bag with
dry ice. The frozen rasher is pulverized in the bag and well
mixed. A 20-g aliquot is accurately weighed into a 500-ml round-
bottomed flask and 60 ml of cold 3N sodium hydroxide are added.
The Polytron homogenizer does an excellent job of disintegrating
the sample in 30 seconds. The probe is rinsed into the flask
with 5-10 ml of distilled water and the flask is connected into
the distillation equipment described previously (6). Normally
four samples are distilled simultaneously. The receivers, which
are 250-ml round-bottomed flasks, are immersed in a dry ice/
acetone bath and vacuum is applied using a water aspirator until
the contents of the flask are distilling well. The vacuum line
is closed and water at 60°-65° C is added to the constant temperature bath up to the neck of the distilling flask. The distillation proceeds for 30 minutes in the closed system. The vacuum
line is opened once after 15 minutes to confirm that the low pressure is being maintained.

The distillate is melted and acidified with concentrated sulfuric acid to pH 1. The liquid is filtered through a glass wool
plug into a 125-ml separatory funnel. The nitrosamines are separated from the acidic solution by extraction with dichloromethane
(3 × 30 ml). The extraction is performed very carefully so that
no aqueous layer is transferred to the dichloromethane. The combined extract is backwashed in a second 125-ml separatory funnel
with 10% sulfuric acid (5 ml) to eliminate the possibility of

amines appearing in the extract. By adding 500-1000 ppm of both
dimethyl amine and pyrrolidine to the distillate it has been
shown that the amines are retained by the sulfuric acid. The
dichloromethane solution is dried over anhydrous sodium sulphate
and anhydrous calcium chloride, filtered into a 250-ml 24/40
Erlenmeyer flask, and evaporated on a hot plate through a Snyder
column until 5 ml of solution remain. The Snyder column is
allowed to drain back into the flask. The cooled solution is
transferred quantitatively with a Pasteur pipette to a 25-ml pear-
shaped flask. A boiling chip is added and a standard 10/14 inner
joint is attached to the flask. The solution is then evaporated
on a hot water bath until the contents of the flask go dry. The
column is rinsed back into the flask with 0.2 ml of glacial
acetic acid. The column is removed and 0.2 ml of 3% hydrogen
bromide in glacial acetic acid is added to the flask. The splitting reaction is allowed to proceed in the stoppered flask for at
least 15 minutes at room temperature but may be left overnight if
necessary. The reaction mixture is transferred to a 25-ml round-
bottomed flask and evaporated carefully and completely to dryness
on a rotary evaporator using a 50°-60° C water bath. The residue
is washed once with about 0.2 ml of hexane. The hexane is
removed with a Pasteur pipette and discarded. The residue is
transferred with small volumes of methanol to a 1-ml glass-
stoppered, calibrated test tube. A total of about 0.5 ml of
methanol is used for the transfer. The solvent is removed by
means of a hot water bath using a gentle stream of nitrogen.

THE FORMATION OF THE NBD AMINE DERIVATIVES

The residue in the tube is dissolved in methanol (40 µl). A
0.05% solution of NBD-Cl in methanol (40 µl) is added, followed
by 0.2 M sodium bicarbonate (20 µl). The tube is stoppered and
immersed just above the liquid level in a 55° C water bath for 1
hour. The solvents are blown off with a stream of nitrogen on a
hot water bath, leaving a dry residue in the tube.

THIN LAYER CHROMATOGRAPHY

The residue is treated with 1:4 dichloromethane-hexane (0.2 ml)
in an ultrasonic bath for about 10 to 15 seconds to dissolve the
derivative. Some residue does not dissolve.
    Each sample (5 µl) is spotted three times on a 20 × 20 cm
Merck silica gel 60 plate, along with a standard solution containing known quantities of NDMA and NPYR that have been split and

derivatized at the same time as the sample. Using 18 tracks, five samples and one standard can be applied to one 20 × 20 cm plate. One set of three tracks is usually reserved for a blank. For spotting the TLC plates a standard 10-μl Hamilton syringe held in a clamp on a burette stand was used. The plate is placed on a lab jack and raised until it just contacts the needle. The solution is applied in small quantities alternating with a gentle air stream to keep the spots compact.

The plate is developed once in a saturated tank for a distance of 12 cm using either 1:1 cyclohexane-ethyl acetate or 33:33:33:1 hexane-chloroform-diethyl ether-acetic acid (10). First the plate is dried by warming gently with a heat gun and then it is sprayed evenly but heavily with a solution of mineral oil in hexane (2:1). This increases the fluorescence intensity of the NBD amines by a factor of about 10 and, in addition, keeps the fluorescence constant for at least a day. This technique was recently described by Uchiyama and Uchiyama (11).

TABLE I. R_f VALUES OF NBD-AMINES ON SILICA GEL 60

Parent Amine	R_f Value	
	System 1[a]	System 2[b]
Dimethylamine	0.16	0.25
Diethylamine	0.35	0.41
Di-N-propylamine	0.58	0.67
Pyrrolidine	0.27	0.35
Piperidine	0.42	0.47
Morpholine	0.19	0.18

[a] Cyclohexane-ethyl acetate (1:1).

[b] Hexane-chloroform-diethyl ether-acetic acid (33:33:33:1).

R_f values obtained with several amine NBD derivatives in the two solvent systems are presented in Table I.

For quantitation of the chromatograms a Vitatron TLD 100 flying spot densitometer was used, although any apparatus capable

of measuring fluorescence on TLC plates can also be used. The
primary filter, an interference line filter, passes light from
360 to 375 nm, with a maximum at 366 nm. The diaphragm between
the plate and the photomultiplier has a diameter of 0.25 mm. The
secondary filter, just before the detector, is also an interference line filter. It exhibits maximum transmission at 546 nm
with a band width of about 5 nm. The combination of filters provides excellent selectivity.

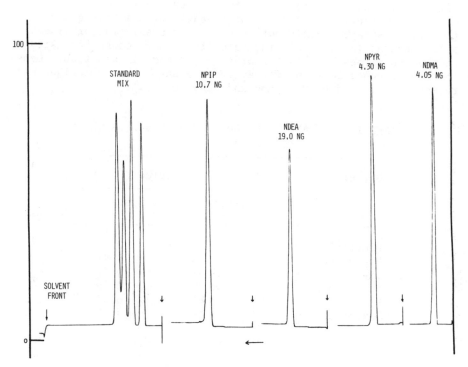

Figure 3.   Scans of standard nitrosamine NBD derivatives.

The scan trace obtained from a series of standard amine-NBD
derivatives is shown in Figure 3. It can be seen that the
responses for the NBD derivative obtained from NDMA and NPYR are
nearly full scale for only about 4 ng of nitrosamine, whereas the
responses for NDEA and NPIP are somewhat lower.

The recovery of NDMA and NPYR added to bacon rasher just before the homogenization step at levels of 5-10 ppb was examined. Twelve samples were analyzed and gave a recovery of 73% ± 5% for each nitrosamine. In other experiments in which the spiking was done at different stages of the analysis we found the major loss of NDMA occurred at the extraction step whereas the major loss of NPYR occurred during vacuum distillation.

TABLE II. A COMPARISON OF NITROSAMINE DETERMINATION
BY TLD AND TEA

| | Nitrosamines (ppb) | | | |
| | NDMA | | NPYR | |
Sample	TEA	TLD	TEA	TLD
174-1C	0.6	5.5	5.8	7.2
174-1D	0.3	2.0	0.7	0.7
174-2C	1.5	2.6	11.1	10.1
174-2D	0.4	1.7	1.0	1.5
174-3C	0.1	1.7	6.3	8.6
174-3D	0.1	0.8	0.4	0.3
174-4C	0.8	1.1	4.1	2.7
174-4D	0.1	0.6	ND[a]	0.2
177-A	6.2	6.5	39.7	37.1
177-B	13.7	12.3	6.7	5.9
181-3A	1.0	1.0	2.7	1.6
181-3B	0.4	0.4	ND	ND
181-4A	1.4	1.1	3.4	2.1
181-4B	0.6	0.4	ND	ND
	r = 0.92		r = 0.994	

[a] Not detected.

Through the cooperation of Dan Brown and Martin Feist at The Hormel Research Laboratories in Austin, Minnesota, the above-described method was compared with the semi-official GC-TEA method that had been applied to our extracts. Several bacon extracts

from 20-g samples were divided in two equal portions before splitting. One half of each extract was analyzed at Hormel Research
Laboratories by the GC-TEA method, which detects unchanged nitrosamines, while the other half was analyzed in our laboratories by
the thin-layer densitometric method applied to the amines obtained
after splitting. The results are presented in Table II. The
agreement in most cases is remarkably good, particularly in the
case of NPYR, where a correlation coefficient of 0.994 was found.
The correlation coefficient for NDMA was only 0.92. This is
probably due to the very low levels found in most cases.

It is felt that this is an excellent method, one that could
be utilized by many laboratories not equipped with a Mass spectrometer or a Thermal Energy Analyzer. Although the method has
only been applied to fried bacon, there is no obvious reason why
it would not work equally well for any other foodstuff.

REFERENCES

1. D. C. Havery, T. Fazio, and J. W. Howard, J. Assoc. Off. Anal.
   Chem. 61, 1374 (1978).
2. N. P. Sen, Food Cosmet. Toxicol. 10, 219 (1972).
3. N. T. Crosby, J. K. Foreman, J. F. Palframan, and R. Sawyer,
   Nature (London) 238, 342 (1972).
4. T. Fazio, R. H. White, and J. W. Howard, J. Assoc. Off. Anal.
   Chem. 54, 1157 (1971).
5. D. H. Fine and D. P. Rounbehler, J. Chromatog. 109, 271 (1975).
6. C. K. Cross, K. R. Barucha, and G. M. Telling, J. Agric. Food
   Chem. 26, 657 (1978).
7. G. M. Telling, T. A. Bryce, D. Hear, D. Osborne, and O. Weti,
   in N-Nitroso Compounds in the Environment, edited by P. Bogovski and E. A. Walker (International Agency for Research on
   Cancer, Lyon, 1974), Scientific Publication No. 9, pp. 12-17.
8. G. Eisenbrand and R. Preussmann, Arzneimmittel-Forsch. 20,
   1513 (1970).
9. H. J. Klimisch and L. Stadler, J. Chromatog. 90, 223 (1974).
10. H. J. Klimisch and L. Stadler, J. Chromatog. 90, 141 (1974).
11. S. Uchiyama and M. Uchiyama, J. Chromatog. 153, 135 (1978).

CHAPTER 20

Metabolism of Pesticide Related Nitrosamines and Nitramines

James E. Oliver and R. H. Smith

Our laboratory has been interested in herbicide-related nitrosamines for several years (1,2), but this interest was intensified by the detection of nitrosamines in two classes of herbicides: acidic products formulated as dialkylamine salts, and dinitroanilines (3-6). In addition, a recent report (7) has shown that certain dinitroaniline herbicides may also be contaminated with N-nitramines. The environmental behavior and metabolism of some of these herbicide contaminants were examined, using thin layer chromatography (TLC) for the metabolism studies. More useful, however, in the isolation of metabolites from soils or microorganism cultures, has been an offspring of TLC, dry-column chromatography. In this chapter a procedure for identifying metabolites of ^{14}C-labeled nitrosamines and nitramines is outlined and structural assignments for several of these metabolites are presented.

Recent interest has focused on the nitrosamine and nitramine impurities associated with the dinitroanilines. There are two types of compounds in the dinitroaniline category, as illustrated in Figure 1. The first consists of tertiary amines, such as trifluralin [α,α,α-trifluoro-2,6-dinitro-N,N-dipropyl-p-toluidine, R = n-propyl, X = 4,trifluoromethyl]. The tertiary amines, when contaminated by nitrosamines, tend to contain the nitroso derivatives of the dialkylamino substituent on the aromatic ring, e.g., N-nitrosodipropylamine (NDPA) in trifluralin. The second category consists of secondary amines, such as pendimethalin [N-(1-ethylpropyl)-3,4-dimethyl-2,6-dinitrobenzenamine, R = 3-pentyl, X = 3,4-dimethyl] and butralin [4-(1,1-dimethylethyl)-N-(1-methylpropyl)-2,6-dinitrobenzenamine, R = 2-butyl, X = 4-t-butyl]. The two examples of secondary amine dinitroaniline herbicides examined contained N-nitroso derivatives of the herbicides (5,6).

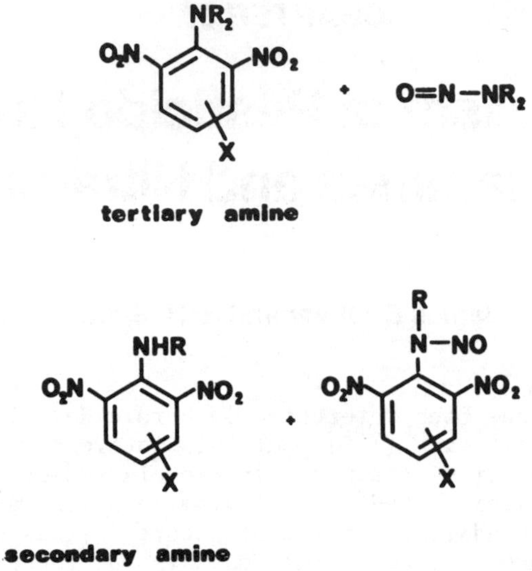

Figure 1. Two types of compounds of the herbicide-related nitrosamines.

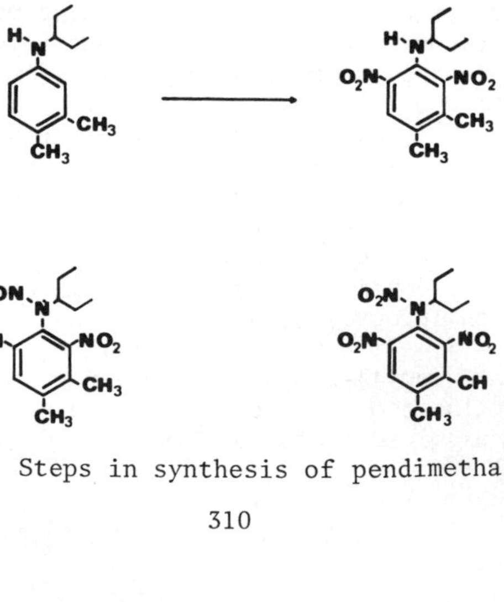

Figure 2. Steps in synthesis of pendimethalin.

310

The final step in the synthesis of pendimethalin involves nitration of the N-(3-pentyl)-xylidine illustrated in Figure 2 (8); presumably both the N-nitroso and N-nitro derivatives are formed by the action of excess nitric acid and/or nitrogen oxides in that step.

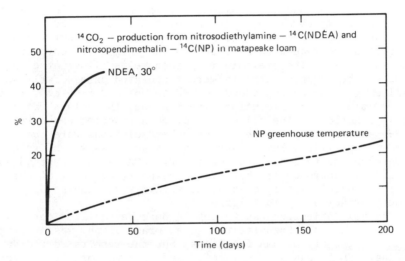

Figure 3. Carbon dioxide production from nitroso derivatives of pendimethalin.

Dinitroanilines are normally present in soil, so one objective was to determine the fates and lifetimes of several nitrosamines in soil. We found that NDPA, nitrosodiethylamine, and nitrosodimethylamine had half-lives in soil of approximately 3 weeks, and were evidently degraded completely to CO_2 (9). In contrast, nitrosopendimethalin I [N-1-ethylpropyl)-N-nitroso-3,4-dimethyl-2,6-dinitrobenzenamine] was considerably more stable in soil and produced CO_2 relatively slowly, as illustrated in Figure 3. In one experiment, soybeans were grown to maturity in a soil amended with various levels of I-[14]C. After the beans were harvested, soil that had been treated with 100 ppb I was extracted with methanol. More than 50% of the added[14]C was recovered, and thin layer chromatography (silica gel F, benzene) and autoradiography indicated that most of this ^{14}C was recovered as unchanged I. A few very faint bands suggested that some metabolism (or decomposition) had occurred, but the amounts of products were too small to permit isolation and characterization.

To study the metabolism of I by a soil microorganism in more detail, radiolabeled I was incubated in a nutrient broth inoculated with a mixture of two uncharacterized Streptomyces species that had been isolated from the same soil used in the soybean experiment (11). This provided a very complex mixture of a few labeled compounds and many unlabeled compounds. Previous experience with column chromatography on related mixtures had been discouraging because of the large number of fractions that rapidly accumulated, so sequential dry-column and thin layer chromatography was tried. The preliminary procedure is illustrated in Figure 4; the organic extracts were evaporated onto a small amount of silica gel, which was added to the top of a 2 × 90 cm silica gel dry column. After elution with benzene, the column was cut into 15 fractions. The silica gel of each fraction was extracted with freshly distilled ether. The ether solutions were concentrated to specific volumes, and aliquots were assayed by scintillation counting to locate radioactivity. One significant advantage of this approach is that often one-half, or more, of the fractions contain little or no radioactivity and can be discarded without further analysis (Figure 5).

The next objective was to obtain pure metabolites for mass spectral analysis. Preparative TLC-autoradiography was attempted on some of the dry column fractions, but the results were disappointing. "Hot" bands were located and scraped; often, however, samples obtained from what appeared to be single spots or bands were still so contaminated with phthalates and other materials that the mass spectra were obscured by extraneous ions. A typical example is shown in Figure 6. The TLC (system A, 3, Table I) is of dry-column origin. A high-pressure liquid chromatogram of the scraped and extracted R_f 0.53 band is shown in the same figure. During the initial HPLC run, each peak was collected and scintillation counted; radioactive components were thus located and subsequently collected. Careful evaporation with a gentle stream of nitrogen then usually provided a sample clean enough for mass spectral analysis. In most subsequent work, dry-column fractions were subjected to HPLC directly without the intermediate preparative TLC, and this two-step sequence, dry-column followed by high pressure liquid chromatography, has now become our standard cleanup procedure.

Figure 6 shows that the R_f 0.53 band from the TLC contained several materials, but only one of them (the peak extending off scale) was labeled. This material was collected; the parent ion in its electron impact mass spectrum appeared at m/e 250 (Figure 7). However, the chemical ionization spectrum (Figure 8)

ISOLATION PROCEDURE
NITROSOPENDIMETHALIN: STREPTOMYCES SP.
METABOLITES

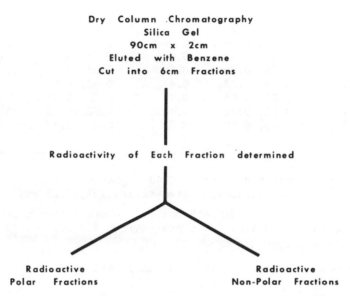

Figure 4. Procedures for isolation of the products.

contained an ion at m/e 281, indicating that the molecular
weight was probably 280 amu. The chemical ionization mass spectrum obtained when deuterium oxide was used as the reagent gas
had m/e 284, indicating that the molecule contained two exchangeable hydrogens. All of these data (summarized in Figure 9) were
consistent with a structure in which a nitro group of 1 was
reduced to an amino group, and a possible structure is shown in
the same figure. (Note that this is one of two possible isomers.)
We were initially reluctant to accept this structure because it
represents a nitrosated o-phenylenediamine. The latter class of
compounds has not been reported previously; o-phenylenediamines,
upon nitrosation by the common reagents, seem to invariably
cyclodehydrate to benzotriazoles (Figure 10). Our skepticism was
temporarily enhanced when we found the metabolite could be

TABLE I. TLC DATA ON PENDIMETHALIN-RELATED COMPOUNDS

Compound[c]	Stationary Phase[a], Solvent[b]			
	A, 1[d]	A, 2	A,3[d]	B, 4[d]
I	0.42, 0.35	0.68	0.61, 0.59	0.46, 0.37
II	0.43, 0.37	0.73	0.60, 0.56	0.44, 0.36
III	0.64, 0.58	0.86	0.72, 0.70	0.29, 0.21
IV	0.39, 0.33	0.66	0.63, 0.59	0.41, 0.32
V	0.36, 0.30	0.55	0.65, 0.61	0.34, 0.26
VI	0.29, 0.23	0.46	0.63, 0.60	0.29
VII	0.04, 0.03	0.12	0.29, 0.26	0.28
VIII	0.13, 0.12	0.40	0.52, 0.49	0.51
IX	0.10, 0.09	0.22	0.46, 0.44	0.45

[a] A: E. Merck precoated silica gel 60 F-254, 0.25 mm.
B: Whatman KC_{18} octadecylsilated silica gel, 0.25 mm.

[b] 1: Toluene; 2: Dichloromethane; 3: Toluene/ethyl acetate/
acetic acid, 60:40:1; 4: Methanol/water, 80:20.

[c] Structures of compounds I-IX are shown in Figure 19.

[d] Two values represent results from duplicate runs.

recovered unchanged after standing several hours in acetic acid
containing a drop of hydrochloric acid. When, however, the
metabolite was heated in toluene with a trace of p-toluenesul-
fonic acid, the anticipated benzotriazole formed rapidly. As
expected, nitrosation of the corresponding o-phenylenediamine
(available from ammonium sulfide reduction of pendimethalin) with
either dinitrogen tetroxide at -78° or aqueous nitrous acid at
room temperature provided the same benzotriazole instantly.
    The proposed structure was confirmed by synthesizing the
metabolite, as illustrated in Figure 11. The stability of this
compound to aqueous acid permitted removal of the pentafluoroben-
zaldehyde-protecting group with no adverse side reactions.
    Similar isolation and mass spectroscopy (where required)
provided identification, or tentative identification, of three
other labeled compounds from the dry column fractions, as illustrated in Figure 12. The benzotriazole and unmetabolized I were
identical to standards. The structure of the final compound

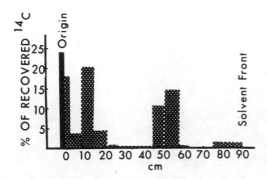

NITROSOPENDIMETHALIN-^{14}C METABOLISM
DRY COLUMN CHROMATOGRAPHY

Figure 5. Diagram showing little activity in column fractions.

(Figure 13) was proposed on the basis of its mass spectrum (16 amu heavier than the nitrosated o-phenylenediamine, three exchangeable hydrogens, and loss of NO--typical of N-nitroso compounds) and its pmr spectrum (disappearance of the signal assigned to the methyl para to the alkylamino group). This metabolite has also recently been synthesized.

The Streptomyces bacteria used in the metabolic study were aerobic soil microorganisms. Our attention was then directed to the metabolism of I under anaerobic conditions. The chamber pictured in Figure 14 was constructed for this purpose. Soil, amended with a little ground hay, was flooded with water; the system was flushed with nitrogen, and then closed. Within 1-2 days the reduction potential measured ca. -300 mV, and the system was then dosed with the compound to be studied (usually 10 ppm with respect to the dry weight of the soil). Preparation consisted of extracting both soil and water, combining the extracts, and submitting them to the dry-column and high-pressure liquid chromatography routine. When I was subjected to these conditions for 14 days, 38% of the initial radioactivity was able to be extracted, and a single metabolite accounted for the bulk of the ^{14}C (Figure 15). The metabolite from anaerobic metabolism was the same as one of the aerobic metabolites. A similar incubation was terminated after 4 days; 51% of the radioactivity was extractable, and in this case too, the same metabolite (along with a little unreacted 1) was all that could be isolated and identified.

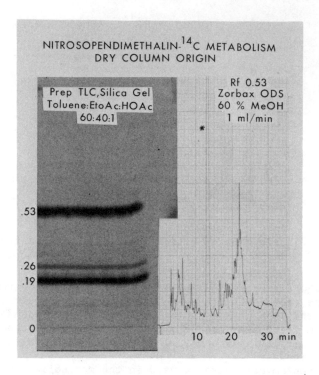

NITROSOPENDIMETHALIN-^{14}C METABOLISM
DRY COLUMN ORIGIN

Prep TLC, Silica Gel
Toluene:EtoAc:HOAc
60:40:1

Rf 0.53
Zorbax ODS
60 % MeOH
1 ml/min

.53

.26
.19

0

10   20   30 min

Figure 6.   Chromatogram of metabolites.

N-Nitropendimethalin [N-(1-ethylpropyl)-3,4-dimethyl-N,2,6-trinitrobenzenamine, II] was incubated anaerobically under the conditions just described. After 20 days, only 7.5% of the radioactivity was recovered. Only 21% of the ^{14}C was extractable in each of two subsequent runs for 4 and 2 days. In the latter run, most of the radioactivity recovered was associated with a single dry-column fraction, and with a single metabolite within that fraction (Figure 16). The benzimidazole structure was deduced from the mass spectrum and confirmed by synthesis (Figure 17).

The nitramine II was also incubated with the aerobic Streptomyces bacteria under the conditions described earlier for I. As with the anaerobic system, recovery of radioactivity was much lower from II than from I, and only one metabolite has been positively identified. The mass spectrum indicated that the metabolite was a hydroxylated derivative of II (gain of 16 amu, one exchangeable hydrogen), and both the fragmentation pattern and

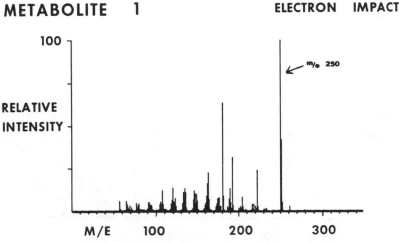

Figure 7. Mass spectrum of metabolite 1 on electron impact.

pmr spectrum indicated that hydroxylation had occurred on one of
the ring methyls, probably the one para to the nitramino function
(Figure 18). This assignment has now been confirmed by synthesis
of the metabolite.

   Although TLC was not extremely useful in the isolation
phases of these studies, it was used, where standards were available, for tentative structural assignments. Table I lists several
compounds of interest and their R_f's on silica gel F with selected
solvent systems and on reverse phase ODS plates with methanol
water. Reproducibility, even on plates eluted simultaneously in
the same chamber, left something to be desired. If one considers
only polarity, one might predict that the compounds moving fastest
on silica gel plates might move slowest on the reverse phase
plates, and vice versa. But the relationships were not that simple. Especially noteworthy was the benzimidazole VII, which was
one of the slowest moving compounds on both stationary phases.

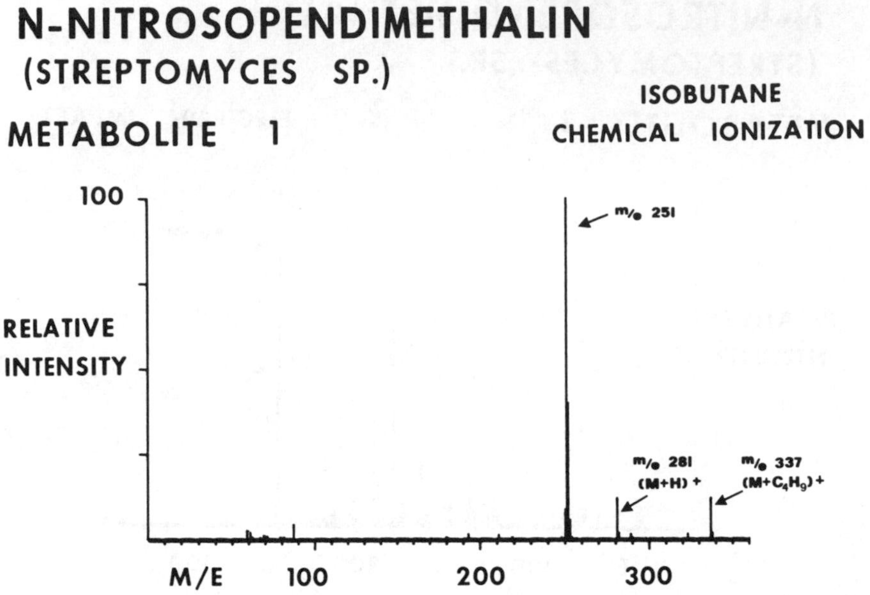

N-NITROSOPENDIMETHALIN
(STREPTOMYCES SP.)
METABOLITE 1
ISOBUTANE
CHEMICAL IONIZATION
100
RELATIVE
INTENSITY
m/e 251
m/e 281
(M+H)+
m/e 337
(M+C4H9)+
M/E
100
200
300

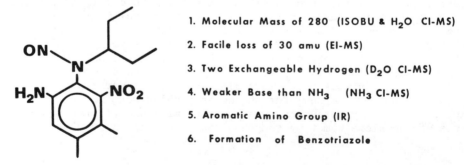

Figure 8. Mass spectrum of metabolite on chemical ionization.

NITROSOPENDIMETHALIN METABOLITE NO 1
ON
N
H2N
NO2
1. Molecular Mass of 280 (ISOBU & H2O CI-MS)
2. Facile loss of 30 amu (EI-MS)
3. Two Exchangeable Hydrogen (D2O CI-MS)
4. Weaker Base than NH3 (NH3 CI-MS)
5. Aromatic Amino Group (IR)
6. Formation of Benzotriazole

Figure 9. Identifying characteristics of metabolite 1.

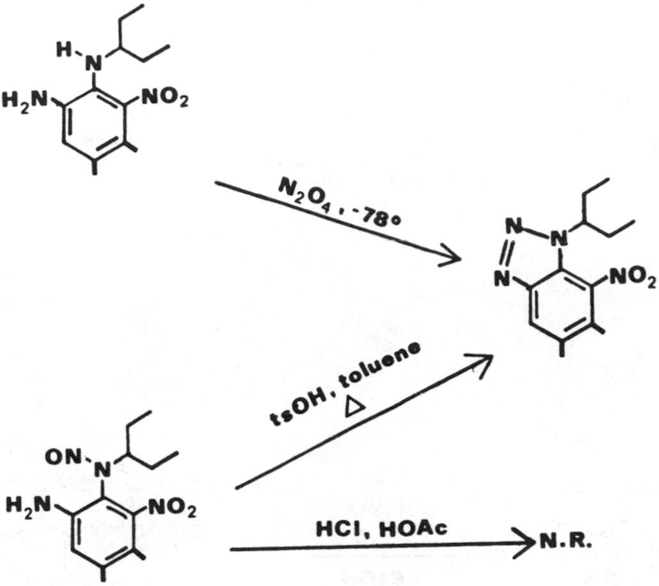

Figure 10.   Reactions involving metabolite 1.

ACKNOWLEDGMENTS

We thank American Cyanamid for a generous gift of ^{14}C-labeled pendimethalin, Arnold Kontson for much technical assistance, Harold Finegold for nmr spectra, and William Lusby for the excellent mass spectral work upon which this work depended on so many occasions.

REFERENCES

1. P. C. Kearney, J. E. Oliver, C. S. Helling, A. R. Isensee, and A. Kontson, J. Agr. Food Chem. 25, 1177 (1977).
2. J. E. Oliver and A. Kontson, Bull. Environ. Contamin. Toxicol. 20, 170 (1978).
3. S. Fan, D. Fine, R. Ross, D. Rounhehler, A. Silvergleid, and L. Song, Am. Chem. Soc. Nat. Meeting, San Francisco, Cal., Sept. 2, 1976.

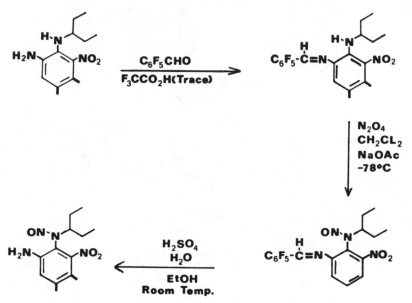

Figure 11.   Proposed structure as synthesized for identification.

NITROSOPENDIMETHALIN-^{14}C METABOLISM
DRY COLUMN CHROMATOGRAPHY

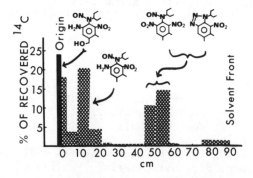

Figure 12.   Fractionation resulting in isolation of three other
metabolites.

NITROSOPENDIMETHALIN   METABOLITE NO 2

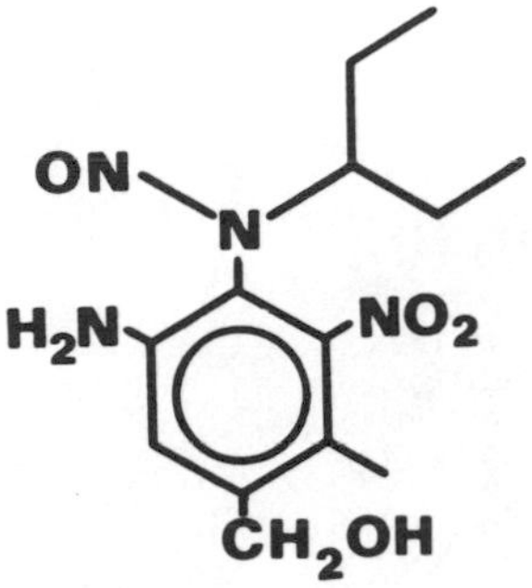

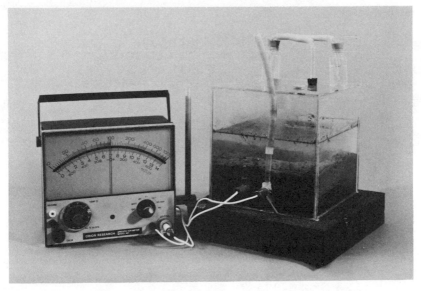

1. Molecular Mass of 296 (ISOBU CI-MS)

2. Facile loss of 30 amu (EI-MS)

3. Three Exchangeable Hydrogen (D O CI-MS)

Figure 13.   Characteristics of metabolite 2.

Figure 14.   Chamber for anaerobic metabolism.

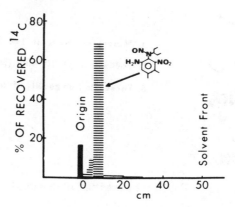

Figure 15. Metabolite resulting from anaerobic modification.

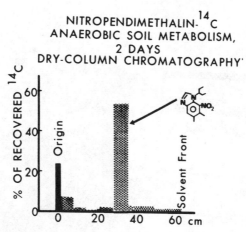

Figure 16. Single metabolite formed by anaerobic metabolism of nitrosopendimethalin for two days.

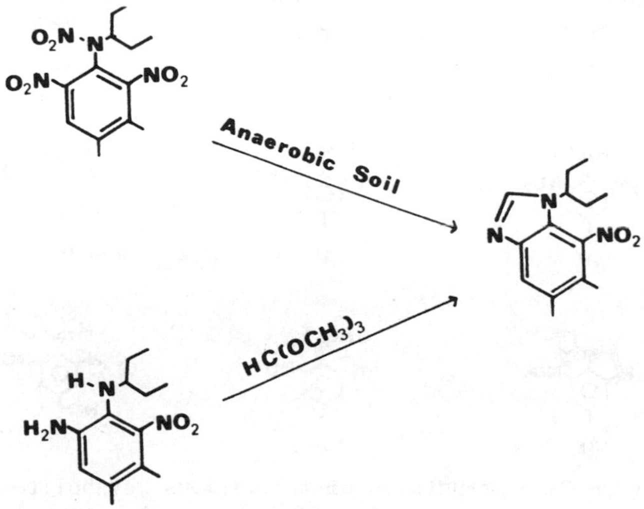

Figure 17. Structure elucidated from mass spectra.

Streptomyces sp.

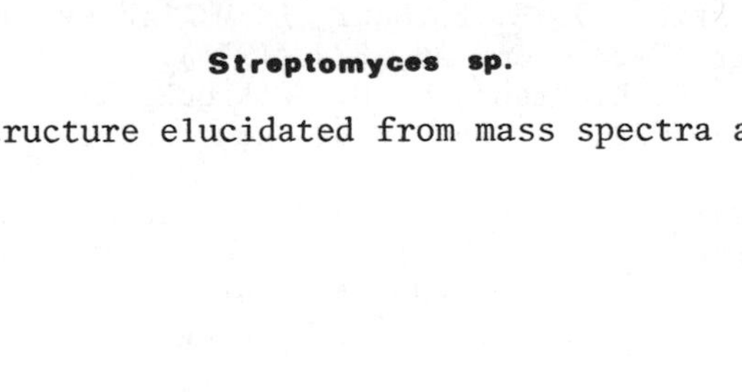

Figure 18. Structure elucidated from mass spectra and synthesis.

323

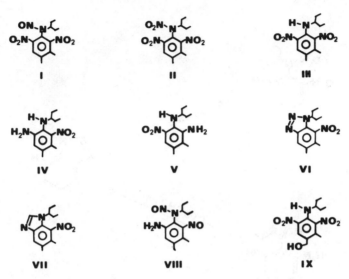

Figure 19.   Structures of the various metabolites.

REFERENCES (continued)

4. R. Ross, J. Morrison, and D. H. Fine, J. Agr. Food Chem. 26, 455 (1978).
5. S. Z. Cohen, G. Zweig, M. W. Law, D. Wright, and W. R. Bontoyan, in Environmental Aspects of N-Nitroso Compounds, edited by E. A. Walker, M. Castegnaro, L. Griciute, and R. E. Lyle (International Agency for Research on Cancer, Lyon, 1978, Scientific Publications No. 19, pp. 333-342.
6. W. R. Bontoyan, M. W. Law, and D. Wright, Jr., Am. Chem. Soc. Nat. Meeting, Anaheim, Cal., March 17, 1978.
7. J. H. Hotchkiss, J. F. Barbour, L. M. Libbey, and R. A. Scanlan, J. Agr. Food Chem. 26, 884 (1978).
8. S. D. Levy, R. E. Diehl, W. H. Gastrock, and L. S. Ross, Ger. Offen. 2,429,958, 16 Jan. 1975; Chem. Abstr. 83, 27865m (1975).
9. J. E. Oliver, P. C. Kearney, and A. Kontson, Am. Chem. Soc. Nat. Meeting, Anaheim, Calif., March 17, 1978.
10. P. C. Kearney, J. E. Oliver, A. Kontson, W. Fiddler, and J. W. Pensebene, Am. Chem. Soc. Nat. Meeting, Anaheim, Cal., March 17, 1978.
11. W. R. Lusby, J. E. Oliver, and P. C. Kearney, Am. Chem. Soc. Nat. Meeting, Miami, Fla., Sept. 14, 1978.

CHAPTER 21

TLC in the Analysis
of the Respiratory Environment

Irwin Schmeltz, Alvin Wenger,
and Dietrich Hoffman

The Naylor Dana Institute is concerned with the presence of
respiratory carcinogens in the environment, with their mechanism
of action, and with reducing their levels. Humans are exposed to
such agents in urban atmospheres, in occupational environments,
in the inhalation of tobacco smoke, and, in general, wherever
combustion products are generated. Such is the nature of industri-
alized society. The ever-increasing incidence of lung cancer in
these societies is a cause of concern (Figure 1) (1). This disease is related to the presence of specific carcinogens in the
environment and human exposure to them. By identifying responsible environmental agents, it is possible, by preventive measures,
to reduce their levels and minimize the risk of developing lung
cancer and related diseases.

Among various pollutants, it has been clearly established
that tobacco smoke is the prime factor for the high incidence of
lung cancer now observed (Figure 2) (2). Urban and occupational
atmospheres may also contribute, although less significantly in
this regard.

The research is generally twofold: the analytical determination of environmental carcinogens (and their sources), and the
evaluation of their biological activity, including mutagenicity,
carcinogenicity, and mode of action (i.e., metabolic activation).
This chapter centers on the analytical aspects of these studies,
with special emphasis on the use of thin layer chromatography
(TLC).

In general, analytical studies begin with the isolation of a
specific compound, or class of compounds, from tobacco "tar" or
urban particulates. Such a procedure, as outlined in Figure 3,
usually involves solvent partition, column chromatography, followed by HPLC, GLC, or GLC/MS of selected fractions. In the
course of the scheme, TLC is usually utilized in order to enrich

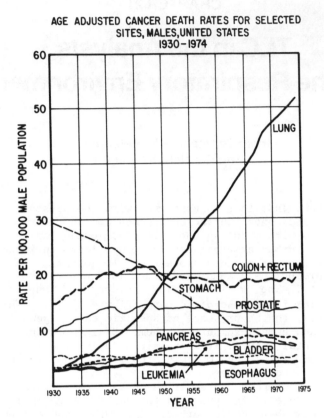

AGE ADJUSTED CANCER DEATH RATES FOR SELECTED
SITES, MALES, UNITED STATES
1930-1974

Figure 1. Age-adjusted cancer death rates in males for selected
sites: United States. 1930-1974 (standardized on the age distribution of the 1940 United States census population). Source:
U.S. National Center for Health Statistics and U.S. Bureau of the
Census. Reproduced by permission (2).

certain fractions with certain constituents or to effect further
purification of isolated compounds. It is emphasized, however,
that TLC is only one of the methods employed in these procedures;
it is usually used in an adjunct fashion, prior to GLC or HPLC,
for purposes of enrichment of desired compounds or fractions.

    Because environmental pollutants comprise a rather diverse
array of chemical compounds, it is necessary to develop analytical
procedures for each compound or class of compounds under study.

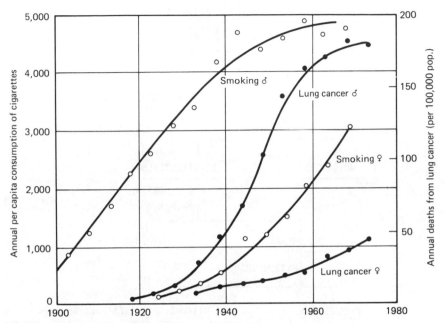

Figure 2. Parallel trends between cigarette consumption and lung cancer in England and Wales (2). Reproduced by permission.

Table I presents only a partial listing of the compounds determined and for which TLC was used.

In the discussion that follows, some examples of the methodology used will be outlined, including the application of TLC.

TLC IN THE DETERMINATION OF BENZO(A)PYRENE (BaP) AND OTHER POLYNUCLEAR AROMATIC HYDROCARBONS (PAH)

In the case of cigarette smoke, the determination of the carcinogen, benzo(a)pyrene (BaP), has always been important, as it is in urban air and other environments. In cigarette smoke, levels of BaP are indicators of the carcinogenic PAH content of the smoke, and they generally correlate with tumorigenicity of the "tar" (3). Its determination has been accomplished in several ways (4), but in any event its presence in very minute amounts amidst numerous

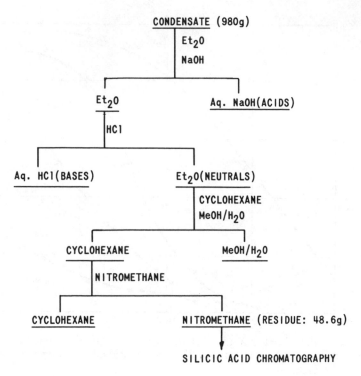

Figure 3.   Typical extraction procedure for cigarette smoke condensate.

other compounds has always required a number of enrichment steps in the analytical scheme, usually involving solvent partition and column chromatography, followed by paper or thin layer chromatography. Quantitation is then achieved by either uv absorption spectroscopy (Figure 4) or spectrophotofluorometry (Figure 5) (5) of the BaP eluted from the paper or sorbent.

When relatively high levels of BaP are present, as for example in pyrolysis products, TLC becomes a rather facile means for determining this compound (6). Following solvent parition and column chromatography, BaP-containing fractions(with ^{14}C-BaP as internal standard) can be subjected to TLC on silica gel; after development with pentane-ether (19:1) and visualization by uv light, it can be extracted with methanol and determined as noted above.

TABLE I. TLC IN THE ANALYSIS OF ENVIRONMENTAL AGENTS

Compound	Substrate	Solvent	Detection
PAH (BaP)	Silica gel	n-pentane/Et_2O (19:1)	uv light
Alkaloids	Silica gel	$CHCl_3$/95% EtOH (9:1)	p-aminobenzoic acid + CNBr
Indoles	Silica gel	$CHCl_3$/HOAc (95:5)	4-dimethylamino-benzaldehyde in HCl + AR
			$SbCl_5$ in CCl_4
			I_2 in $CHCl_3$
N-bases	Silica gel	ϕH/MeOH (95:5)	
Fatty acid Aziridines	Silica gel	Hexane/ϕ-H (1:1)	50% H_2SO_4 + charring
Hydrazines (as azines or hydrazones)	Silica gel, alumina	ϕ-H/hexane (1:1)	uv light (R_f 0.28 for hydrazine)

Others:

N-nitrosodiethanolamine
Various ^{14}C-compounds
^{14}C-tobacco isolates
Myristicin and related compounds
Quinolines
PAH metabolites

More current methods for resolving mixtures of environmental PAH including BaP involve the use of HPLC (Figure 6). However, in the case of cigarette smoke, such methods, especially for use in the determination of BaP, need further development. Fluorescence detectors, in conjunction with HPLC, show some promise in this regard.

A very recent study illustrates the utility of TLC in conjunction with GLC for determining polycyclic aromatic hydrocarbons (PAH) in airborne particulate matters (7). In this case, a

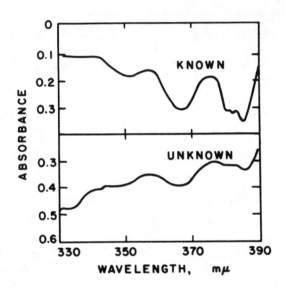

Figure 4. uv absorption spectrum for benzo(a)pyrene.

cyclohexane extract of the latter is chromatographed on a 20%
acetylated-cellulose-coated plate with a mobile phase consisting
of 1-propanol-acetone-water (2:1:1). The PAH separate into four
distinct regions of the plate, which are removed, examined, and
quantitated by GLC (Figure 7 and Table II). Among the reported
advantages of this method is its ability to separate BaP from
BeP, and benzo(k)- from benzo(b)fluoranthene. In addition, aza-
arenes migrate as a group on the TLC plate, appearing near the
solvent front.

In a study on the origin of components in tobacco smoke,
cigarettes with certain ^{14}C-tobacco constituents were prepared to
see if, among other things, their constituents gave rise to BaP
during tobacco combustion (8). The cigarettes were smoked, and
the collected "tar" was extracted in the usual way (9) to give a
PAH fraction. The latter was subjected to TLC radio-scanning
techniques. As can be seen (Figure 9), a radioactive zone corresponding to BaP was observed, indicative of the pyrosynthetic

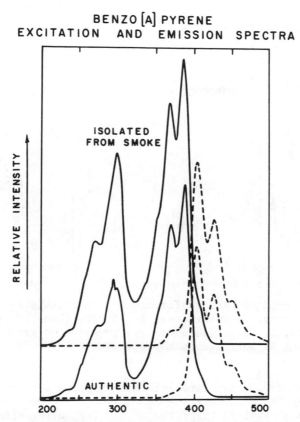

Figure 5. Excitation and emission spectra for benzo(a)pyrene.

formation of this compound during actual tobacco combustion, from
neophytodiene, a terpenoid hydrocarbon.

PAH METABOLITES

Metabolism of carcinogens, such as BaP, has been the subject of
much study because it is believed that many carcinogens require
metabolic activation to proximate and/or ultimate forms in order
to exert their biological effects (10). Prior to the demonstra-
tion of the efficacy of HPLC techniques to establish metabolite

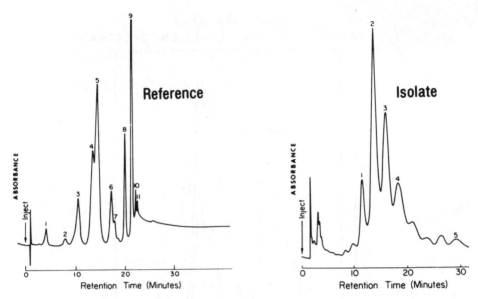

Figure 6.   HPLC of PAH.   Reference: 1. Naphthalene; 3. Biphenyl;
4. Phenanthrene; 5. Anthracene; 6. Fluoroanthene; 7. Pyrene;
8. Chrysene; 9. BaP.   Isolate: 1. Pyrene; 2. 4-Methylpyrene;
4. Chrysene; 5. BaP.   On Permaphase ETH.

profiles of BaP, metabolites of this compound were readily
resolved by TLC procedures.

In such studies, tritiated or ^{14}C-BaP, along with cold carrier, is metabolized by various microsomal fractions (or in vivo).
The ethyl acetate soluble metabolites are then applied to TLC
plates (silica), which are developed with successively applied
mobile phases, first nonpolar and then polar (11,12). Zones containing metabolites are visualized in uv light, and metabolites
are quantitated by radioisotopic techniques. Examples of such
TLC separations, in conjunction with radioscanning, are shown in
Figures 9 and 10. One study measured the extent of metabolism of
BaP by microsomal fractions taken from the placentas of smokers
and nonsmokers (13). Greater degree of metabolism (Table III)
was observed in the case of microsomal fractions obtained from
smokers. This observation suggests that smokers have an enhanced
ability (i.e., induced) to metabolically activate the carcinogen
BaP, which is present in cigarette smoke. Nonsmokers, in contrast,
possess less of this enzyme-inducing capability.

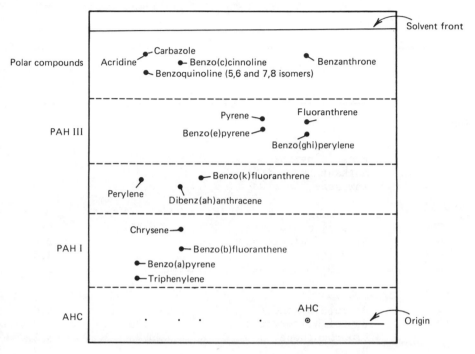

Figure 7.   TLC of PAH using acetylated-cellulose coated plates (7).

BaP, which is present in cigarette smoke. Nonsmokers, in contrast, possess less of this enzyme-inducing capability.

Of course, although TLC can be used to monitor BaP metabolism, HPLC techniques have provided very characteristic and reproducible metabolic profiles (Figure 11).

OTHER CARCINOGENS

As noted earlier, TLC has also been used in analytical schemes developed for the determination of environmental carcinogens such as hydrazines (15), quinolines (16), and certain N-nitrosamines (17).   These will be discussed below.

TABLE II. R_f VALUES OF SOME POLYCYCLIC AROMATIC HYDROCARBONS
            SEPARATED BY TLC ON ACETYLATED-CELLULOSE-COATED PLATES
            (7)

Compound No.	Compound	R_f	R_B
Band I			
1	Triphenylene	0.16	0.8
2	Benzo[a]pyrene	0.23	1.0
3	Anthanthrene	0.31	1.4
4	Chrysene	0.36	1.6
Band II			
5	Benz[a]anthracene	0.57	2.5
6	Dibenz[a,h[anthracene	0.59	2.6
7	Perylene	0.58	2.5
8	7,12-Dimethylbenz[a]anthracene	0.66	2.9
Band III			
9	Benzo[e]pyrene	0.70	3.0
10	Benzo[ghi]perylene	0.70	3.0
11	Fluoroanthene	0.73	3.2
12	Pyrene	0.73	3.2
Polar Compounds			
13	Benzanthrone	0.85	3.7
14	Benzo[h]quinoline	0.88	3.8
15	Benzo[f]quinoline	0.89	3.9
16	Benzo[c]cinnoline	0.92	4.0
17	Acridine	0.96	4.2
18	Carbazole	0.97	4.2

N-Nitrosodiethanolamine

N-nitrosodiethanolamine (NDELA) was found in tobacco. Its presence
can be traced to the use of the herbicide MH-30 on tobacco,
especially in formulations containing diethanolamine (17).

     This nitrosamine is rather polar, and extremely water-solu-
ble. The overall method for its isolation and quantitation is

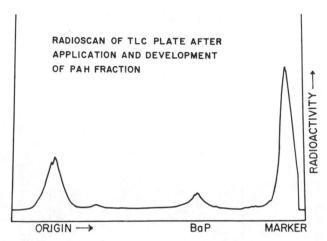

Figure 8. Radioscan of TLC plate after development, showing ^{14}C-benzo(a)pyrene. (8).

shown in Figure 12. It involves solvent partition, column chromatography, and GLC/MS. However, prior to GLC, the appropriate
fraction off the liquid column, identified by use of a ^{14}C-inter-
nal standard, is further enriched by TLC. Here, TLC is used for
obtaining an enriched sample of either NDELA or its silylated
derivative. The latter is amenable to TLC on silica gel with
benzene-ether (1:1) and visualized with I_2. After removal from
the sorbent, it is determined by GLC. In this manner, up to 173
ppb of NDELA was discovered in tobacco treated with certain formulations of MH-30.

Hydrazines

Hydrazines are another class of compound that are potential
environmental carcinogens (18). Their isolation is initiated by
immediate derivatization with pentafluorobenzaldehyde (Figure 14).
Again, this is followed by solvent partition and enrichment by
TLC and then by GLC (15,17). In this case, and in the case of
NDELA, TLC is not the crux of the analytical procedure used, but
represents an important step in providing enriched fractions of
the compounds sought.

TABLE III. YIELDS OF ETHYL ACETATE EXTRACTABLE METABOLITES OF BaP PRODUCED BY PLACENTAL MICROSOME OF 10 NONSMOKERS AND SIX SMOKERS

Metabolite	Nonsmokers[a]	2 Cigarettes/Day		10-20 Cigarettes/Day		20 Cigarettes/Day	
Origin[b]	1.9 ± 0.7	7.3	7.5	30.5	12.3	27.7	9.3
9,10-diol	0.8 ± 0.3	4.5	3.6	10.6	6.6	18.7	15.5
7,8-diol	8.0 ± 0.3	22.5	28.2	36.2	43.9	106.5	24.6
4,5-diol	1.0 ± 0.6	5.0	3.5	9.3	8.2	14.4	8.3
3-OH-BaP and other phenols of BaP	3.4 ± 1.7	132.3	39.4	287.6	102.9	194.5	63.6
Quinones	15.5 ± 3.0	37.4	34.3	37.4	29.9	102.0	19.3
Totals	18.4 ± 4.9	209.0	116.5	401.6	203.8	436.8	140.6

[a] Metabolite yields are the mean of 10 subjects $\pm$ standard deviations.
[b] Polar metabolites at and near the origin of the thin layer chromatogram.

336

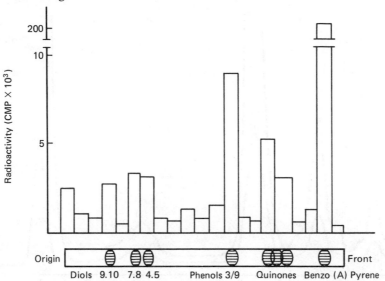

Figure 9. Metabolism of ^{3}H-BaP by human liver homogenate from a 41-year-old man with stomach carcinoma without hepatic metastases. Unlabeled reference (donated by H. V. Gelboin, Bethesda, Maryland) compounds were cochromatographed with extracts of incubation mixtures, located by fluorescence under UV light and the profile of radioactivity was determined by liquid scintillation counting (11).

Quinolines

Quinolines and other aza-arenes are readily separable by TLC (Figure 14). These N-containing analogs of PAH have become rather important lately as potential environmental mutagens and carcinogens. Quinoline, for example, has been shown to be hepatocarcinogenic in mice and rats (19). Other quinolines, in addition to the parent compound, have also been shown to be mutagenic in the Ames Salmonella typhimurium/microsome assay (20).
        The TLC separations for quinolines on silica gel are rather analogous to the separations obtained on HPLC normal phase columns (Figure 15). However, reverse phase columns are satisfactory as well. In addition, they are quite useful for the separation of quinoline metabolites generated by microsomal preparations (21).

TLC-Radioscanning Techniques in Studies on Tobacco

In another application, TLC was used in conjunction with radio-scanning to isolate radioactive constituents of tobacco grown in

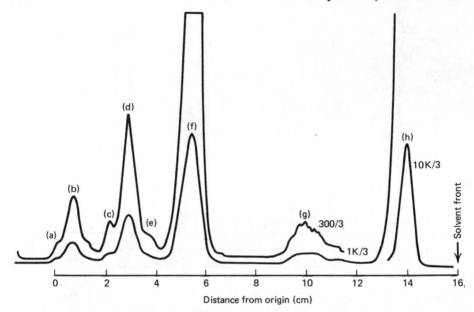

Figure 10. Radioscan of a TLC plate on which ^{3}H-BaP and its metabolites produced in an incubation system with placental micro-somes of a cigarette smoker were separated. The incubation sys-tem, extraction and separation of ^{3}H-BaP and its metabolites were carried out as described in the text. The numbers 300/3, 1K/3 and 10K/3 represent the scanning range (counts per second)/time con-stant (seconds.   Peak (a), origin of TLC; peak (b), unidentified polar metabolites of ^{3}H-BaP; peak (c), 9,10-diol; peak (d), 7,8-diol; peak (e), 4,5-diol; peak (f), 3-OH-BaP and other phenols of BaP; peak (g),quinones of BaP; peah (h), unmetabolized ^{3}H-BaP (12,13).

a $^{14}CO_2$ atmosphere (8). Ethanol extracts of the radioactive tobacco underwent preliminary fractionation on a Sephadex LH-20 column (Figure 16). By monitoring the effluents by GLC/MS, it was possible to characterize fractions containing neophytadiene phyto-sterols, amyrins, and fatty acids. However, in order to further enrich the radioactive compounds in the column effluents for tracer studies in cigarettes, the LH-20 effluents were applied to TLC (on silica gel). By parallel radioscanning techniques, it was then possible to isolate pure (>97%) compounds and estimate their activities. Examples of TLC scans thus obtained are shown in Fig-ures 17-20. Radioactive compounds isolated in this manner were put into cigarettes. The cigarettes were smoked and the label was

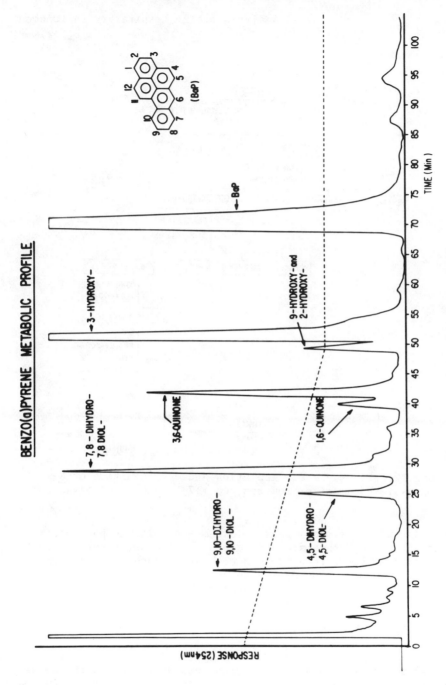

Figure 11. HPLC of BaP metabolites.

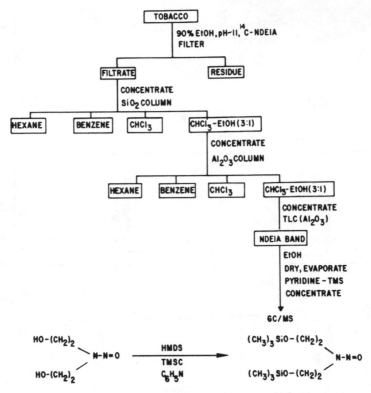

Figure 12.   Procedure for isolation and determination of N-nitroso-diethanolamine from tobacco (17).

traced into the smoke stream. In this way, it was possible to demonstrate the conversion of several leaf constituents to the carcinogenic BaP during tobacco combustion, and, as shown in Figure 21, the conversion of the tobacco alkaloid nicotine to a variety of other smoke constituents (22).

SUMMARY

In the foregoing, aspects of the analytical determinations of a number of environmental carcinogens were discussed, with the use of TLC in these determinations being emphasized. Also discussed

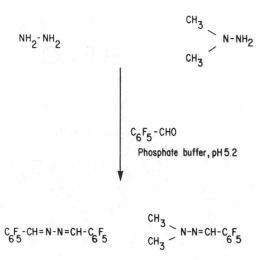

Figure 13. Derivatization of hydrazines prior to TLC (15).

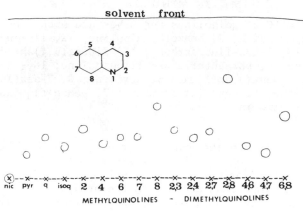

Figure 14. TLC of quinolines.

341

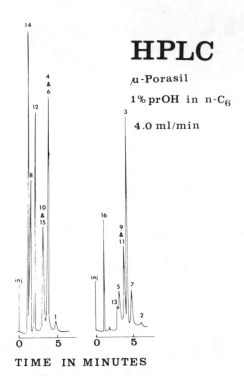

TIME IN MINUTES

Figure 15.   HPLC of quinolines and other aza-arenes on μ Porasil.
14, Benzo(c)acridine; 8, benzo(b)quinoline; 12, 11H-indeno(1,2b)-
quinoline; 10, 1-aza-fluoranthene; 15, dibenz(a,j)acridine; 4,
4-azafluorene; 6, phenanthridine; 1, quinoline; 16, dibenz(a,h)-
acridine; 13, benz(a)acridine; 5, acridine; 9, 7-azafluoranthene;
11, 4-azapyrene; 3, benzo(c)cinnoline; 7, benzo(f)quinoline; 2,
2,6-dimethylquinoline.

was the use of TLC in the analysis of metabolites of benzo(a)-
pyrene. Although many of the analyses require GLC, or spectroscopic techniques for purposes of final quantitation, TLC, nevertheless, is an important adjunct in the analytical schemes
described, serving to obtain fractions enriched with the com-
pound(s) being sought.

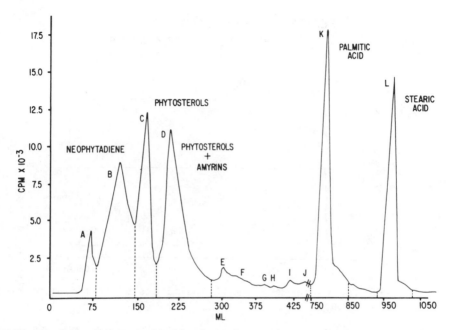

Figure 16.   Tobacco lipids fractionated on Sephadex LH-20 (8).

ACKNOWLEDGMENT

This work was supported by the National Cancer Institute under
Contract No. 1-CP-55666.

REFERENCES

1. C. Silverberg, Cancer J. Clin. 27, 26 (1977).
2. M. G. M. Curnen, Bull. N. Y. Acad. Med. 54, 349 (1978).
3. E. L. Wynder and D. Hoffmann, Tobacco and Tobacco Smoke.
   Studies in Experimental Carcinogenesis (Academic Press, New
   York, 1967).
4. I. Schmeltz, K. D. Brunnemann, and D. Hoffmann, in Detection
   and Prevention of Cancer, edited by E. Nieburgs (Marcel
   Dekker, New York, 1977), part 1, vol. 2, pp. 1973-1992.
5. I. Schmeltz, R. L. Stedman, and W. J. Chamberlain, Anal. Chem.
   36, 2499 (1964).

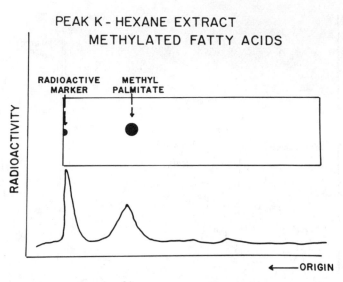

Figure 17. Radioscan of ^{14}C-methylpalmitate after isolation from radioactive tobacco.

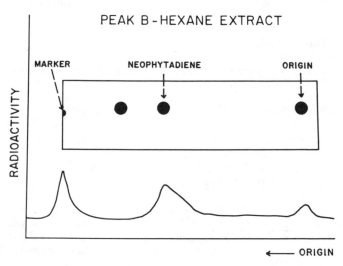

Figure 18. Radioscan of ^{14}C-neophytodiene isolated from radioactive tobacco (8).

344

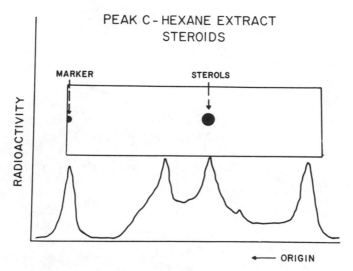

Figure 19.   Radioscan of ^{14}C-phytosterols (8).

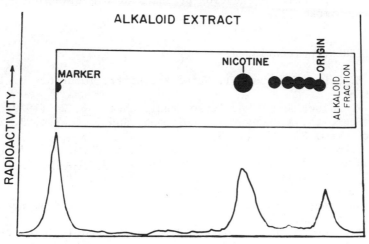

Figure 20.   Radioscan of ^{14}C-nicotine (8).

EFFECTS OF NICOTINE ON SMOKE COMPOSISTION

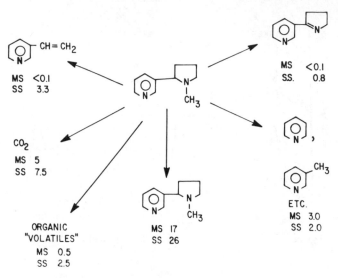

Figure 21. Thermal degradation products of nicotine that appear in tobacco smoke (22).

REFERENCES (continued)

6. W. S. Scholthauer and I. Schmeltz, Beitr. Tabakforsch. $\underline{4}$, 176 (1968).
7. J. M. Daisey and M. A. Leyko, Anal. Chem. $\underline{51}$, 24 (1979).
8. I. Schmeltz, A. Wenger, D. Hoffmann, and T. C. Tso, J. Agric. Food Chem. $\underline{26}$, 234 (1978).
9. I. Schmeltz, J. Tosk, D. Hoffmann, Anal. Chem. 48, 645 (1976).
10. R. Freudenthal and P. W. Jones, Eds., Carcinogenesis, A Comprehensive Survey (Raven Press, New York, 1976).
11. O. Pelkonen, op. cit., pp. 9-21
12. I. Y. Wang, R. E. Rasmussen, N. L. Petrakis, and A. C. Wang, op. cit., pp. 77-89.
13. I. Y. Wang, R. E. Rasmussen, R. Creasey, and T. T. Crocker, Life Sci. $\underline{20}$, 1265 (1977).
14. I. Schmeltz, J. Tosk, and G. M. Williams, Cancer Lett. $\underline{5}$, 81 (1978).

15. Y. Y. Liu, I. Schmeltz, and D. Hoffmann, Anal. Chem. <u>46</u>, 885 (1974).
16. M. Dong and D. C. Locke, J. Chromatog. Sci. <u>15</u>, 32 (1977).
17. I. Schmeltz, S. Abidi, and D. Hoffmann, Cancer Lett. <u>2</u>, 125 (1977).
18. I. Schmeltz, D. Hoffmann, and B. Toth, Proc. FDA Symp. on Structural Correlates of Carcinogenesis and Mutagenesis. A Guide to Testing Priorities (1977), pp. 172-178.
19. Y. Schinohara, T. Ogiso, M. Hananouchi, K. Nakaniski, T. Yoshimura, and N. Ito, Gann <u>68</u>, 785 (1977).
20. M. Dong, I. Schmeltz, E. LaVoie, and D. Hoffmann, in Ref. 10.
21. I. Schmeltz, K. D. Brunnemann, and D. Hoffmann, in <u>Proc. Newly Identified Pollutants</u>, AIChE Meeting, Philadelphia, June 1978, in press.
22. I. Schmeltz, A. Wenger, D. Hoffmann, and T. C. Tso, J. Agric. Food Chem., in press.

CHAPTER 22

Quantitation of Polycyclic Aromatic Hydrocarbons

R. E. Allen and L. J. Deutsch

Most of the methods described in the literature for the determination of benzo(a)pyrene (BaP) and other polycyclic aromatic hydrocarbons (PAH) in matrices such as cigarette smoke condensate, coal tar, pitch, and air particulates are very complex and time-consuming, which makes them impractical for routine analysis. The complexity of these methods results primarily from the fact that these matrices are exceedingly complex containing many thousands of compounds and that PAH usually occur at very low concentrations in these matrices. In this chapter are described methods that have been specifically formulated for the routine determination of PAH in cigarette smoke condensate and that are most likely suitable for other applications.

Numerous papers (1-21) describe procedures for obtaining PAH-enriched fractions prior to quantitation of PAH. These procedures include liquid-liquid partitioning, acid-base extractions, and liquid, gel, thin layer, or paper chromatography. Once suitable PAH-enriched fractions have been obtained, highly refined analytical separations have been used for quantitation. Thin layer (12, 16) and paper chromatography (1,16) have been successfully used in conjunction with ultraviolet or fluorescence spectroscopy for the determination of single compounds such as BaP. Where the quantitation of several PAH was desired, classical liquid chromatography (4,5) and high-performance liquid chromatography (19,22-24) have been used; however, because its powers of resolution are greater, gas chromatography (2,3,6-8,11,14,17,20,25,26) has been most commonly used. Packed, capillary, and support-coated open tubular columns have been used but no one GC column has been found to provide adequate resolution of the numerous PAH isomers of general interest.

In the first of the two methods described herein, BaP is determined by thin layer chromatography using a semiautomatic plate scanner to obtain direct fluorescence measurements for a partially acetylated cellulose plate. Liquid-liquid partitioning and liquid chromatography are necessarily used prior to TLC to obtain a BaP-enriched fraction.

In the second method a quantitative and qualitative survey of PAH in cigarette smoke condensate is obtained. Again it is necessary to use liquid-liquid partitioning and column chromatography (dextran gel) prior to TLC, but quantitation is obtained subsequent to TLC by packed-column gas chromatography. Using this method, at least 16 PAH have been identified and quantitative results have been obtained for phenanthrene-anthracene, 3-methylphenanthrene, fluorene, pyrene, benzfluoroanthene, and benzpyrene.

EXPERIMENTAL

All of the solvents discussed in this chapter were spectroscopic grade, except 2-propanol, which was electronic grade and was not further purified. Quantitative TLC was done on Analtech Uniplates (20% acetylated cellulose, 250 microns thick) and preparative TLC on Mallinckrodt Chrom-AR Sheet (70% silicic acid/30% glass fiber, 650 microns thick). Gel chromatography was performed using Pharmacia Sephadex LH-20, liquid chromatography using Fisher silica gel Grade 922, and gas chromatography using Alltech Associates Dexsil 300 on Chromosorb W. PAH standards were obtained from Analabs and Nanogen.

Fluorescence measurements were made using a Perkin-Elmer Model MPF-44A Spectrophotofluorimeter equipped with a semiautomatic plate scanning accessory. All GC work was performed on a Perkin-Elmer Model 3920 gas chromatograph linked to a Perkin-Elmer PEP-2 data system. For the determination of BaP, cigarettes were smoked on a Filamatic Vial Filler while for the determination of PAH, cigarettes were smoked on a Mason 20-port Rotary Smoking Machine.

PROCEDURE

Determination of BaP

A general flow chart for the determination of BaP is shown in Figure 1.

The smoke from 25 cigarettes was collected in a cryogenic trap and the entire amount of condensate dissolved in methanol:

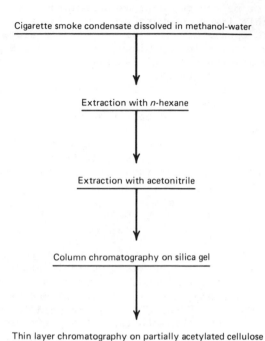

Figure 1.   Schema for analysis of material from cigarette smoke.

water (6:1). All cigarettes were conditioned at least 48 hours
at 74 ± 2° F and 60 ± 2% RH before they were weight-selected and
smoked to a butt length of 23 mm for 72-mm filtered cigarettes or
30 mm for 85-mm unfiltered cigarettes. The cigarettes were
smoked according to the internationally recognized standards of
one 35 ml puff of 2 seconds duration, once per minute.

The aqueous solution was extracted with n-hexane (5 × 30 ml)
and the volume of hexane was evaporated to about 50 ml and
extracted with acetonitrile (5 × 30 ml). The acetonitrile fraction was evaporated just to dryness and the residue was dissolved
in 3 ml hexane. Except where noted, all evaporation of solvent
was done in a rotary evaporator at 35 ± 5° C under a vacuum supplied by a water aspirator. Figure 2 shows a flow chart of the
extractions.

The hexane solution was transferred to the head of a silica
gel column (12 mm id × 25 cm). The silica gel was prewashed with
methanol and activated at 180° C and 27 in. Hg for 16 hours prior

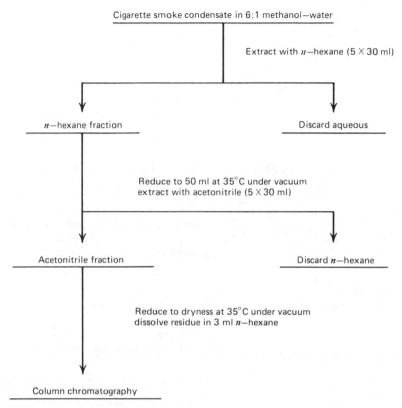

Cigarette smoke condensate in 6:1 methanol—water

Extract with n—hexane (5 × 30 ml)

n—hexane fraction   Discard aqueous

Reduce to 50 ml at 35°C under vacuum
extract with acetonitrile (5 × 30 ml)

Acetonitrile fraction   Discard n—hexane

Reduce to dryness at 35°C under vacuum
dissolve residue in 3 ml n—hexane

Column chromatography

Figure 2. Flow diagram for extraction of smoke condensates.

to use. A 220-ml portion of hexane was passed through the column, after which time elution was begun with hexane-benzene (95:5). The fraction containing BaP (as determined using standard BaP solutions) was collected and evaporated just to dryness. The details of the column chromatography are summarized in Figure 3.

The residue was dissolved in 1.0 ml hexane and 1.0-μl aliquots were spotted on a 20 × 20 cm cellulose (20% acetylated) TLC plate used as received from the supplier. In addition, 1.0-μl aliquots of standard BaP solutions (with concentrations bracketing the sample solutions) were spotted to obtain a calibration curve. The plate was developed in a rectangular TLC tank for 1 hour with methanol-ethyl ether-water (4:4:1). The R_f value for BaP was 0.35. The details of the TLC are summarized in Figure 4.

Column dimensions: 12 mm id × 25 cm
Stationary phase: Activated silica gel
Mobile phase: I. n-hexane
                       II. 95:5 n-hexane:benzene
Flow rate: 2-3 ml/min

Discard all n-hexane (220 ml)
Collect n-hexane-benzene fraction containing BaP
Reduce collected volume to dryness at 35°C under
    vacuum
Dissolve residue in 1.0 ml n-hexane

Figure 3.   Column chromatography parameters for extracts of
cigarette smoke.

Stationary phase: Cellulose plate (20% acetylated)
Plate dimensions: 20 cm × 20 cm × 250 μ thick
Mobile phase: 4:4:1 methanol-ethyl ether-water

Develop 1 hour
Air dry
Read fluorescence directly λ_{ex} = 300 nm

                               λ_{em} = 407 (maximum)

R_f for BaP = 0.35

Figure 4.   Parameters for thin layer chromatography of the column
fractions.

The developed plate was air dried and scanned with the excitation wavelength at 300 nm and the emission wavelength at 407 nm.
Other instrument parameters were set to give the maximum signal
for the spot corresponding to the most concentrated BaP standard
solution.   Using the X-Y adjustment controls, the center of each
BaP spot was located and an emission spectrum obtained at the
fixed excitation wavelength.   Typical emission spectra are shown
in Figure 5.   A calibration curve was prepared by plotting maximum fluorescence versus nanograms BaP/ml.   The concentration of
BaP in each sample was taken from the calibration curve and
divided by the number of cigarettes smoked (usually 25) to convert the results to BaP delivery per cigarette.
     The BaP recovery was determined for the procedure by analyz-
ing duplicate samples, one of which had a known amount of BaP

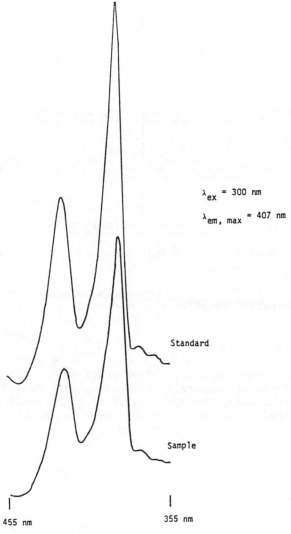

Figure 5. Emission spectra of material separated by thin layer
chromatography.

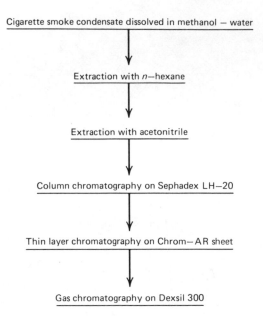

Figure 6.   Schema for use of methanol-water extracts of cigarette smoke.

added approximately equal to the amount of BaP in the unadulterated sample.

    Determination of PAH

In Figure 6 is shown a general flow chart of the procedure. For this analysis large numbers of cigarettes were smoked in a rotary smoking machine using the same preparation and smoking protocol described above. The smoke was collected in impaction traps. The entire quantity of condensate for each sample was dissolved in methanol-water (6:1). An aliquot of the condensate solution was dried in a rotary evaporated, weighed, and the residue redissolved in methanol-water. PAH deliveries per cigarette were obtained by multiplying the PAH deliveries per gram by the independently determined dry tar deliveries per cigarette.
    The condensate was redissolved in methanol-water (6:1) and the solution was extracted by the same scheme as described in the BaP procedure (Figure 2) except that the final residue was dissolved in 1 ml of tetrahydrofuran (THF).

The THF solution was transferred to the head of a column of Sephadex LH-20 (11 mm id × 30 cm) and eluted with 2-propanol at 0.35 ml/min under a slight hydrostatic pressure. The fraction of eluent containing 3-6 fused rings was collected and evaporated just to dryness. Figure 7 summarizes the details of the gel chromatography.

Column dimensions: 11 mm id × 30 cm
Stationary phase: Sephadex LH-20
Mobile phase: 2-propanol
Flow rate   0.35 ml/min

Discard first 25 ml
Collect next 65 ml
Reduce collected volume to dryness at
    35° C under vacuum
Dissolve residue in 0.3 ml tetrahydrofuran

Figure 7. Column chromatography of methanol-water extract on Sephadex LH-20.

The residue was dissolved in 0.3 ml THF and the solution was streaked on a Chrom-AR sheet (20 × 20 cm). The Chrom-AR was activated at 100° C under vacuum at least 4 hours prior to use. The sheet was developed with 9:1 n-hexane-benzene for 30 minutes, after which time the strip containing the PAH fraction was excised and extracted with methanol (5 × 30 ml). The methanol solution was evaporated to about 1 ml, transferred to a conical vial, and evaporated to dryness using a stream of nitrogen. Figure 8 summarizes the details of the preparative TLC.

The residue was dissolved in 100 µl THF and the solution was analyzed by GC using the conditions summarized in Figure 9. Individual PAH were identified by GC retention times of pure PAH and mass spectrometry. Quantitation was accomplished by computerized integration or measurement of peak heights followed by reference to calibration curves determined with standard PAH solutions. Although many PAH were detected, quantitative results were reported only for those compounds whose GC peaks were shown by mass spectrometry to be of the order of 95% pure and for which suitable degrees of recovery could be determined.

The degree of recovery for each PAH was determined in a manner similar to that described for BaP.

Stationary phase: Chrom-AR sheet (70% silicic acid - 30%
                           glass fiber)
Sheet dimensions: 20 cm × 20 cm × 650 μ thick
Mobile phase: 9:1 n-hexane-benzene

Develop 30 minutes
Cut out band containing PAH
Extract band with methanol (5 × 30 ml)
Reduce to dryness at 35° C under vacuum
Dissolve residue in 0.100 ml tetrahydrofuran

Figure 8.   Preparative thin layer chromatography of the methanol-
water extracts.

Column dimensions: 0.125 in. × 20 ft. stainless steel
Column packing: 60/80 mesh Chromosorb W - 5% Dexsil 300
Flow rate: 40 cm^3/min helium
Detectors: Dual flame ionization
Detector temperature: 300° C
Injector temperature: 250° C
Injection size: 5 μl
Oven program: Initially 165° C for 8 min
                         Programmed to 295° C at 1°/min
                         Final hold at 295° C for 64 min

Figure 9.   Conditions for gas chromatography of material isolated
by TLC.

RESULTS AND DISCUSSION

An indication of the precision of the BaP procedure is shown in
Table I.   The sample was an 85-mm unfiltered cigarette.   The
deliveries have been corrected for loss during analysis by using
the average recovery shown.
    A gas chromatogram of a mixture of PAH standards is shown in
Figure 10.   Certain PAH isomers could not be resolved with the
Dexsil 300 column; therefore, for the purposes of quantitation,
benzpyrene representing the sum of "a" and "e" are reported.
Phenanthrene and anthracene are also reported as a single result.
Typical chromatograms depicting the PAH content of condensates
from filtered flue-cured tobacco cigarettes are shown in Figure
11.
    An indication of the precision of the PAH procedure can be
seen in Table II.   The sample was a 72-mm filtered cigarette.

TABLE I.   PRECISION OF THE BaP PROCEDURE

BaP delivery (ng/cig)   22, 17, 16, 18, 19,
                             18, 18, 19, 22, 22

                             $\overline{X}$ = 19.1
                             S = 2.2
                             CV = 11.5%

BaP recovery (%)   80, 86, 83, 83, 81,
                             84, 88, 83, 81, 85

                             $\overline{X}$ = 83
                             S = 2.5
                             CV = 3.0%

The deliveries have been corrected for the loss during analysis by using the average recoveries shown.

PAH deliveries from six different flue-cured tobacco cigarettes of various constructions are shown in Table III.

The comparison of the deliveries obtained using the BaP procedure and the benzpyrene deliveries obtained using the PAH procedure is shown in Table IV. The results for Samples A, B, and C are in excellent agreement by the two methods whereas the results for Samples D, E, and F are higher by the PAH method. The higher values are most likely caused by compounds unresolved from the benzpyrene isomers by the GC column.

REFERENCES

1. H. J. Davis, L. A. Lee, and T. R. Davidson, Anal. Chem. **38**, 1752 (1966).
2. E. W. Robb, G. C. Guvernator III, M. D. Edmonds, and A. Bauley, Beitr. z. Tabakforsch. **3**, 278 (1965).
3. C. I. Ayres and R. E. Thornton, Beitr. z. Tabakforsch. **3**, 285 (1965).
4. L. Dubois, A. Zdrojewski, C. Baker, and J. L. Monkman, J. Air Pollution Control Assn. **17**, 818 (1967).
5. G. E. Moore, R. S. Thomas, and J. L. Monkman, J. Chromatog. **26**, 456 (1967).
6. N. Carugno and S. Rossi, J. Gas Chromatog. **5**, 103 (1967).
7. A. Zane, Tobacco Sci. **12**, 54 (1968).

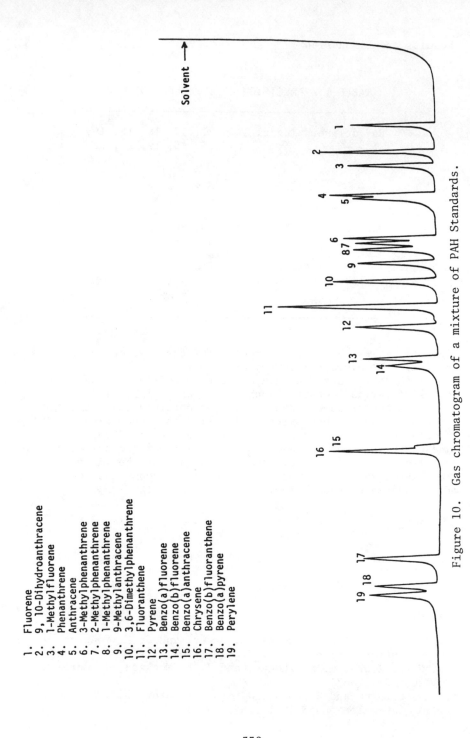

1. Fluorene
2. 9,10-Dihydroanthracene
3. 1-Methylfluorene
4. Phenanthrene
5. Anthracene
6. 3-Methylphenanthrene
7. 2-Methylphenanthrene
8. 1-Methylphenanthrene
9. 9-Methylanthracene
10. 3,6-Dimethylphenanthrene
11. Fluoranthene
12. Pyrene
13. Benzo(a)fluorene
14. Benzo(b)fluorene
15. Benzo(a)anthracene
16. Chrysene
17. Benzo(b)fluoranthene
18. Benzo(a)pyrene
19. Perylene

Figure 10. Gas chromatogram of a mixture of PAH Standards.

358

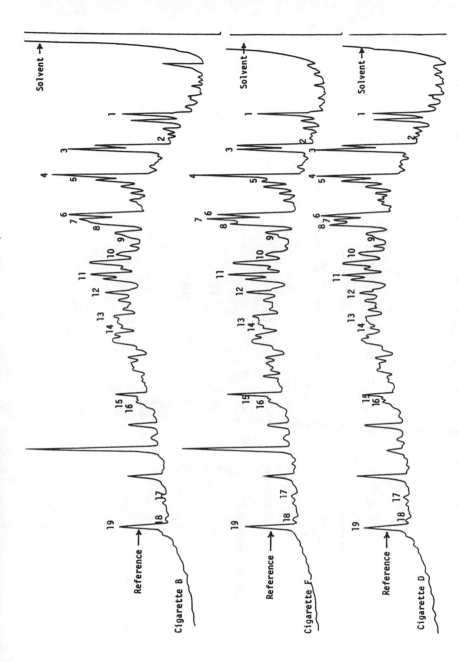

Figure 11. Gas chromatogram of PAH content of cigarette smoke.

TABLE II. PRECISION OF THE PAH PROCEDURE

	Average Recovery	Deliveries (ng/cigarette)					
		1	2	3	4	$\overline{X}$	S
Phenanthrene + anthracene	60	362	401	448	343	389	46
3-Methylphenanthrene	71	251	246	276	218	248	24
Fluoranthene	70	154	172	162	138	157	14
Pyrene	67	130	107	103	87	107	18
Benzfluoranthene	87	22	26	26	26	25	2
Benzpyrene	64	18	20	16	12	17	3

360

TABLE III. PAH DELIVERIES FROM FLUE-CURED TOBACCO CIGARETTES

	Deliveries (ng/cigarette)					
	A	B	C	D	E	F
Phenanthrene + anthracene	360	560	390	420	520	430
3-Methylphenanthrene	200	300	250	240	270	230
Fluoranthrene	130	160	160	140	160	140
Pyrene	91	110	110	88	90	99
Benzfluoranthene	21	24	25	24	24	22
Benzpyrene	20	26	17	26	30	34

361

TABLE IV.   COMPARISON OF BENZPYRENE DELIVERIES

Cigarette	BaP Method Benzo(a)pyrene (ng/cig)	PAH Method Benzpyrene (ng/cig)
A	21	20
B	22	26
C	19	17
D	18	26
E	20	30
F	21	34

REFERENCES (continued)

8. J. H. David, Anal. Chem. 40, 1583 (1968).
9. R. L. Stedman, R. L. Miller, L. Lakritz, and W. J. Chamberlain, Chem. Ind. 394 (March 23, 1968).
10. A. P. Swain, J. E. Cooper, R. L. Stedman, and F. G. Bock, Beitr. z. Tabakforsch. 5, 109 (1969).
11. H. J. Davis, Talanta 16, 621 (1969).
12. L. E. Stromberg and G. Widmark, J. Chromatog. 47, 27 (1970).
13. D. Hoffman and E. L. Wynder, Cancer 27, 848 (1971).
14. G. Grimer and H. Nohnke, Z. Anal. Chem. 261, 310 (1972).
15. R. Gladen, Chromatographia 5, 236 (1972).
16. E. T. Oakley, L. F. Johnson, and H. M. Stahr, Tobacco Sci. 16, 19 (1972).
17. R. C. Lao, R. S. Thomas, H. Oja, and L. Dubois, Anal. Chem. 45, 908 (1973).
18. H. J. Klimisch, Z. Anal. Chem. 264, 275 (1973).
19. D. B. Walters, W. J. Chamberlain, M. E. Snook, and O. T. Chortyk, Anal. Chim. Acta 73, 194 (1974).
20. M. E. Snook, W. J. Chamberlain, R. F. Severson, and O. T. Chortyk, Anal. Chem. 47, 1155 (1975).
21. B. B. Chakraborty, British-American Tobacco Co. Ltd., in press.
22. B. L. Karger, M. Martin, J. Loheac, and G. Guiochou, Anal. Chem. 45, 496 (1973).
23. R. E. Jentoft and T. H. Gouw, Anal. Chem. 40, 1787 (1968).
24. H. J. Klimisch, Anal. Chem. 45, 1960 (1973).
25. K. Bahtia, Anal. Chem. 43, 609 (1971).
26. D. A. Lane, H. K. Moe, and M. Katz, Anal. Chem. 45, 1776 (1973).

CHAPTER 23

The Characterization of Hazardous Wastes by Thin Layer Chromatography

R. D. Stephens and James J. Chan

A paper on the characterization of hazardous waste would have never been included in a proceeding such as this five years ago. However, since that time the impact of toxic and otherwise hazardous industrial waste products has become an important public and governmental concern.

The first major step toward solution of this problem was taken in late 1976 with the passage of Public Law 94-580, the Resource Conservation and Recovery Act. This far-reaching law mandates the U.S. Environmental Protection Agency to develop regulations for minimum standards for those who generate, transport, and dispose of hazardous waste. In the words of Douglas Castle, U.S. EPA Administrator, "We do not underestimate the complexity and difficulty of our proposed regulations. Rather they reflect the large amounts of hazardous wastes generated in our diverse society. The regulation will affect a large number of industries . . . as well as laboratories, pesticide applicators, and waste transporters. The agency estimates that approximately 270,000 waste-generating facilities, 10,000 waste transporters, and many thousands of waste disposal and processing facilities will be regulated."

In the two years since the passage of RCRA, EPA has been working to develop these regulations, and on December 18, 1978, the core of the regulations was proposed in the Federal Register.

The phrase "core of the regulations" is used because the December 18 Federal Register covers what is called "Subtitle C" of RCRA. Subtitle C is directed toward the identification and characterization of hazardous wastes. The regulation outlines biological and chemical testing methodologies by which these complex industrial and other toxic waste products may be examined to evaluate their potential hazard.

This highlights a key point in the regulation and management of hazardous waste. Such activities must be founded on an adequate understanding of waste composition. The generation of waste composition data by the laboratory is in many cases a formidable problem. However, if proper procedures are to be developed to assure safe handling, transportation, recycling, or disposal of waste, as well as assessment of public health and environment impacts, waste chemical composition must be known.

Attempts to develop a variety of methodologies to attack this problem are under way. These include many of the modern instrumental techniques such as gas chromatography, HPLC, atomic absorption, x-ray fluorescence, and GC/MS. In addition to these techniques, thin layer chromatography was found to be well suited to waste analysis, and it has certain definite advantages over some of the more "equipment intensive" methods. Many laboratories that must provide data on waste composition do not have the financial resources to obtain expensive sophisticated instrumentation. TLC in these laboratories is an ideal alternative.

Because waste analysis is of essentially all products and by-products of industry, it is necessarily complex. The capability of TLC is only beginning to be realized, but even our rudimentary methods have proved useful.

A scheme of analysis that uses multiple TLC systems is proposed as an approach to this type of problem. This is a screening procedure that may also be confirmatory or may direct the analyst to other analytical methods for confirmation. In this scheme, a bank of TLC systems is set up by the laboratory and kept ready for use when needed.

Each system is designed to detect a number of specific chemicals belonging to a class of chemical compounds frequently found in hazardous wastes. For example, a system for toxic metals includes the capability of detecting mercury, lead, cadmium, and the other priority pollutant metals; the system for chlorinated pesticides can detect p,p' DDT, dieldrin, lindane, etc.; a system for organophosphorus pesticides can detect malathion, diazinon, methyl parathion, etc. Systems for other types of compounds, such as phenols and amines, are also prepared by selecting the proper combination of sorbent mobile phase, and visualization agent.

As an example, a sample suspected of containing cadmium is quickly screened with the system calibrated specifically for toxic metals in order to confirm its presence or absence. A sample with no history is screened with the entire bank of TLC systems. A positive response in any system directs the analyst to do further study using methods with greater specificity. This may include quantitative TLC or other methods, such as gas

chromatography or infrared spectrometry. If negative findings
are obtained, many classes of compounds have been eliminated as
being present above the detection limit. The analyst can then
begin to search for other classes of compounds not tested in this
scheme.

Thin layer chromatography is fast, sensitive, and economical,
and requires minimal equipment and personnel training. The
screening procedure can be easily adapted for use by mobile laboratories in the field.

Our laboratory has developed several TLC systems and is now
expanding this number of systems for routine screening. Systems
for toxic metals, chlorinated pesticides, and organophosphorus
pesticides have been calibrated and tested with spiked samples in
matrices frequently encountered in hazardous wastes. The detection levels on the TLC plate range from approximately 0.5 to 2.0
μg, varying with the compound tested. One to five microliters of
1 μg/μl solutions are applied. Samples found to be below the sensitivity of the test are pretreated by extraction and concentration for retesting. The use of proper solvents for extraction
enhances the specificity of the test and isolates the compound of
interest from interfering materials.

Solid samples are extracted with organic solvents or digested with acid and adjusted to a suitable volume. Liquid waste samples are spotted directly on the plate or subjected to pretreat-
ment by solvent partition or column absorption cleanup. Another
technique for purification is selective precipitation of the compounds of interest. Some of these techniques are illustrated in
the descriptions of the TLC systems below.

In developing each system, numerous mobile phases and chromogenic reagents were tested. Those combinations that gave the
best resolution and reproducibility were selected.

TOXIC METALS SYSTEM

This system can detect the following metals of the hydrogen sulfide group at detection levels indicated in Table I.

The standard solutions of the metals were applied as chloride
or nitrate salt solutions onto microcrystalline cellulose layers
(0.1 mm thick) (Polygram Cell 400) and developed in n-butanol
saturated with 3N hydrochloric acid for approximately 2 hours.
The plate was then sprayed with a 2% ammonium sulfide solution to
yield a chromatogram with the colored spots on a white background
indicated in Table II.

A 10-ml sample of galvanizing waste was spiked with the
above six metals to a concentration of 200 ppm. The metals were

TABLE I.   DETECTION LIMITS FOR METALS

Metal	Microgram
Mercury	1.5
Cadmium	1.2
Bismuth	1.0
Arsenic	1.9
Lead	1.6
Copper	0.7

TABLE II.   TLC CHARACTERISTICS FOR METALS

Metal	Spot Color	R_f
Mercury	Gray	0.92
Cadmium	Yellow	0.85
Bismuth	Brown	0.69
Arsenic	Yellow	0.61
Lead	Brown	0.31
Copper	Yellow	0.23

isolated by adjusting the solution to pH 3 and by precipitating
the metals with 0.5% thioacetamide solution as the sulfides. The
precipitate was centrifuged, washed, and dissolved in nitric acid
and evaporated to a small volume. This solution and a solution
of unspiked galvanizing waste sample, treated in the same manner,
were analyzed. All metals in this spiked sample were recovered
and readily detectable.

CHLORINATED PESTICIDES SYSTEM

The method described below will detect the following nine compounds at the detection levels indicated in Table III.
     Standard pesticide solutions in ethyl acetate were spotted
on Silica Gel 60 plates. The chromatogram was developed in

TABLE III. DETECTION LIMITS FOR PESTICIDES

Pesticide	Microgram
Aldrin	1.0
BHC	0.4
p,p' DDT	1.0
Dieldrin	1.0
Endrin	1.0
Heptachlor	0.7
Lindane	1.0
Methoxychlor	0.9
Mirex	0.9

heptane-acetone (98:2) for 50 minutes. The chromatogram was sprayed with a 0.5% diphenylamine in 95% ethyl alcohol, then irradiated under an ultraviolet germicidal lamp for 20 minutes. Colored spots appeared on the plate with a beige background as indicated in Table IV.

Most compounds are well resolved, except for overlapping of lindane, dieldrin, and the major spot of BHC at an R_f of 0.21. These three compounds can be differentiated as follows:

Observation	Pesticide
One green (R_f .21)	Dieldrin
Three violet (R_f .35, .21, .06)	BHC
One violet (R_f .21, none at .35 and .06)	Lindane

If the spot color at R_f .21 is not clearly distinctive, gas chromatographic analysis of the scraped spot from an unsprayed plate can show whether dieldrin, lindane, or both are present.

A 3-ml sample of waste crankcase oil, spiked with the above nine pesticides to a level of 220 ppm, was analyzed. The sample was first extracted with petroleum ether/acetonitrile. The extracted pesticides were then partitioned into hexane, concentrated, and analyzed. All pesticides were recovered and readily detectable. An unspiked sample was carried through the entire procedure. No spots were detectable.

TABLE IV. TLC CHARACTERISTICS OF PESTICIDES

Pesticide	Spot Color	R_f	
Mirex	Violet	0.86	
Aldrin	Blue	0.77	
Heptachlor	Violet	0.69	
p,p' DDT	Violet	0.52	
BHC	Violet	0.35	(minor spot)
Endrin	Green	0.28	
BHC	Violet	0.21	(major spot)
Lindane	Violet	0.21	
Dieldrin	Green	0.21	
Methoxychlor	Violet	0.14	
BHC	Violet	0.06	(minor spot)

ORGANOPHOSPHORUS PESTICIDE SYSTEM

This system can detect the following six pesticides at the detection levels.

Pesticide	Micrograms
Diazinon	0.5
Ronnel	1.0
Ethion	0.5
Malathion	1.0
Methyl parathion	0.5
Guthion	0.5

The standard solutions were spotted on high-performance TLC silica gel 60 plates (Mackery-Nagel) containing a fluorescent dye. The plate was developed in a mixture of hexane-acetone (5:1) to a distance of 5 cm. Development time required 7 minutes. The chromatogram was viewed under short wavelength ultraviolet light for diazinon, methyl parathion, and guthion. The plate was then sprayed with a solution of 0.3% N,2,6-trichloro-p-benzoquinone imine in cyclohexane. On heating on the surface of a hot plate at 200° C for 5 minutes, dull red spots appeared in a beige background as listed in Table V.

Diazinon did not develop color spot, but may be detectable under uv before spraying. Malathion has two red spots, at R_f's of

TABLE V.   TLC CHARACTERISTICS OF PESTICIDES

Pesticide	Viewed under uv	Sprayed with Reagent	R_f
Diazinon	Quenched fluorescence		0.58
Ronnel		Red	0.46
Ethion		Gray-red	0.42
Malathion		Red	0.24, 0.18
Methyl parathion	Quenched fluorescence	Red	0.24
Guthion	Quenched fluorescence	Gray-red	0.12

0.24 and 0.18. Methyl parathion quenches fluorescence and develops a red spot at an R_f of 0.24.

A 5-g sample of soil with added pesticides to 200 ppm was
extracted with 50 ml of hexane/acetone, 41/51. Sodium chloride
solution was added to the mixture to shift the pesticide to the
hexane layer. The hexane was reduced to a suitable volume for
analyses. All pesticides were detectable. No spots were visible
on the blank soil treated in the same manner.

Future work in this area can include plans for more extensive use of:

1. High-performance thin layer chromatography plates to
   increase sensitivity and reduce analysis time.
2. Plates with preabsorbent layers to enable direct application of some samples without prior extraction.
3. Reverse phase chromatography for better resolution of
   certain types of compounds.
4. Methods for rapid quantitation of detected materials.
5. Use of two-dimensional development.

ACKNOWLEDGMENT

The authors gratefully acknowledge the financial support of this
work by the U.S. Environmental Protection Agency under research
grant number R804692, Richard A. Carnes, Project Officer.

CHAPTER 24

Safe Use and Disposal of Hazardous Chemicals in the Laboratory

J. R. Songer and D. T. Braymen

Chemicals which pose a threat to life and property may be hazardous by virtue of their toxicity, flammability, or explosiveness. Hazardous chemicals are found in the laboratory in a wide variety of solid, liquid, or gaseous forms. Frequently chemical solutions are both toxic and flammable. In some cases, toxic chemicals are suspended in flammable solvents which are, to a greater or lesser degree, toxic also. The major emphasis of this presentation is the use of safe methods in handling and disposal of organic solvents alone or in combination with other toxic chemicals.

Flammable liquids are those having a flash point below 140° F and a vapor pressure not exceeding 40 pounds per square inch absolute at 100° F. Combustible liquids have a flash point at or above 140° F. Liquid compressed gases are flammable liquids with a vapor pressure above 40 pounds per square inch absolute at 100° F. When flammable liquids are handled or stored, care must be exercised to assure that explosive or flammable mixtures with air do not occur. If, in the course of work, it is necessary to create explosive or flammable mixtures, all sources of ignition must be eliminated. Protection against static electricity when transferring flammable liquids should be provided by bonding and grounding.

Ventilated cabinets (Figure 1) should be used for bulk storage of flammable chemicals. Unventilated cabinets (Figure 2) constructed of 1-inch thick plywood for storage of working amounts of solvents are OSHA-approved.

If refrigerators are used for the storage of flammable liquids, they must be explosion-safe (Figure 3). Many conventional refrigerators can be made explosion-safe by relocating the thermostat on the outside and removing all other sources of ignition from the box.

Figure 1. Ventilated flammable-liquid storage cabinet. The vent
line connected at the bottom of the cabinet collects vapors that
are heavier than air. For vapors lighter than air the vent line
can be located near the top on the opposite end.

     All work with flammable liquids should be conducted in areas
with controlled ventilation. Laboratory hoods (Figure 4) should
be used when larger quantities of liquids are involved. If the
quantity of liquid is relatively small, high-velocity spot ventilation can be used. High-velocity spot ventilation systems are

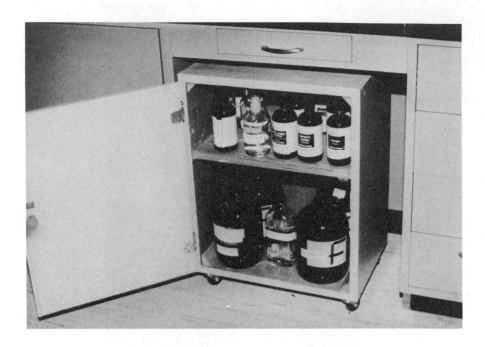

Figure 2. Unventilated flammable-liquid storage cabinet. Cabinet
is constructed of 1-in. thick plywood and painted with a fire-
retardant paint.

described in Industrial Ventilation (1), and an example is shown
in Figure 5. The velocity of air moving through hood openings
should be at least 100 feet per minute (fpm). The velocity
through the opening of a spot ventilator should be 2,000 fpm.
Exhaust fans for hoods or spot ventilation systems should be
located on the roof of the building.

Gastight Class III Biological Safety Cabinets (Figure 6) provide the highest level of containment available. These units
should be used when working with highly hazardous pathogenic microorganisms or very toxic or carcinogenic chemicals in studies that
afford a high risk of respiratory exposure.

All organic solvents are toxic, and appropriate safeguards
must be maintained when used. Threshold limit values (TLV) as
recommended by the American Conference of Governmental Industrial
Hygienists are given in parts of vapor or gas per million parts of

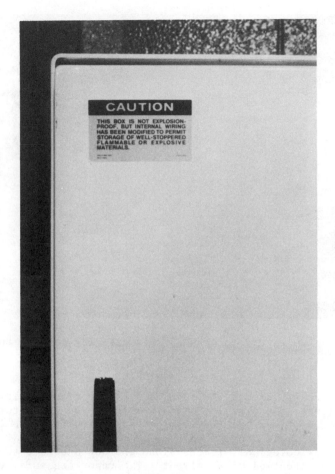

Figure 3. Explosion-safe refrigerator. All sources of ignition have been removed from the storage area.

air by volume at 25° C and 760 mm Hg pressure. These values are obviously for respiratory exposure. The TLV values should be used as guides in the control of health hazards and should not be regarded as fine lines between safe and dangerous concentrations.

Laboratory hoods and high-velocity spot ventilation systems are used for the control of toxic organic solvent vapors. Ventilation requirements necessary to meet the TLV for organic

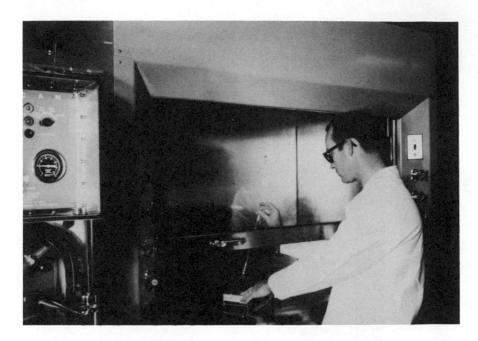

Figure 4.   Laboratory fume hood.   Air velocity through the hood opening should be at least 100 fpm.

solvents greatly exceed the requirements for flammability control. Table I lists some organic solvents used in the laboratory and the number of times the TLV must be exceeded before the lower flammability limit is reached.

The toxicological effects alone are not adequate to assess the hazard potential of a solvent. The vapor pressure, ventilation, and the manner of use will determine the concentration in air.

The vapor-hazard ratio is one approach to a numerical comparison of potential hazard under a given set of conditions. This number is the ratio of the equilibrium vapor concentration at 25° C to the TLV (ppm/ppm). The lower the ratio the lower the potential hazard. As an example, methylene chloride with a TLV of 500 ppm might be judged safer than 1,1,1,trichloroethane, which has a TLV of 350 ppm, if judged on TLV alone. The latter solvent should be safer when the potential for vaporization is taken into account. The vapor-hazard ratio for methylene chloride is 1080,

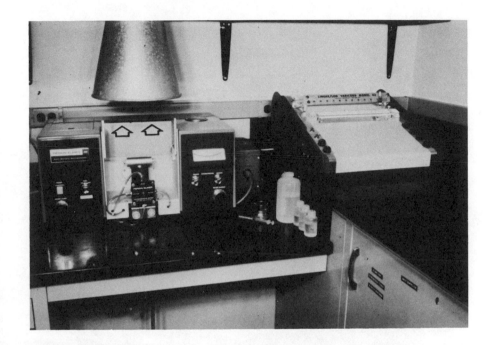

Figure 5. High-velocity spot ventilation. Air velocity through the ventilator opening should be 2,000 fpm.

whereas the vapor-hazard ratio for 1,1,1,trichloroethane is only 489. Table II is a list of some organic solvents and their threshold limit values and vapor hazard ratios.

Care should be exercised to prevent skin contact with organic solvents. Appropriate gloves and protective clothing should be worn when handling solvents. The eyes should also be protected when working with hazardous chemicals.

COLLECTION AND DISPOSAL OF HAZARDOUS CHEMICAL WASTE

Identification of Chemical Waste

In laboratory operations it is imperative that people generating hazardous chemical waste properly segregate and identify these wastes. In a typical laboratory, time does not permit extensive

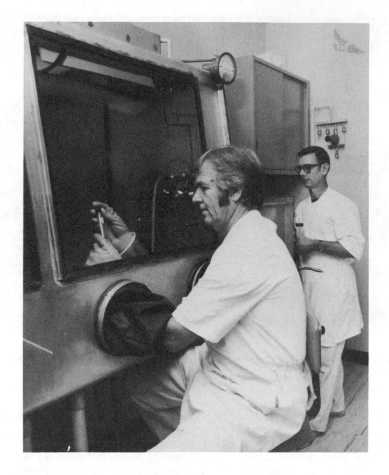

Figure 6. Class III biological safety cabinet. This cabinet is gastight, operates at a negative pressure, the supply and exhaust air is filtered through HEPA filters, and sewage is sterilized.

examination of waste for disposal. Unidentified waste chemicals remain the responsibility of the investigator.

Robert Stephens and James Chan, in Chapter 23, Application of TLC to Characterize Hazardous Waste, consider the problem of identification of unknowns in hazardous waste.

TABLE I.   THRESHOLD LIMIT VALUES FOR VARIOUS SOLVENTS

Organic Solvent	Factor by which TLV is raised to the lower explosive limit
Acetonitrile	2000
Pyridine	3600
Benzene	560
Ethyl ether	46
Methyl alcohol	300
Methyl ethyl ketone	100
Dioxane	200
Ethyl acetate	55
Acetone	30
Toluene	70
Isopropyl alcohol	58
Propyl alcohol	140
Butyl alcohol	140
Ethyl alcohol	33
Stoddard solvent	16

Selecting Disposal Methods

A key for handling hazardous waste problems was developed by
Ghassemi et al. (2), and is shown in Figure 7.

Once a chemical has been identified and determined to be
hazardous, several options for disposal can be considered. The
best option is to reuse or recycle the chemical. If this is not
a viable option, disposal in a hazardous waste management facil-
ity should be the next consideration.

The geographic distribution of hazardous waste management
facilities in continental United States is shown in Figure 8 (3).
The geographic distribution of heavy metal processors, solvent
reclaimers, petroleum re-refiners, and industrial waste information clearinghouses is shown in Figure 9 (3).

Hazardous waste disposal sites are concentrated in the north-
eastern and southwestern parts of the United States. Lack of
access to these sites,or other reasons for being unable to use
them, necessitates the exploration of other methods.

An in-house treatment and/or disposal system can be developed if space, personnel, and equipment are available.

TABLE II.   CHARACTERISTICS FOR HAZARDOUS SOLVENTS

Substance	Vapor Hazard	Threshold Limit (ppm by vol.)
Acetonitrile	11,200	20
Chloroform	4,970	50
	(9,940)	
Pyridine	5,260	5
Benzene	5,000	25
Ethyl ether	1,380	400
Methylene chloride	1,080	500
Methyl alcohol	820	200
Methyl ethyl ketone (2-butanone)	625	200
Dioxane	490	100
Ethyl acetate	303	400
Acetone	290	1,000
Toluene	184	200
Isopropyl alcohol	140	400
Propyl alcohol	140	400
Butyl alcohol	92	100
Ethyl alcohol	76	1,000
Stoddard solvent	7	500

Disposal in municipal incinerators or disposal in sanitary landfills are possible options for some chemicals.

There are a few chemical compounds for which there are no approved methods of disposal. The only alternative available for these compounds is storage until a disposal method is developed.

COLLECTION AND DISPOSAL OF HAZARDOUS CHEMICAL WASTE AT THE NATIONAL ANIMAL DISEASE CENTER (NADC)

The volume and nature of chemical waste generated at the NADC and the fact that no hazardous waste management facilities are located in Iowa favored the development of an "in-house" disposal system.

For the purpose of disposal, waste chemicals were grouped as follows: (1) acids, (2) alkalies, (3) flammable liquids, water

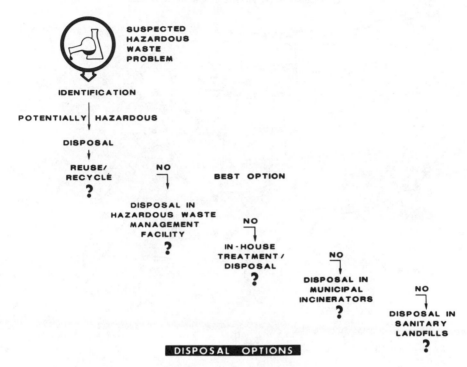

Figure 7. Hazardous waste chemical disposal options from
Ghassemi (2).

immiscible, (4) flammable liquids, water miscible, (5) flammable
solids, (6) toxic nonflammable chemicals, and (7) nontoxic nonflammable chemicals.

Waste chemicals must be identified by the user at the time
they are collected; using this information, the best method of
disposal can be selected.

Chemicals are sometimes received in unopened containers in
perfectly usable condition. This occurs when a particular phase
of work is completed and excess chemicals are left. Before
attempting to dispose of them, every possibility of reusing these
chemicals should first be exhausted. For example, embedding fluid
discontinued by one project can be quite satisfactory for another.

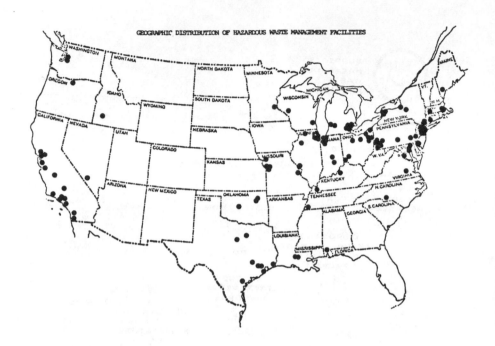

Figure 8. Geographic distribution of hazardous waste management facilities. Data from Straus (3).

Recycling used chemicals is the preferred method of disposal. As metallic mercury presently costs $50.00 per pound, distillation is not only the preferred method of disposal--it is also the most economical.

Waste chemicals are collected in appropriate containers and transported to the safety unit. Organic solvents are collected in nonmetallic waste solvent cans equipped with flash arrestors and self-closing lids. Concentrated liquid acids and alkalies in glass jugs are transported in safety carriers.

Acids and alkalies are neutralized and mixed with copious amounts of water, and are flushed down the drain at the time of their collection.

Chemicals to be disposed of by incineration are transported to the chemical storage building and held. This building is maintained at 55° F and is equipped with an automatic CO_2 fire extinguishing system. When approximately 50 gallons of organic solvents have accumulated, a burn is scheduled.

To determine fetal lung maturity, amniotic fluid is obtained by transabdominal amniocentesis using ultrasound locali- zation of the fetus. Fetal age at amniocentesis is determined by maternal dates, biparietal diameter, and physical examination for gestational age.

Five milliliters of amniotic fluid is centrifuged at 800 × g for 5 minutes to remove cellular debris. Lipids are extracted using the technique of Bligh and Dyer (38) and the extracted lipids are exposed to acetone at 0° C. The fraction precipitated by cold acetone is plated to 20 × 20 cm Pyrex plate coated with silica gel H containing 5% ammonium sulfate. The fraction solu- ble in cold acetone is also plated but run only in the first dimension. Authentic phospholipid standards of known quantity PC, PI, and PG (Supelco Co.) are applied in the plate margins and co-chromatographed with our samples. The first phase is devel- oped for 10 cm in chloroform-methanol-water-glacial acetic acid (65:25:4:8). After the first phase, the plate is dried in room air, placed in an oven at 70° C for 10 minutes, and then placed in the second phase at 90° clockwise from the first run. After developing through the second phase (THF-methylal-methanol- ammonium hydroxide) the plate is charred on a hot plate at 280° C. Upon heating, the ammonium sulfate in the silica forms sulfuric acid, which chars the phospholipid spots and, using reflectance densitometry or phosphorus analysis, makes comparison of densi- ties possible.

By comparing the relative amounts, expressed as percentages, of acetone precipitated to acetone-soluble PC, and sphingomyelin to acetone-precipitated PC, PI, and PG, it has been possible, since July 1, 1978, to predict fetal lung maturity in 24 infants without false positives or negatives. By scheduling pregnancy intervention only after the appearance of PG and when the soluble- to-precipitable ratio of PC is 50%, the ability to predict the optimal timing of interruption in an uncomplicated pregnancy is enhanced. This methodology is being explored in pregnancies com- plicated by several factors known to influence lung maturation in utero. Recent experience with pregnancies complicated by diabe- tes A-C has shown that if delivery is delayed until the appearance of PG (mean gestational age 38 weeks, 2 days, regardless of L/S alone, it should be performed when warranted by antenatal assess- ment of fetal maturity. Unpublished observations indicate that phosphatidylglycerol is not present in meconium, vaginal secre- tions, or blood; thus amniotic fluid samples "contaminated" with these fluids do not interfere with the two-dimensional thin layer chromatographic separation of phosphatidylglycerol, as is the case with the undimensional L/S ratio.

With expanded knowledge of lung development the role of quantitative thin layer chromatographic analysis must now continue to move from the research laboratory to the clinical laboratory in order to provide optimal perinatal health care for timing delivery.

REFERENCES

1. L. Gluck, M. Kulovich, R. C. Borer, P. H. Brenner, G. G. Anderson, and W. N. Spellacy, Am. J. Obstet. Gynecol. 109, 440 (1971).
2. M. J. Bryson, J. A. Gabert, and M. A. Stenchever, Am. J. Obstet. Gynecol. 114, 208 (1972).
3. B. D. Kulkarni, J. Bienauz, L. Burd, and A. Scommegna, Obstet. Gynecol. 40, 173 (1972).
4. C. R. Whitfield, W. H. Chan, and W. B. Sproule, Br. Med. J. 2, 85 (1972).
5. W. H. Spellacy and W. C. Buhi, Obstet. Gynecol. 39, 852 (1972).
6. J. Nakamura, J. F. Roux, E. G. Brown, and A. Sweet, Am. J. Obstet. Gynecol. 113, 363 (1972).
7. L. Gluck and M. Kulovich, Am. J. Obstet. Gynecol. 115, 539 (1973).
8. C. J. Dewhurst, A. M. Dunham, D. R. Harvey, and C. E. Parkinson, Lancet 1, 1475 (1973).
9. P. Harding, F. Possmayer, K. Milne, N. T. Jaco, and J. N. Walters, Am. J. Obstet. Gynecol. 115, 298 (1973).
10. I. R. Donald, R. K. Freeman, U. Goebelsmann, W. H. Chan, and R. M. Nakamura, Am. J. Obstet. Gynecol. 115, 547 (1973).
11. J. C. Hobbins, W. Brock, L. Speroff, G. G. Anderson, and B. Caldwell, Obstet. Gynecol. 39, 660 (1972).
12. T. I. Wagstaff and D. R. Bromhaus, J. Obstet. Gynecol. Br. Commonw. 80, 412 (1973).
13. J. A. Lemons and R. B. Jaffe, Am. J. Obstet. Gynecol. 115, 233 (1973).
14. T. Lindback, T. Frantz, J. Skjaeraasen, and S. Graven, Acta Obstet. Gynecol. Scand. 53, 219 (1974).
15. J. S. Meyer, E. H. Coch, G. M. Goldman, and G. Kessler, South. Med. J. 67, 431 (1974).
16. J. C. L. Merola, L. M. Johnson, R. J. Bolognese, and S. L. Corson, Am. J. Obstet. Gynecol. 119, 243 (1973).
17. T. K. Mukherjee, B. K. Rajegowda, L. Glass, J. Auerbach, and H. Evans, Am. J. Obstet. Gynecol. 119, 648 (1974).
18. W. B. Sproule, M. E. Greene, and C. P. Whitfield, Am. J. Obstet. Gynecol. 119, 653 (1974).

19. J. L. Duhring and S. A. Thompson, Am. J. Obstet. Gynecol. 121, 218 (1975).
20. R. C. Keniston, M. L. Pernolls, N. R. Buist, M. Lyon, and J. R. Swanson, Am. J. Obstet. Gynecol. 121, 324 (1975).
21. E. B. Olson, S. N. Granen, and R. D. Zachman, Am. J. Obstet. Gynecol. 121, 324 (1975).
22. D. Harven, C. E. Parkinson, and S. Campbell, Lancet 1, 42 (1975).
23. P. M. Farrell and M. E. Avery, Rev. Resp. Dis. III, 657 (1975).
24. G. W. Dahlenburg, F. I. R. Martin, P. R. Jeffrey, Br. J. Obstet. Gynecol. 84, 294 (1977).
25. I. D. Frantz and M. F. Epstein, Sem. Perinatol. 2, 347 (1978).
26. T. I. Wagstaff, G. A. Whyley, and G. Freedman, Ann. Clin. Biochem. II, 24 (1974).
27. S. A. Olowe and A. Akinkugke, Pediatrics 62, 38 (1978).
28. C. F. Baxter, G. Rouser, and G. Simon, Lipids 4, 243 (1967).
29. W. R. Harlan, J. H. Margraf, and S. L. Said, Am. J. Physiol. 211, 855 (1966).
30. S. A. Rooney, P. M. Canavan, and E. K. Motoyama, Biochim. Biophys. Acta 300, 56 (1974).
31. T. A. Merritt, M. Obladen, and L. Gluck, in Perinatal Physiology, edited by U. Stave (Plenum, New York, 1978), p. 106.
32. G. Arvidson, H. Ekelund, and B. Astedt, Acta Obstet. Gynecol. Scand. 51, 71 (1972).
33. M. Hallman, B. Feldman, E. Kirkpatrick, and L. Gluck, Pediatr. Res. 11, 714 (1977).
34. M. Obladen, Eur. J. Pediatr. 128, 129 (1978).
35. M. Hallman, M. V. Kulovich, E. Kirkpatrick, R. G. Sugarman, and L. Gluck, Am. J. Obstet. Gynecol. 125, 613 (1976).
36. M. Kulovich, M. Hallman, and L. Gluck, Am. J. Obstet. Gynecol., in press.
37. B. S. Saunders, M. Kulovich, and L. Gluck, Perinatology 5, 231 (1978).
38. E. G. Bligh and W. J. Dyer, Can. J. Biochem. Physiol. 37, 391 (1959).

CHAPTER 26

Metabolism of B-Aminopropionitrile (BAPN) During Wound Healing

Joseph H. Fleisher and M. Chvapil

A promising method for controlling the physical properties of newly synthesized scar tissue in animal models of human disease and in healing wounds of humans is interference with covalent cross linking of collagen by administration of a lathyrogen such as β-aminopropionitrile (BAPN) (1-3).

However, discrepancies were noted between clinical data and the predicted biochemical reactions from the dose of BAPN administered. These observations have raised questions concerning the mechanisms for metabolic disposition of BAPN.

The only previous study of the metabolism of BAPN in man was reported by Keiser and Sjoerdsma (4). They found that BAPN disappeared rapidly from human plasma following oral administration. The disappearance might have been caused, in part, by enzymatic degradation to the nonlathyrogenic metabolite, cyanoacetic acid (CAA) (5), by a monoamine oxidase present in human serum (6). This possibility was examined by incubating BAPN with human serum. An earlier finding by McEwen et al. (7) that BAPN was also a competitive inhibitor for rabbit serum monoamine oxidase prompted study of the effect of BAPN on the ability of human serum to deaminate benzylamine, a classical substrate for monoamine oxidase activity.

Sensitive methods for the determination of BAPN and CAA in biological fluids following administration of the lathyrogen would contribute to more effective dosage with BAPN as well as to the interpretation of future clinical studies.

In this regard, the colorimetric determination of BAPN with ninhydrin requires separation from other nitrogenous compounds that also respond to this reagent (8). Recently Wilk et al. (9) estimated BAPN by thin layer chromatography (TLC). Their method is based upon extraction of the free base into butanol at alkaline pH, back extraction into the aqueous phase under acid conditions, lyophilization, and TLC in butanol-water-acetic acid (5:4:1). The spots obtained were eluted and measured spectrophotometrically. They reported only 5% recovery by their procedure.

The present method for the estimation of BAPN differs greatly from that used by Wilk et al. (9). Solvent extraction was eliminated avoiding major losses. TLC in acetone, 1 M NH_4OH (9:1), yielded excellent separation between BAPN and amino acids. Direct densitometry of the TLC plate yielded satisfactory quantitation (10).

RECOMMENDED PROCEDURES

Materials

β-Aminopropionitrile was purchased as the fumarate salt from Aldrich Chemical Co., Milwaukee, Wisconsin. Ninhydrin aerosol spray* and cyanoacetic acid were purchased from Brinkmann Instruments, Westbury, New York, and Eastman Kodak Co., Rochester, New York, respectively. Plastic thin layer plates (Macherey Nagel & Co.) precoated with silical gel (250 μm) without gypsum were obtained from Brinkman Instruments, Westbury, New York. ^{14}C-Nitrile labeled BAPN (2.3 mCi/nmole) and 2^{-14}C-CAA (55 mCi/nmole) were obtained from Hoffman LaRoche, Inc.** and ICN Nuclear Division respectively.

Chromatography

Standard solutions of BAPN were made in 80% ethanol. A 10-μl Hamilton syringe was used to apply uniform microliter aliquots of standard and experimental solutions to the same TLC plate. The TLC plate was not heat activated before sample application. The chromatogram was developed by ascending chromatography for 10 cm in glass tanks linked with filter paper and equilibrated with 100 ml of acetone, 1 M NH_4OH (9:1). The plate was air dried briefly, then heated at 50°-60° C for 30 minutes to remove ammonia before spraying with ninhydrin. The plate was again reheated at 85°-95° C for up to 10 minutes. The chromatogram was scanned immediately with a Beckman Model R-110 Microzone densitometer after cutting the plate into strips suitable for this instrument. The resulting

---

* Ninhydrin spray reagent, 0.1% in isopropanol, EM Reagent, Catalog No. 6758, gave optimal visualization and low background.
** We are grateful to the Hoffman-LaRoche Co. of Nutley, New Jersey for a gift of the ^{14}C-BAPN.

tracings were carefully cut out with scissors and weighed on an analytical balance.

Analysis of Urinary BAPN

Recent treatment of a 16-year-old female with massive keloids by surgical excision of the keloids and induction of lathyrism with orally administered BAPN* enabled the metabolism of this compound to be studied in a human being. After partial excision of the keloids, one tablet of BAPN fumarate containing 250 mg of free base was administered at 6-hour intervals each day for 21 days. Urine was collected from day 1 through day 14, at which time urine collection was stopped although BAPN dosage was continued. The volume of each urine specimen was measured, a portion was centrifuged, and a 4-ml aliquot of the supernatant was lyophilized. The lyophilized residue was extracted with 2.0 ml of cold 60% ethanol; the extract was allowed to stand 30 minutes in ice and then centrifuged at 2000 rpm for 5 minutes. The supernatant was retained and subjected to TLC as described in the section "Chromatography."

Assay of CAA in Urine

Aliquots (0.2 ml) of centrifuged urine corresponding to the samples used in the BAPN study were diluted to 1.0 ml with H_2O. Sixty percent H_2SO_4 (0.2 ml) was added, followed by 10 ml of ethyl acetate saturated with H_2O. Samples were extracted for 10 minutes on a mechanical shaker, centrifuged at 2000 rpm for 5 minutes, and 5 ml of supernatant was transferred to graduated tubes containing 0.5 ml of 0.1 M acetate buffer at pH 5.0. The contents were shaken and the ethyl acetate evaporated under N_2 at 55°-65° C. All samples were brought to 10 ml with H_2O. Control specimens of urine (prior to BAPN treatment) with and without 100 μg of CAA were processed concurrently as a standard and specimen blank, respectively. CAA in the samples was determined colorimetrically according to Sievert et al. (11).

RESULTS

Evaluation of Chromatography

Under the conditions described, the spots resulting from the application of BAPN moved 6 cm, yielding an R_f of 0.60.

---

* Research protocol approved by Clinical Research Unit operating under Public Health Service GCRC Grant 1-M01-RR00714.

The weight of the tracings obtained by scanning with the densitometer was linear from 0.1 to 0.7 µg of BAPN free base applied in identical microliter volumes to the same plate (Figure 1).

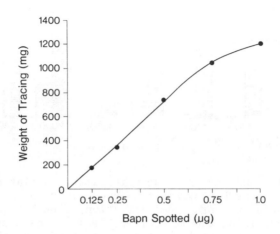

Figure 1. Densitometry of BAPN following thin layer chromatography (see text, Ref. 10).

The reproducibility of the weights of tracings resulting from spotting six separate 5-µl samples of BAPN containing 0.5 µg of free base/sample on the same plate was 0.0874 ± 0.0018 g. This yielded a coefficient of variation approximately equal to 2.1%.

Recovery of BAPN and CAA During the First Day of Drug Administration

Figure 2 shows unchanged BAPN appearing in the patient's urine within an hour of oral administration of 250 mg (3.68 mM) of BAPN at noon. A moderate increase was noted in the specimen collected at 2 hours after drug administration, followed by a marked decline in 6 hours to less than one-fourth the amount of BAPN found in the 1-hour sample. Administration of a second dose of BAPN at 6 P.M. resulted in a marked increase in urinary excretion of BAPN in the sample collected 2 hours later.

The CAA in the first post-BAPN sample collected at 1 P.M. approximated 2.2% of the BAPN excreted. Figure 2 shows a gradual

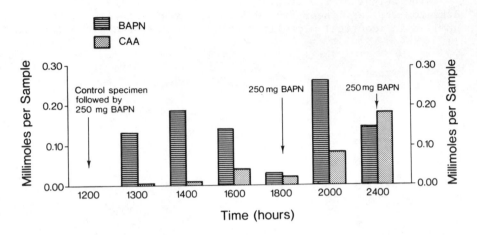

Figure 2.   BAPN and CAA in human urine after initiating oral dosage with BAPN.   The histograms show the urinary output of BAPN
and CAA following the oral administration of BAPN fumarate tablets (each containing 250 mg of free base) at the times indicated.
Urine specimens were obtained at 1, 2, 4, 6, 8, and 12 hours following the initial dosage with BAPN.

increase in the relative proportion of CAA to BAPN in the urine.
The specimen collected at midnight of the first day was the first
to show more CAA than BAPN.

      Recovery of BAPN and CAA in Human Urine During Daily Oral
      Dosage with BAPN

Figure 3 shows the recovery of BAPN to approximate a steady state
by the second day of dosage.   Urinary BAPN averaged 16.3 + 0.6
(SEM) percent of the daily dose administered from days 2 to 14
inclusive.
      Urinary CAA output exceeded that of BAPN by the second day
of dosage.   Recovery of CAA continued to increase, reaching a
rough plateau at the seventh day.   Thereafter it averaged 48.1 +
2.5% (SEM) of the daily dose adminstered from the seventh through
the fourteenth days of urine collection.*

---

* Day 9 was omitted from Figure 3 because of incomplete urine
  collection.

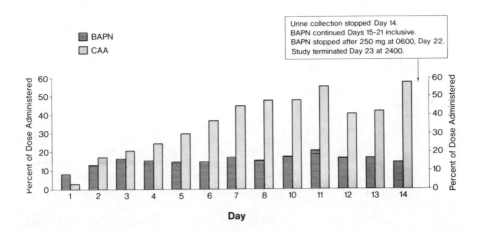

Figure 3. BAPN and CAA in human urine during daily dosage with
BAPN. The histograms show the urinary output of BAPN or CAA in
percentages of the daily dose of orally administered BAPN fumarate (equivalent to 250 mg of free base every six hours) with
urine collection at time of drug administration.

BAPN and CAA Content of Human Urine Following Cessation of
BAPN Dosage

Figure 4 shows urinary BAPN to undergo a rapid decrease in specimens collected at 4 and at 7 hours after the last dose of BAPN.
BAPN was not detected (ND) in any subsequent urine specimens. In
contrast, Figure 4 shows that urinary CAA, although variable, was
consistently present in all specimens collected during the final
two days of the study.

Participation of Serum in Conversion of BAPN to CAA

The possibility that serum might participate in conversion of BAPN
to CAA was studied in vitro. Experimental and control mixtures in
screw-cap culture tubes were incubated in a shaker-water bath at
37° C for 3 hours. The experimental tubes contained 0.2 ml of the
patient's serum, 0.2 ml of 0.2 M phosphate buffer at pH 7.4, and
0.1 ml H_2O. The tubes were gassed with O_2 for 30 seconds and then
incubated with 20 µl of ^{14}C-BAPN (carrier-free, 40,787 dpm; initial concentration of BAPN, 0.02 mM). A second sample of serum

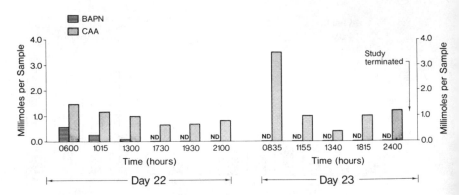

Figure 4.   BAPN and CAA in human urine after stopping dosage with
BAPN.   The histograms show the urinary output of BAPN and CAA in
millimoles per specimen following oral administration of the
final tablet of BAPN fumarate (250 mg of free base) at 6 A.M. on
day 22.   Urine collection at times indicated on days 22 and 23.

was incubated with 20 µl of ^{14}C-CAA (carrier-free; 7,442 dpm).
Control tubes contained water in place of serum.   Following incubation, 20 µl of unlabeledBAPN (as free base) and CAA were added
to experimental and control tubes to yield 0.01 M concentrations
of unlabeled carrier. This was followed by adding 0.5 ml of H_2O
and 0.2 ml of 60% H_2SO_4.   Water-saturated diethyl ether, 10 ml,
was added to extract the CAA.   After shaking and centrifugation,
9 ml of the ether supernatant was added to scintillation vials
and dried under N_2.   The residue was made to 1.0 ml with H_2O, 10
ml of Aquasol added, and the mixture was counted in a liquid scintillation counter.   Results are given in Table I.
    Table I shows that ^{14}C-CAA added to either a water control
or the reaction mixture containing serum is almost completely
recovered from an acidified reaction mixture by extraction with
diethyl ether.   In contrast, the samples containing serum and
^{14}C-BAPN showed virtually no ^{14}C-BAPN-derived radioactivity in
the ether extract.

Effect of BAPN on Amine Oxidase Activity of Human Serum

Screw-cap culture tubes containing 0.2 ml of serum and 0.2 ml of
0.2 M phosphate buffer at pH 7.4 received either 20 µl of BAPN,
to yield concentrations of 10^{-2} M, 2.5×10^{-3} M, and

TABLE I. RECOVERY OF ^{14}C-BAPN and ^{14}C-CAA FROM CONTROL AND SERUM[a] CONTAINING ASSAY MIXTURES

Assay Mixture	^{14}C-BAPN Added (dpm)	^{14}C-CAA Added (dpm)	dpm Recovered in Ether Layer[b]	Recovery (%)
Water (control)	40,787	-----	68	0.2
Water (control)	-----	7,442	7,334	98.0
Serum (experimental)	40,787	-----	47	0.1
Serum (experimental)	-----	7,442	6,400	86.0

a Serum (0.2 ml) or water (controls) were incubated with carrier-free ^{14}C-BAPN or ^{14}C-CAA at 37° C for 3 hours in mixtures containing 0.2 ml of 0.2 M phosphate buffer at pH 7.4

b Twenty microliters of unlabeled BAPN and CAA were added to 0.01 M concentration, followed by 0.5 ml H_2O and 0.2 ml of 60% H_2SO_4 prior to extraction with ether.

0.625×10^{-3} M, or 20 µl of H_2O as a control activity in the absence of BAPN. Reaction mixtures were incubated for 1 hour at 37° C. The samples were chilled in ice, 0.1 ml of 4×10^{-3} M benzylamine was added, and the tubes were gassed with O_2 for 30 seconds. The samples were then incubated in the water-shaker bath at 37° C for 3 hours. A nonenzymic control was identical, except that the substrate, benzylamine, was not added until the end of the incubation.

Following incubation, 0.1 ml of 4 N perchloric acid was added, followed by 1.5 ml of cyclohexane. The samples were shaken at room temperature for 5 minutes and centrifuged at 2,000 rpm.

The absorbency of the cyclohexane extract from the nonenzymic control and the various assay tubes was measured at 242 nm with a spectrophotometer set at 0 absorbency with spectrophotometric grade cyclohexane. The percent inhibitions corresponding to the various concentrations of BAPN added to serum are given in Table II.

Table II shows that BAPN can inhibit the amino oxidase activity of serum toward benzylamine as a substrate. A curve of percent inhibition against concentration permits the interpolation of a value for the concentration of BAPN that causes 50% inhibition. The value, designated I_{50}, was found to be 2×10^{-3} M.

DISCUSSION

The TLC procedure described for determination of BAPN offers greater sensitivity than that described by Wilk et al. (9) because it eliminates losses of BAPN due to extraction of the compound into butanol from biological material and back extraction into acid. The direct densitometric reading of the spots due to BAPN following visualization by ninhydrin offers a further saving of time as well as satisfactory quantitation when compared with a co-chromatographed standard. The choice of acetone, 1 M NH_4OH (9:1), as a solvent for development permitted excellent resolution between BAPN (R_f = 0.6) and amino acids. The latter showed little or no mobility because of their very low solubility in acetone (12) and relatively high polarity compared to BAPN.

Eighty-six percent of isotopically labeled CAA added to serum was recovered in the assay (Table I). The use of known quantities of CAA, added as an internal standard to control serum or urine undergoing the same protocol, is recommended to correct for the relatively small loss in this procedure.

This study showed that urinary excretion of unchanged BAPN began shortly after oral administration and decreased rapidly

TABLE II. EFFECT OF BAPN ON BENZYLAMINE OXIDASE ACTIVITY
OF HUMAN SERUM[a]

Sample	Optical Density	Inhibition (%)[b]
Control activity (no BAPN)	0.148	0
0.625×10^{-3} M BAPN	0.126	26.2
2.5×10^{-3} M BAPN	0.100	57.2
10^{-2} M BAPN	0.073	89.3
Nonenzymatic control[c]	0.064	----

[a] Serum (0.2 ml) was preincubated with the concentrations
of BAPN listed at 37° C for 1 hour prior to incubation
with 0.1 ml of 4×10^{-3} M benzylamine at 37° C for 3
hours.

[b] Calculated from the absorbence of the control and BAPN-
treated samples after correcting for the nonenzymatic
absorbence.

[c] The nonenzymic control contained all components, but
benzylamine was not added until the end of the incubation.

until a second dose of the drug was given (Figure 2). BAPN was
discontinued on day 22, and it was not detectable in the urine
specimens taken more than 7 hours after the last dose (Figure 4).
These observations imply that unchanged BAPN is rapidly removed
from the circulation, as was suggested by the rapid disappearance
of BAPN from human plasma noted by Keiser and Sjoerdsma (4). The
possibility that the plasma concentrations of BAPN undergo this
decrease due to degradation by a serum monoamine oxidase would
appear to be excluded by the finding that no ^{14}C-CAA was formed
from ^{14}C-BAPN in a 3-hour incubation with the patient's serum
(Table I); indeed, the addition of BAPN to the serum inhibited
the ability of the latter to deaminate the classical monoamine
oxidase substrate, benzylamine (Table II).

The formation of CAA represents a pathway for metabolic
inactivation because CAA is devoid of lathyrogenic properties (5).
It is also worth noting that CAA with a pK of 2.45 (13) would
exist as a polar anion at physiological pH and be susceptible to

glomerular filtration. The rapid excretion of directly injected
CAA in animal studies is consistent with this mechanism (14).

The relative lag in the appearance of urinary CAA compared
to that of BAPN (Figure 2) and the gradual increase of the metabolite in the urine during the first seven days of BAPN administration (Figures 2 and 3), as well as the prolonged excretion of
BAPN-derived CAA during the 48 hours after the cessation of BAPN
dosage (Figure 4), suggests that some of the BAPN enters the tissues, is metabolized to CAA, and is subsequently excreted as the
metabolite.

SUMMARY

Thin layer chromatography of β-aminopropionitrile (BAPN) in acetone and 1 M ammonium hydroxide (9:1) allowed separation of that
compound from amino acids present in physiological fluids without
prior solvent extraction. Direct densitometry of the spots
obtained with ninhydrin yielded satisfactory quantitation of the
β-aminopropionitrile present.

Human serum did not convert BAPN to the deaminated, nonlathyrogenic metabolite, cyanoacetic acid (CAA). Instead, its enzymic
activity for oxidizing benzylamine was inhibited by BAPN
(I_{50}= 2 × 10^{-3} M).

Two hundred fifty milligrams of BAPN administered orally at
6-hour intervals each day for 21 days resulted in the appearance
of BAPN in the urine within 1 hour after the initial dose and an
average daily recovery of 16% of the total dose. Urinary CAA
appeared more slowly than BAPN and increased to approximately
three times that of urinary BAPN. After BAPN was discontinued,
prolonged excretion of BAPN-derived CAA occurred.

ACKNOWLEDGMENTS

The kindness of Academic Press, Inc., 111 Fifth Avenue, New York,
N. Y., for permission to reproduce Figure 1 and relevant text
from Analytical Biochemistry 74, 254 (1976), is grateful acknowl-
edged.

We are also grateful to C. V. Mosby Co. of 11830 Westline
Industrial Drive, St. Louis, Mo., for permission to reproduce
Figures 2-4 and relevant text from Clinical Pharmacology and Ther-
apeutics 23, 520 (1978).

This research was supported in part by National Institutes of
Health Grant AM 18706.

REFERENCES

1. R. C. Siegel, S. R. Pinnell, and G. R. Martin, Biochemistry 9, 4486 (1970).
2. M. L. Tanzer and J. Gross, J. Exper. Med. 119, 275 (1964).
3. E. E. Peacock, Jr., and J. W. Madden, Surgery 66, 215 (1969).
4. H. R. Keiser and A. Sjoerdsma, Clin. Pharmacol. Ther. 8, 593 (1967).
5. J. J. Lalich, Science 128, 206 (1958).
6. C. M. McEwen, Jr., and G. D. Cohen, J. Lab. Clin. Med. 62, 766 (1963).
7. C. M. McEwen, Jr., K. T. Cullen, and A. J. Sober, J. Biol. Chem. 241, 4544 (1966).
8. E. G. Young, in Comprehensive Biochemistry, edited by M. Florkin and E. H. Stolz (Elsevier, New York, 1963), vol. 7, p. 35.
9. A. L. Wilk, C. T. King, E. A. Horigan, and A. J. Steffek, Teratology 5, 41 (1972).
10. J. H. Fleisher, K. Brendel, M. Chvapil, and E. E. Peacock, Jr., Anal. Biochem. 74, 254 (1976).
11. H. W. Sievert, S. H. Lipton, and F. M. Strong, Arch. Biochem. Biophys. 86, 311 (1960).
12. E. G. Cohen and J. Edsall, Proteins, Amino Acids and Peptides as Ions and Dipolar Ions (Reinhold, New York, 1943), pp. 201 and 203.
13. R. C. Weast, Handbook of Chemistry and Physics (CRC Press, Cleveland, Ohio, 1976), Ed. 56, Sect. D, p. 150.
14. J. H. Fleisher, A. J. Arem, M. Chvapil, and E. E. Peacock, Jr., Proc. Soc. Exp. Biol. Med. 152, 469 (1976).

CHAPTER 27

Gangliosides from Prenatal and Postnatal Rat Brain: Early Developmental Changes Exceed Regional Variation

Carol C. Irwin and L. N. Irwin

The gangliosides are a group of glycosphingolipids that occur in greater abundance and complexity in the vertebrate nervous system than in any other tissue. The complexity of gangliosides is due primarily to variation in the number of carbohydrate moieties attached to the sphingolipid backbone, especially the number and sites of attachment of sialic acid residues. These are mainly responsible for the variable polarity of individual gangliosides, which forms the basis for their separation by thin layer chromatography. For some purposes, other methods have been used in separating and analyzing the internal constituents of gangliosides, such as analysis by gas-liquid chromatography of $\underline{N}$-acetyl and $\underline{N}$-glycolyl variants of neuraminic acid (1). In addition, recent reports have indicated that monosialogangliosides may be effectively separated and analyzed by high-performance liquid chromatography (2). However, thin layer chromatography remains virtually the only method that effectively allows good separation of multisialogangliosides in the complex mixtures found in most nervous tissue. Recently, high-performance TLC plates have made possible even better separations, particularly of minor gangliosides.

The role that gangliosides play in neural function is unknown, although several possibilities have been postulated. They may have a structural role, as building blocks for the plasma membrane, or an excitatory role, as in cation binding or as contributors to the fixed charge of the cell surface; they may serve as receptors for neurotransmitters or hormones or as informational transducers (4).

Developmentally, the normal accretion and turnover of gangliosides is clearly critical, as shown by the severe neurological consequences of impaired ganglioside metabolism (e.g.,

G_{M1}-gangliosidosis, Tay-Sachs disease). Much remains to be
learned about the normal development of ganglioside metabolism,
including the time course of ganglioside maturation in relation
to specific neuroanatomical features, cell types, stages of differentiation, etc. The developmental appearance of gangliosides
in mammalian brain has been studied almost exclusively at postnatal stages, with minimal regard given to possible regional variations. Presented in this chapter are data on ganglioside patterns from several rat brain regions, ranging from a late prenatal
age to adult, as well as early fetal patterns from whole brain.
Our major findings are, first, that a simplified ganglioside pattern predominates at early fetal stages but changes dramatically
by the time of birth to a nearly "adult" pattern; and, secondly,
that maturation to the adult ganglioside pattern correlates with
the sequence of neurological maturation of specific brain regions.

MATERIALS AND METHODS

Brains of Sprague Dawley rats (both sexes) from 1 to 60 days of
age, as well as 2 days prenatal (gestational age 19), were dissected into medulla, corpora quadrigemina, diencephalon, basal
ganglia, parietal cortex, and frontal cortex. In addition, whole
brains were obtained from fetal rats at 14 to 20 days gestation.
Brain tissue was quickly removed, frozen on dry ice, and lyophilized to a constant dry weight. Gangliosides were extracted by
the method of Irwin and Irwin (5), a procedure designed for small
amounts of tissue (0.5-15 mg dry weight). Tissue samples were
heated at 60° C for 5 minutes with 1 ml chloroform-methanol (2:1,
v/v), and the total chloroform-methanol extract was passed
through a silicic acid column (Unisil, 100-200 mesh, Clarkson
Chemical Co., Williamsport, Pa.). The bulk of nonganglioside
lipid was eluted with either chloroform-methanol (2:1) (Method A)
or chloroform-methanol-water (65:25:4) (Method B). The ganglioside fraction was eluted with chloroform-methanol-water (10:10:3).
    Gangliosides were quantified by the thiobarbituric acid (TBA)
assay for sialic acid (6) following hydrolysis for 2 hours in 0.1
$\underline{N}$ sulfuric acid at 80° C. In some cases, the TBA assay was
scaled down (all reagents added to the ganglioside hydrolysate
were 1/5 the usual volume), which gave a linear response in the
range of 1.5-4.6 µg sialic acid. For thin layer chromatography,
ganglioside aliquots were taken to dryness, resuspended in 40 µl
chloroform-methanol (1:1), and spotted on thin layer plates
coated with a 250-micron layer of silica gel G (20 × 20 cm Merck
plates, Brinkman, Wesbury, N. Y.). The plates were developed
with chloroform-methanol-0.25% $CaCl_2$ (65:35:8) and in most cases

visualized with resorcinol reagent (7). For comparison of various chemical methods of quantitation, chromatograms were visualized with iodine and individual spots were scraped into test
tubes for assay by the modified TBA method. Alternatively, the
plate was sprayed lightly with modified resorcinol reagent (9)
and the spots were scraped into tubes and assayed by a resorcinol
method (10). Ganglioside patterns were also quantified by scanning with a Kontes Chromaflex K-495000 densitometer.

RESULTS AND DISCUSSION

In order to determine the reproducibility of TLC patterns, gangliosides from 1 day and adult rat brain regions were isolated
separately in triplicate, separated by TLC, and quantified by
densitometry (Table I). Variance is very low (coefficient of
variation < 2.5% for the four major gangliosides), indicating
that this densitometric procedure possesses a high degree of
reproducibility. These results illustrate the applicability of
relatively inexpensive instrumentation (the Kontes Chromaflex
K-495000 scanner) to screening large numbers of ganglioside samples. A high level of precision has already been shown using
densitometric quantitation of highly purified gangliosides (3).
Here, the rapid isolation method yields gangliosides of sufficient purity for the rapid analysis and quantitation of a large
number (20 or more) of small samples (in the range 0.5-15 mg
dry weight of tissue) in one or two days. In comparing densitometric and chemical methods of quantitation on identical sets of
triplicate samples (data not shown), the precision of the time-
consuming TBA method was found to be only slightly better than
that of densitometry. The quantitation of individual bands by
resorcinol assay gave a higher coefficient of variation (12%) and
the sensitivity of this assay was only 1/3 that of the modified
TBA procedure. The rapidity, simplicity, and reproducibility of
densitometry thus make it the best method for processing large
numbers of ganglioside samples.
    Qualitative and quantitative studies were undertaken in a
developmental and regional analysis of gangliosides from rat
brain. Initially, tissues were obtained from rats at -2 (19 days
gestational), 1, 6, 13 days, and adult (210 days). The patterns
are quite similar from late fetal age to adult. Certain general
trends can be seen, as in the uniform increase in the uniform
increase in ganglioside MI with increasing age in all brain
regions. This reflects the increase in myelination that takes
place during neural development, MI being a predominant ganglioside present in myelin (11). Another major ganglioside, D1a,

TABLE I. DISTRIBUTION OF GANGLIOSIDES IN FOREBRAIN, DIENCEPHALON, AND MEDULLA OF 1 DAY AND ADULT RATS[a]

Major Ganglioside Component	Forebrain		Diencephalon		Medulla	
	1 Day	Adult	1 Day	Adult	1 Day	Adult
M1	14.0 ± 1.0	24.2 ± 1.3	27.6 ± 0.6	35.8 ± 1.7	24.1 ± 2.3	23.0 ± 0.4
D3	15.0 ± 1.4	7.0 ± 0.6	15.2 ± 0.7	9.2 ± 0.5	15.0 ± 1.6	12.5 ± 0.9
D1a	30.4 ± 1.8	34.0 ± 0.6	30.6 ± 0.6	24.6 ± 1.5	31.5 ± 3.3	17.1 ± 0.7
D1b	14.8 ± 0.6	13.6 ± 0.4	12.8 ± 1.3	10.9 ± 1.3	13.0 ± 0.8	16.8 ± 0.8
ALG	3.4 ± 0.5	4.2 ± 0.4		2.8 ± 0.2	1.8 ± 0.2	4.3 ± 0.6
T1b	14.4 ± 1.3	11.2 ± 1.0	11.0 ± 0.8	13.4 ± 1.5	10.2 ± 0.8	18.5 ± 1.1
Q	8.1 ± 0.6	5.9 ± 0.7	2.6 ± 0.3	3.2 ± 0.4	5.2 ± 0.3	7.8 ± 0.8

[a] Gangliosides were extracted by Method B, assayed by the modified TBA procedure, and separated by TLC as described under "Materials and Methods." Lane widths were 5-8 mm and gangliosides were spotted on the basis of 1 µg ganglioside sialic acid/mm. The ganglioside pattern was quantified by scanning densitometry of resorcinol-positive bands, and expressed as percent (mean ± s.e.m. for four replications) ganglioside sialic acid in each band of the total ganglioside sialic acid in the sample.

increases in all brain regions at early ages. Among the minor
gangliosides at these ages is a band that migrates between MI and
D1a and decreases with development in all regions studied. The
ganglioside that has been reported to migrate at this position in
the mobile phase used is disialolactosyl ceramide, D3 (12-14).
Another minor ganglioside, which migrates between GD1b and GT1b,
appears in greater abundance in the adult rat brain. This band
corresponds to an alkalilabile ganglioside reported previously to
occur in mouse brain (15).

Analysis of multiple regions from a single rat brain has
revealed that, despite a general pattern of ganglioside maturation common to all regions, the time course of maturation is
region-specific. For example, ganglioside D1a reaches an apparent maximum at different ages in the medulla (day 1), diencephalon (day 6), and frontal cortex (day 13), remaining relatively
constant thereafter or decreasing at a later age (Figure 1, Table
I). Similar differences in the timing of increase in M1 and
decrease in D3 were also noted in several regions. These changes
generally correlate with the developmental sequence of neural differentiation in each region. The medulla differentiates earlier
than the frontal cortex, hence the "mature" ganglioside pattern
appears earlier in the medulla than in the frontal cortex. Later
changes in ganglioside pattern presumably reflect the unique mor-
phological properties of each fully differentiated region. Note
the relatively higher content of M1 and lower level of D1a in the
heavily myelinated medulla than in the neurophil-rich frontal cortex of the 210-day-old rat (Table I).

Having observed significant changes in the ganglioside pattern during the perinatal period, a determination of the earliest
definitive ganglioside pattern of prenatal brain tissue was sought.
Accordingly, gangliosides were isolated from whole brain of fetal
rats. Table II indicates the quantitative distribution of gangliosides from brain at 15 and 20 days of gestation. As observed
earlier (Figure 1), by 19 days of gestation (approximately 2 days
prenatal), the four major gangliosides of adult rat brain (M1,
D1a, D1b, T1b) are present in proportions that correspond rather
closely with the "adult" composition. At 15 days of gestation,
however, the pattern is much simpler, consisting of a double band
migrating between M1 and D1a in the mobile phase. This double
band has also been reported in the D3 position in smaller amounts
in older animals and presumably reflects a difference in the fatty
acid portion of the molecule (3). Significant amounts of M3 are
also seen at the early fetal stages. The double bands in this
region are likewise thought to reflect differences in the fatty
acid composition of the ceramide moiety (Denise Chou, personal

TABLE II.   DISTRIBUTION OF GANGLIOSIDE SIALIC ACID
IN FETAL RAT BRAIN[a]

Major Ganglioside Component	Age (Days Gestation)	
	15	20
M3	21.7 + 3.0	4.4 + 1.0
M2	6.0 + 3.6	2.5 + 1.5
M1	9.8 + 3.3	14.0 + 0.4
D3	55.2 + 7.3	11.2 + 1.5
D1a	0.5 + 0.4	32.5 + 3.3
D1b	1.8 + 0.8	11.4 + 1.0
T1b	4.6 + 2.1	21.5 + 3.5
Q	<1	1.4 + 0.6

[a] Gangliosides were extracted, separated, visualized, and quantified as described in "Materials and Methods" and in Table I. Ganglioside composition is expressed as percent composition (mean + s.e.m. for 3-5 determinations).

communication). Gangliosides M3 and D3 together account for 77% of the total resorcinol-positive stain at 15 days of gestation, but only 19% at 20 days of gestation (Table II). Comparable combined values for the four major gangliosides of postnatal brain-- M1, D1a, D1b, and T1b--are 18% at 15 days and 79% at 20 days of gestation. Thus, ganglioside metabolism shifts from production of the smaller molecules M3 and D3 to synthesis of a full ganglioside pattern as neural differentiation proceeds.

TLC analysis has shown the alterations in ganglioside metabolism that occur developmentally, in the form of a dramatic change in pattern between 15 and 20 days of gestation, and more subtle changes later in development (postnatally) in several brain regions. The significance of the major change in fetal pattern is unknown. The shift from a pattern characterized by gangliosides M3 and D3 to the "mature" pattern dominated by M1, D1a, D1b, and T1b corresponds roughly with the end of cellular proliferation (division) and proceeds most cellular differentiation and synaptogenesis. A rapid method for isolating gangliosides from small amounts of tissues, combined with quantitation by densitometry, should greatly simplify the developmental, regional, subregional, and subcellular analysis of ganglioside patterns employed in assessing the role of gangliosides in neural differentiation.

REFERENCES

1. R. K. Yu and R. W. Ledeen, J. Lipid Res. 11, 506 (1970).
2. E. Bremer, S. Gross, and R. H. McCluer, Fed. Proc. 37, 1645 (1978).
3. S. Ando, N. Chang, and R. K. Yu, Anal. Biochem. 89, 437 (1978).
4. L. N. Irwin, in Reviews of Neuroscience, edited by S. Ehrenpreis and I. J. Kopin (Raven Press, New York, 1974), pp. 137-179.
5. C. C. Irwin and L. N. Irwin, Anal. Biochem., in press.
6. D. Aminoff, Biochem. J. 81, 384 (1961).
7. L. Svennerholm, Biochim. Biophys. Acta 24, 604 (1957).
8. L. Svennerholm, J. Neurochem. 10, 613 (1963).
9. A. J. Yates and D. Thompson, J. Lipid Res. 18, 660 (1977).
10. V. H. MacMillan and J. R. Wherrett, J. Neurochem. 16, 1621 (1969).
11. K. Suzuki, S. E. Poduslo, and W. T. Norton, Biochim. Biophys. Acta 144, 375 (1967).
12. H. Dreyfuss, P. F. Urban, S. Edel-Harth, and P. Mandel, J. Neurochem. 25, 245 (1975).
13. P. F. Urban, S. Edel-Harth, and H. L. Dreyfus, Exp. Eye Res. 20, 397 (1975).
14. T. N. Seyfried, E. J. Weber, and R. K. Yu, Lipids 12, 979 (1977).
15. N. Baumann, S. Sonnino, R. Ghidoni, M. L. Harpin, and G. Tettamanti, C.R. Acad. Sci. Paris D. 286, 1807 (1978).

ACKNOWLEDGMENT

This work was supported by NSF Grant BNS 77-20575.

CHAPTER 28

Composition of Brain Gangliosides Determined by Quantitative HPTLC

Robert K. Yu

Gangliosides are a family of sialic acid-containing glycosphingo-lipids that consist of a hydrophilic oligosaccharide portion gly-cosidically linked to a hydrophobic ceramide moiety. They are generally found in highest concentrations in nervous tissues and in reduced amounts in many extraneural tissues and body fluids (for review, see Ledeen and Yu [1-3]). Since their discovery some 30 years ago by Klenk (4,5) nearly 40 different ganglioside species have been isolated and characterized (3). In the brains of most higher vertebrates, the usual major gangliosides are G_{M1} G_{D1a}, G_{D1b}, and G_{T1b} (nomenclature of Svennerholm [6]). Excep-tions are the primate and avian brains, which contain an additional ganglioside, sialosylgalactosylceramide (G_{M4}) as a major compon-ent (1,7-9). On the other hand, the fish brain generally contains a preponderance of tetra- and penta-sialogangliosides (10-12). In addition to the major gangliosides, most brain tissues also con-tain a wide variety of minor gangliosides. At least 18 different ganglioside species have been isolated from brain tissues, and the number is likely to grow with the advent of improved methodologies in ganglioside resolution.

Because of its simplicity and excellent resolving power, silica gel thin layer chromatography (TLC) has been extensively employed as a universal tool for separating complex ganglioside mixtures into individual components. The individual gangliosides can then be directly visualized by spraying the plate with resor-cinol-HCl reagent of Svennerholm (13). This reagent specifically reacts with sialic acid-containing molecules. Quantification of individual gangliosides separated by TLC has traditionally been achieved by the widely used method of Suzuki (14). The method involves scraping the silica gel from each ganglioside zone (visu-alized by nondestructive means) from the plate, followed by the colorimetric determination of the ganglioside sialic acid in each fraction. However, the method generally requires relatively large amounts of gangliosides (at least 40 µg of ganglioside sialic acid) for a single determination, and its use for small-sized samples is therefore limited. In addition, the detection and quantification of minor gangliosides are often quite diffi-cult (MacMillan and Wherrett [15]). An alternative to the Suzuki method employs densitometry. Because it is simple, fast, and sensitive, it has already been applied to the analysis of ganglio-sides in several laboratories with some success (16-19).

Recently an improved TLC technique, termed high-performance thin layer chromatography (HPTLC), was developed (see review by Zlatkis and Kaiser [20]). The method employs plates coated with a thin, uniform layer of fine adsorbent (silica gel) that exhibits enhanced resolution for a wide variety of substances. In addition, the uniformity of the coating material allows the plates best suited for photometric densitometry. This chapter describes the application of HPTLC and densitometry for the analysis of several representative brain ganglioside mixtures from higher vertebrates. Evidence is presented that sialosylgalactosyl ceramide (G_{M4}) is specifically localized in the oligodendroglial cell and its plasma membrane (myelin) in the brain.

MATERIALS AND METHODS

Brain Specimens

Human brains were obtained at autopsy from adult patients who died without any known neurological diseases. The gray and white mat-ter of the human brain and the white matter of adult chimpanzee, monkey, pig, bovine, sheep, and chicken brains were analyzed. Human myelin samples were prepared from the white matter by the method of Norton and Poduslo (21). Human oligodendroglia and neu-rons were isolated from the white matter and gray matter, respec-tively, by Dr. K. Iqbal using the methods of Iqbal et al. (22) and Iqbal and Tellez-Nagel (23).

Ganglioside Isolation and Purification

Gangliosides were isolated from the above specimens using the method of Lebeen and Yu (1). Quantitative and qualitative analy-ses of ganglioside sialic acid were carried out using the gas-liquid chromatographic method of Yu and Ledeen (24). With the exception of bovine, sheep, and pig brain white matter ganglio-sides that contained about 1% of N-glycolylneuraminic acid, the major sialic acid present was N-acetylneuraminic acid.

Although the gangliosides prepared by the above procedure
were sufficiently pure for TLC analysis, small amounts of interfering contaminants were frequently encountered in preparations
where the ganglioside concentrations were extremely low. These
ganglioside preparations were further purified by applying to a
small DEAE-Sephadex column, followed by mild base treatment and
desalting with a Sephadex G-50 column. This entire procedure,
described in detail elsewhere (9,25), was also useful in destroy-
ing possible artifacts arising from esterification of the ganglioside sialic acid after prolonged storage in solvents containing
chloroform and methanol (Yu, unpublished observation). Authentic
individual gangliosides, prepared from adult human brain as
described previously (26), were used for TLC reference standards
and for the estimation of relative detector response. The ganglioside samples were dissolved in appropriate volumes of chloroform-methanol (1:1) for spotting.

Thin Layer Chromatography

Ganglioside patterns were examined on HPTLC (10 × 20 cm) plates
precoated with a 0.2-mm layer of silica gel 60 (E. Merck, Darmstadt, Germany, or EM Laboratories, Inc., Elmsford, N. Y.). The
plates were activated by heating at 100° C for 10-20 minutes and
were then cooled to room temperature under desiccation before use.
For routine analysis, an aliquot of ganglioside solution containing 5-15 µg of lipid-bound sialic acid was spotted as a 7-mm
streak with a 10-µl Hamilton syringe. Two different developing
solvent systems were found to be most effective: (A) chloroform-
methanol-water (55:45:10, v/v/v) containing 0.02% (w/v) $CaCl_2$ ·
$2H_2O$, and (B) chloroform-methanol-2.5 N ammonium hydroxide
(65:35:8, v/v/v). After the plate was developed in a well-satu-
rated tank lined inside with Whatman No. 1 paper (usually 3-3.5
hours), the plate was dried over a stream of warm air. The gangliosides were made visible by spraying the plate with resorcinol-
HCl reagent as described by Svennerholm (13). The plate was then
placed, face down, on a clear glass plate preheated at 95 ± 2° C
on an aluminum heater block. After 5 minutes, the covered plate
was reversed and the heating was continued for 30 minutes at the
same temperature. Under the conditions described, all gangliosides appeared as deep purple bands. The plate can be stored in
a refrigerator for several months without appreciable decolorization of the bands. Storage can be prolonged if excess HCl vapor
is removed. This is accomplished by simply lifting the cover
plate briefly at the end of heating.

Densitometry

The chromatograms were scanned with a Transdyne RFT scanning densitometer (Transidyne General Co., Ann Arbor, Mich.). The scanning mechanism was modified to provide a wide range of scanning
speeds by attaching a Motomatic motor generator (ElectroCraft Co.,
Hopkins, Mich.). For routine measurements, the scanning speed
was set at 0.65 cm/min. The absorbance was measured at 580 nm,
the absorption maximum of the sialic acid chromophore, in a
transmittance mode with a slit of 0.3 × 2 mm. Peak areas were
measured with a Hewlett-Packard digital integrator (Model 3380A,
Hewlett-Packard, Avondale, Pa.).

RESULTS AND DISCUSSION

Thin Layer Chromatography

Because the ganglioside patterns of most brain tissues are rather
complex, it is important to achieve high resolution of individual
ganglioside components with minimal overlapping on a thin layer
plate. A number of homemade and commercially available precoated
plates were tried, with HPTLC plates consistently offering the
best resolution and superior reproducibility. Although, in most
cases, the best resolution was achieved with the full-length (20
cm) plates, excellent resolution could still be achieved with the
half-length plates (10 × 10 cm). The advantage in the latter
case is that the developing time could be reduced to about 1 hour
or less.
     Successful separation of a complex ganglioside mixture also
depends upon the mobile phase. A variety of different mobile
phases have been used in many laboratories (3,19,27-29). These
include chloroform-methanol-water (60:40:9, v/v), chloroform-
methanol-2.5 N ammonium hydroxide (60:40:9, v/v), n-propanol-
water (7:3, v/v), n-butanol-pyridine-water (7:3, v/v), and n-
propanol-water-concentrated ammonium hydroxide (6:2:1, v/v).
Van den Eijnden (28) and Zanetta et al. (19) have reported that
addition of small amounts of inorganic salts can enhance resolution. Although any of these solvent systems can be useful in
separating gangliosides, the neutral mobile phase A and the basic
mobile phase B consistently offered the most satisfactory separation. This was particularly true in the case of solvent A (Figure 1A). For this reason, solvent A was chosen as a general
mobile phase for the densitometry determination of ganglioside
compositions on HPTLC plates. This system could be made more
polar by adjusting the solvent ratios to chloroform-methanol-

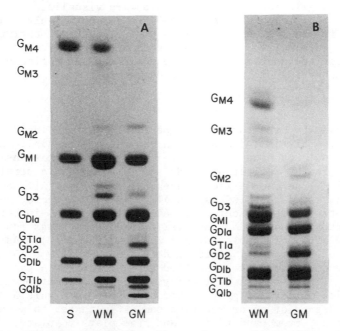

Figure 1. Thin layer chromatograms of gangliosides from human white matter (WM), gray matter (GM), and standards (S). The standards consisted of 0.5 µg (as sialic acid) each of G_{M4}, G_{M1}, G_{D1a}, and G_{D1b}, and 0.25 µg (as sialic acid) of G_{T1b}. The plate on the left was developed with solvent system A, and the plate on the right with solvent system B.

water (50:45:10, or 45:45:10) containing 0.02% $CaCl_2 \cdot 2H_2O$. This modification was particularly effective in separating fish brain gangliosides, which contained large amounts of highly polar tetra- and penta-sialogangliosides (12).

Because gangliosides are acidic in nature, their TLC mobilities are expected to be strongly affected by the pH of the mobile phase. Thus G_{D3} was located below G_{M1} using the neutral solvent A, but migrated ahead of G_{M1} with the ammonia-containing solvent B (Figure 1B). The focuse-containing G_{D1b} (Fucose-G_{D1b}) was located between G_{D1b} and T_{T1b} with solvent A, but it migrated between G_{T1b} and G_{Q1b} with solvent B (30). Hence, the use of both solvents A and B is necessary for the correct assignment of various ganglioside bands.

The ganglioside bands on a plate were visualized by spraying with the resorcinol-HCl reagent (13), a reagent highly sensitive and specific to sialic acid. This was followed by heating at 95° C for 30 minutes. This heating condition was found to be most effective in producing purple bands without noticeable non-specific charring of the ganglioside molecule.

Densitometry

The scanning by densitometry was performed with an incident light of 580 nm. This is the absorption maximum of the sialic acid chromophore. Scanning proved more sensitive in the transmittance mode than in the reflectance mode and it also gave a straight calibration line over a wide concentration range. All the gangliosides tested revealed essentially the same detector response based on their sialic contents. A linear relationship between the detector response and the sialic acid content was found in the range of 0.5-10 nmole (0.15 - 3.0 µg) (Figure 2). As little as 30 ng of sialic acid could be detected. Usually 5-15 µg of lipid-bound sialic acid is spotted for a total brain ganglioside mixture. Typical scans of total ganglioside chromatograms from human white matter and gray matter are shown in Figure 3.

The degree of color reaction could fluctuate, to a very small extent, from band to band and from plate to plate. This was attributed to the unevenness in spraying. This problem was

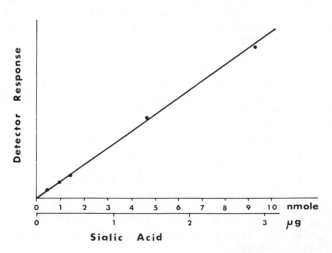

Figure 2.   Detector response versus ganglioside sialic acid content.

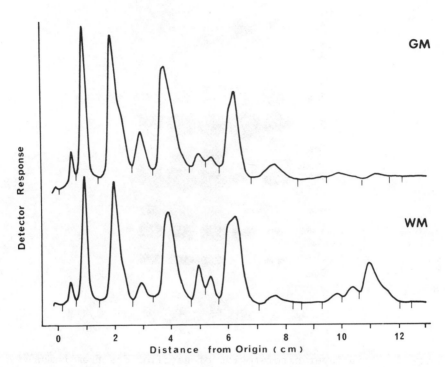

Figure 3. Densitograms of gangliosides from human white matter (WM) and gray matter (GM).

corrected by applying known amounts of standards close to the test samples on the same plate. The standard samples were then used to calibrate the color intensities of the corresponding individual gangliosides (Figure 1A).

Ando et al. compared densitometry to the colorimetric method of Suzuki (25). Excellent agreement was obtained between the two methods. However, the variations obtained with Suzuki's method were generally higher than those obtained by densitometry. This could result from cross contamination between bands caused by scrapping or by uncertainty in the detection of faint bands.

Ganglioside Distribution

The "scan" of the chromatogram of gangliosides of human brain gray matter, white matter, myelin, oligodendroglia, and neurons

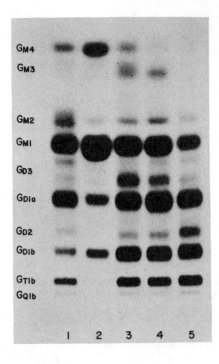

Figure 4. Thin layer chromatogram of gangliosides from human
white matter (1), myelin (2), oligodendroglia (3), neurons (4),
and gray matter (5). The developing solvent system was A.

is shown in Figure 4; and that of the gangliosides of the brain
white matter of chimpanzee, monkey, chicken, bovine, sheep, and
pig is shown in Figure 5. The ganglioside compositions, determined by densitometry and expressed as the percent distribution
of ganglioside sialic acid, are listed in Table I. It is appar-
ent that all cerebral tissues contain G_{M1}, G_{D1a}, G_{D1b}, and G_{T1b}
as the predominant ganglioside species. About 70% of the total
ganglioside sialic acid could be ascribed to these four species.
However, the white matter of primate and avian brains contained
an additional ganglioside, G_{M4}, as the second or third most
abundant ganglioside on a molar basis (24). The extremely low
concentration of G_{M4} in the brains of other animal species indicates that its occurrence in animal brains is highly species specific. It is interesting to note that G_{M4} always appears as
double bands because both normal and α-hydroxy fatty acids are

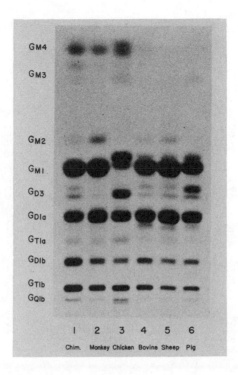

Figure 5.   Thin layer chromatogram of white matter gangliosides.
The developing solvent system was A.

present (1). G_{M1} ganglioside of chicken, bovine, sheep, and pig
samples also appears as doublets. The basis for their separation
is, however, due largely to difference in fatty acid chain lengths
(S. Ando and R. K. Yu, unpublished results).

   In addition to the major gangliosides, most of the known
minor brain gangliosides as well as several gangliosides of
unknown structures could be clearly separated by HPTLC and are
easily quantified with the aid of densitometry. Among the minor
gangliosides, G_{M3} and G_{D3} again always appear as doublets on TLC,
due most likely to differences in their lipophilic constituents.

   In order to gain some insight as to the cellular and subcellular origin of the structurally unique ganglioside, G_{M4}, a comparative study on the ganglioside compositions of human myelin,
oligodendroglia, and neurons was performed (9). It has previously

TABLE I. PERCENT DISTRIBUTION OF GANGLIOSIDE SIALIC ACID

	Human WM	Chimp WM	Monkey WM	Chick WM	Bovine WM	Sheep WM	Pig WM	Human GM	Human Myelin	Human Oligos.	Human Neurons
Sialic Acid Content	275^a	374^a	300^a	620^a	326^a	342^a	406^a	875^a	2000^b	350^b	1300^b
G_{M4}	8.6	8.8	9.0	10.8	2.7	0.6	0.7	1.5	18.4	5.9	0.9
G_{M3}	4.8	4.6	2.6	6.9	2.6	1.1	3.4	2.7	1.6	8.1	4.1
G_{M2}	2.5	2.5	6.0	2.7	2.3	4.0	1.0	4.1	6.2	5.7	4.5
G_{M1}	21.6	24.7	30.0	18.9	31.0	34.1	32.9	14.9	47.8	20.1	22.6
G_{D3}	8.8	6.6	2.8	13.1	6.0	8.5	14.8	5.4	2.8	11.9	7.2
G_{D1a}	16.6	15.4	21.4	16.4	24.1	25.1	22.7	21.7	8.2	16.4	21.5
G_{D1a}-GalNAc	1.1	1.7	1.9	---	---	---	---	0.4	---	---	---
G_{D1a}-NeuGc	---	---	---	---	5.1	4.5	3.5	---	---	---	---
G_{T1a}	2.2	2.5	2.5	5.1	2.6	3.9	2.1	1.8	---	2.7	1.3
G_{D2}	3.1	2.6	2.0	1.3	1.4	1.1	---	8.0	1.4	3.3	3.9
G_{D1b}	16.9	14.8	9.6	9.1	10.5	8.5	8.2	18.2	11.2	15.0	20.7

$X_2,0.8^c$ $X_4,1.6$

G_{T1b}	11.1	12.0	10.0	9.8	9.1	6.7	6.7	16.3	2.2	9.3	11.3
G_{Q1b}	2.7	3.8	2.2	4.9	1.2	1.9	2.4	5.0	0.2	1.9	2.2

$X_1, 1.0$ $X_3, 0.6$

[a] μg lipid-bound sialic acid per g fresh tissue.

[b] μg sialic acid per g protein.

[c] X_1, X_2, X_3, X_4, unidentified gangliosides.

423

been shown that the pattern of human myelin is rather simple in
that G_{M4} and G_{M1} are its predominant ganglioside species (7,30).
Since myelin is generally considered to be an extension of the
oligodendroglial plasma membrane, G_{M4} should, therefore, be an
intrinsic component of the oligodendroglial cell body. This is
clearly demonstrated in Figure 4 and Table I. It is present only
in extremely low concentrations in human gray matter and neurons
(Figure 4 and Table I), and it is totally absent in demyelinated
plaques of multiple sclerosis brains (24). The results thus indicate that G_{M4} cannot be a constituent of neurons, astrocytes, or
neuronal processes and it can serve as a unique marker for human
myelin and oligodendroglia. G_{M4} and G_{M1} are heavily concentrated
in human myelin, which constitutes the bulk of white matter.
This accounts for the relatively high concentrations of these two
gangliosides in the latter source.

ACKNOWLEDGMENTS

This work was performed with the collaboration of Dr. S. Ando,
Dr. K. Iqbal, and Dr. N. C. Chang. Financial assistance was provided by a USPHS grant (NS 11853) and a grant from the Kroc Foundation.

REFERENCES

1. R. W. Ledeen and R. K. Yu, in Lysosomes and Storage Diseases,
   edited by H. G. Hers and F. VanHoff (Academic Press, New York,
   1973), pp. 105-214.
2. R. W. Ledeen and R. K. Yu, in Glycolipid Methodology, edited
   by L. L. Witting (American Oil Chemists' Society, Champaign,
   Ill., 1976), pp. 187-214.
3. R. W. Ledeen and R. K. Yu, in Research Methods in Neurochemistry, edited by N. Marks and R. Rodnight (Plenum Press, New
   York, 1978), vol. iv, chap. 12, pp. 371-409.
4. E. Klenk, Hoppe-Seyler's Z. Physiol. Chem. 235, 24 (1935).
5. E. Klenk, Hoppe-Seyler's Z. Physiol. Chem. 262, 128 (1935).
6. L. Svennerholm, J. Neurochem. 10, 613 (1963).
7. F. B. Cochran, R. K. Yu, and R. W. Ledeen, Proc. Int. Soc.
   Neurochem. 6, 540 (1977).
8. K. Ueno, S. Ando, and R. K. Yu, J. Lipid Res. 19, 863 (1978).
9. R. K. Yu and K. Iqbal, J. Neurochem., in press.
10. N. F. Avrova, J. Neurochem. 18, 667 (1971).
11. I. Ishizuka, M. Kloppenburg, and H. Wiegandt, Biochim. Bio-
    phys. Acta 210, 299 (1970).

12. R. K. Yu and S. Ando, Trans. Amer. Soc. Neurochem. 9, 135 (1978).
13. L. Svennerholm, Biochim. Biophys. Acta 24, 604 (1957).
14. K. Suzuki, Life Sci. 3, 1227 (1964).
15. V. H. MacMillan and J. R. Wherrett, J. Neurochem. 16, 1621 (1969).
16. K. Sandhoff, K. Harzer, and H. Jatzkewitz, Hoppe-Seyler's Z. Physiol. Chem. 349, 283 (1968).
17. R. O. Brady, C. Borek, and R. M. Bradley, J. Biol. Chem. 244, 6552 (1969).
18. F. Smid and J. Reinisova, J. Chromatog. 86, 200 (1973).
19. J.-P. Zanetta, F. Vifiello, and J. Robert, J. Chromatog. 137, 481 (1977).
20. A. Zlatkis and R. E. Kaiser, in HPTLC High Performance Thin-Layer Chromatography (Elsevier, Amsterdam, 1977).
21. W. T. Norton and S. E. Poduslo, J. Neurochem. 21, 749 (1973).
22. K. Iqbal and I. Tellez-Nagel, Brain Res. 45, 296 (1972).
23. K. Iqbal, I. Grundke-Iqbal, and H. M. Wisniewski, J. Neurochem. 28, 707 (1977).
24. R. K. Yu, R. W. Ledeen, and L. F. Eng, J. Neurochem. 23, 169 (1974).
25. S. Ando, N.-C. Chang, and R. K. Yu, Anal. Biochem. 89, 437 (1978).
26. S. Ando and R. K. Yu, J. Biol. Chem. 252, 6247 (1977).
27. R. J. Penick, M. H. Meisler, and R. H. McCluer, Biochim. Biophys. Acta 116, 279 (1966).
28. D. H. Van den Eijnden, Hoppe-Seyler's Z. Physiol. Chem. 352, 1601 (1971).
29. K. Eberlein and G. Gercken, J. Chromatog. 106, 425 (1975).
30. S. Ando and R. K. Yu, Proc. Int. Meetings Glyconjug., 1977, p. 1.
31. R. W. Ledeen, R. K. Yu, and L. F. Eng, J. Neurochem. 21, 829 (1973).

CHAPTER 29

Quantitative TLC for the Study of Regulation of Rat Leukocyte HMG-CoA Reductase

N. L. Young and Victor W. Rodwell

Among the "risk factors" for degenerative heart disease identified by the American Heart Association is an elevated level of cholesterol in low-density lipoproteins (LDL). The disease familial hypercholesterolemia (FH) illustrates in striking and tragic fashion this correlation between plasma cholesterol and cardiovascular disease. FH, an autosomal, dominant genetic disorder, afflicts an estimated 0.1-0.2% of the population (1). FH heterozygotes exhibit fasting plasma cholesterol levels of 350-500 mg/dl, develop cholesterol deposits in arteries and other tissues, and generally cause fatal heart disease in young or early middle age. FH homogygotes exhibit fasting blood cholesterol levels of 700-1000 mg/dl, develop cholesterol deposits in early life, exhibit symptoms of coronary heart disease in childhood, and almost invariably cause death prior to age 30 (2,3). Apart from its rapid onset, the clinical presentation in FH closely resembles that seen in later-life degenerative cardiovascular disease in otherwise normal subjects. For this reason, FH is widely regarded among physicians as the best available model for the degenerative process that ultimately kills well over half the population worldwide.

Circulating plasma cholesterol arises from two sources: absorption of cholesterol and cholesterol precursors present in the diet, and de novo synthesis. Humans and other animals can synthesize cholesterol from the amphibolic intermediate acetyl-CoA. Because liver tissue converts all foodstuffs to acetyl-CoA, a portion of which is converted to cholesterol, mere restriction or elimination of cholesterol from the diet does not by itself markedly lower normal plasma cholesterol levels. Although restriction of dietary cholesterol intake does lower plasma cholesterol, it is actually far less effective than might be supposed, because

decreased intake is partially compensated for by increased biosynthesis. This is explained below.

The biosynthesis of cholesterol in human and animal tissues
is subject to precise and generally effective control or regulation by the end-product of the biosynthetic process--cholesterol.
Although as long ago as 1933 cholesterol was inferred to regulate
the rate of its own biosynthesis (4), only in 1950 did Gould and
Taylor (5) show conclusively that fasting or the ingestion of
dietary cholesterol extensively depresses cholesterol biosynthesis in animal models. Somewhat later, Gould and Popják (6),
Bucher et al. (7), and Siperstein and Guest (8) localized the
site at which dietary cholesterol regulates its biosynthesis as
prior to mevalonate, and by inference, subsequent to 3-hydroxy-3-
methylglutaryl-coenzyme A (HMG-CoA). Conversion of HMG-CoA to
mevalonate is catalyzed by the enzyme HMG-CoA reductase
(EC 1.1.1.34) (Figure 1).

In the ensuing decades, an impressive body of evidence has
been accumulated, indicating that, under most physiological conditions, HMG-CoA reductase catalyzes the rate-limiting, key regulated reaction of mammalian cholesterogenesis and is the primary
locus for feedback control of cholesterogenesis by cholesterol
(9). The interest of physicians in HMG-CoA reductase is best
illustrated with reference to FH. FH homozygotes fail to respond
in normal fashion to increased cholesterol intake by decreasing
cholesterol biosynthesis. Although increasing cholesterol intake
decreased cholesterogenesis in four adult FH heterozygotes and
controls (10), cholesterogenesis was unaffected in three homozyg-
ous children (11).

This apparent defect in FH has been shown by Brown, Goldstein, and their collaborators to be located at the plasma membrane, where serum lipoproteins carrying cholesterol are not normally transported into the cell. Using cultured fibroblasts from
normal and FH subjects, these investigators demonstrated that
reductase of FH fibroblasts is not inhibited by cholesterol in
lipoproteins added to the culture medium because added cholesterol
fails to reach its intracellular site of action.

In a genetically heterogeneous population of human subjects,
serum cholesterol levels range from "low risk" to levels characteristic of FH. The responsiveness of HMG-CoA reductase, and
hence of cholesterol biosynthesis, to regulation by dietary and
other factors may also exhibit wide variations. Although this is
suggested by the available data for FH subjects, it is not known
for large-scale populations of "normal" individuals. This lack
of knowledge reflects the undesirable features of any large-scale
investigation of hepatic HMG-CoA reductase in living subjects.
Similar information from small groups has been obtained by sterol

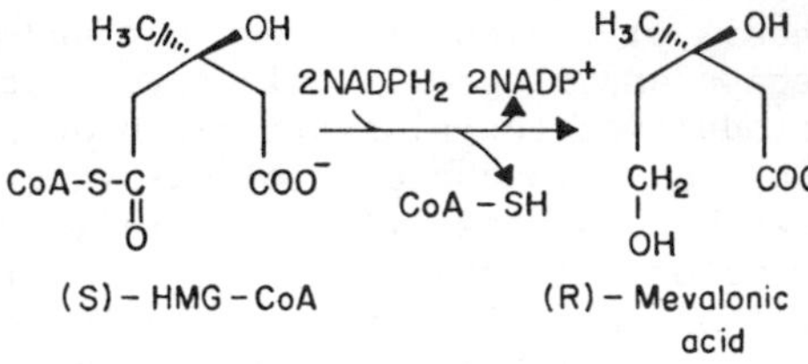

Figure 1.   The HMG-CoA reductase reaction.

balance studies, turnover measurements after injection of isotopically labeled cholesterol or its precursors (12) or assay of
liver biopsy tissue (13). However, such techniques are time-
consuming, expensive, or potentially hazardous to subjects, and
have not been widely employed on large populations of normal subjects. Clearly needed are safe, rapid, and inexpensive techniques
for studying the regulation of human cholesterol synthesis in vivo
under physiological conditions as well as in response to therapy.
     Assay of HMG-CoA reductase activity in freshly isolated leu-
kocytes might fulfill these criteria. In the rat liver, the rate
of cholesterol synthesis generally is directly proportional to
HMG-CoA reductase activity (14) and, in cultured human leukocytes,
reductase activity and cholesterogenesis vary in parallel and are
subject to regulation by cholesterol in the culture medium (15,
16). To provide clinically useful information, the regulatory
characteristics of the HMG-CoA reductase of freshly isolated human
leukocytes must reflect those of the corresponding human liver and
intestinal reductase. This could not, however, be addressed in
normal human subjects because of the aforementioned difficulties
that preclude obtaining comprehensive data on human hepatic and
intestinal reductase. It was therefore first necessary to examine
the laboratory rat, the animal model in which the regulation of
HMG-CoA reductase of liver, intestines, and certain other tissues
has been most extensively studied (9). It was sought to determine
whether reductase activity could be detected in freshly isolated
rat leukocytes, whether its activity was subject to metabolic
regulation by factors that affect hepatic reductase, and whether
the direction and magnitude of changes in leukocyte reductase
activity resembled those of hepatic HMG-CoA reductase activity.

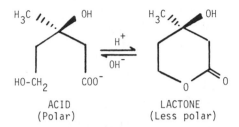

Figure 2.   Interconversion of mevalonic acid and mevalonolactone.

EXPERIMENTAL PROCEDURES

Materials

[3-^{14}C]HMG and [5-^{3}H] mevalonic acid (dibenzyethylene diammonium salt) were from New England Nuclear Corp., Boston, Mass.   The salt was converted to mevalonic acid lactone by passing it through a short column of Dowex 50-X8 (H$^+$ form) (17).   To convert the lactone to the free acid, a slight excess of KOH was added (pH about 8).   After incubation at 37° C for 30 minutes, the pH was adjusted to 7.4 and the solution was stored at 5° C.

R,S-[3-^{14}C]HMG-CoA (specific activity 22.8 Ci/mole, prepared as described by Nordstrom et al. (18), was purified before use by descending paper chromatography in 1-butanol:glacial acetic acid: H$_2$O (5:2:3 v/v) on sheets of Whatman No. 3-mm paper that had been previously washed, first with the above solvent, then with 1.0 mM HCl.   R$_f$ values were HMG-CoA = 0.27, CoA = 0.34, and HMG = 0.77. The region R$_f$ 0.14-0.36 was eluted by descending development with 1.0 mM HCl, diluted with water to an HMG-CoA concentration of 1.0 mM, and stored at -20° C.   All other chemicals were from previously listed sources (19).

Silica gel TLC sheets, cellulose TLC sheets, and sandwich (S) TLC development chambers were from Eastman Organic Chemicals, Rochester, N. Y.

Buffer I contained 250 mM NaCl, 5.0 mM Na$_2$EDTA, 5.0 mM dithiothreitol, and 50 mM K$_x$PO$_4$, pH 7.5.

Animals

Wistar strain rats, 300-400 g, housed in a room with alternate 12-hour periods of light and dark to standardize the diurnal rhythm in hepatic HMG-CoA reductase, were fed a chow diet ad libitum.

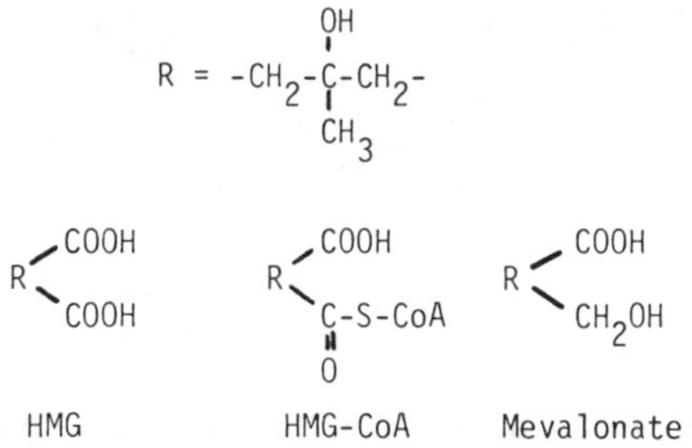

Figure 3. Products to be resolved by the standard TLC system, shown as derivatives of t-butanol to emphasize polarity differences.

Isolation of Leukocytes

From rats anesthetized with ether, approximately 10 ml of blood were drawn by cardiac puncture into a polypropylene syringe containing 1.0 ml of 50 mM EDTA in 50 mM K_xPO_4, pH 6.5, as anticoagulant. The leukocyte fraction was isolated, frozen, a cell-free extract was prepared and fractionated by high-speed centrifugation, and a "microsomal" suspension* was prepared as described by Young and Rodwell (19).**

Assay of HMG-CoA Reductase Activity in Microsomal Suspensions

A 100-μl portion of microsomal suspension (about 500 μg protein) was added to 50 μl of a substrate-cofactor solution in Buffer I (see Materials) to give the following final concentrations and

---

  * Microsomes are here defined as material sedimenting at field strengths of 8,000-113,000 x g. Although not functionally analogous to liver microsomes, these preparations contain HMG-CoA reductase.

** Isolation of rat leukocytes by centrifugation in Ficol-Hypaque offers advantages over the original isolation technique (19). Fewer granulocytes are recovered and the yield of lymphocytes is improved. Since the latter contain most of the reductase activity, higher specific activity reductase preparations are obtained using this procedure (N. L. Young, unpublished observations).

amounts: 20 μM [3-^{14}C]HMG-CoA, 20 nM (4 × 10^4 dpm) [5-^{3}H] mevalonic acid (internal standard, 3.0 mM NADP$^+$, 30 mM glucose-6-
phosphate (International Enzyme Units = pmole mevalonate formed
per minutes at 37° C), and 0.3 E.U. of glucose-6-dehydrogenase.
The mixture was incubated at 37° C for 90 minutes. Kyro-EOB,
25 μl of 0.2% solution in 6 N HCl, was then added. The detergent
facilitated subsequent application of samples to TLC sheets.
Samples were incubated for 15 minutes at 37° C to lactonize mevalonic acid and centrifuged to sediment protein. The supernatant
liquids were then subjected to TLC to separate mevalonolactone
from HMG-CoA, HMG, and other possible polar side-products.

In the analytical portion of this work, advantage is taken of
the ability of mevalonic acid (the product of the HMG-CoA reductase reaction) to exist both in a polar and less polar form (Figure 2). As is frequently the case, the resolving power of the
TLC systems employed are best understood in terms of the relative
polarities of the sample components (Figure 3).

Thin Layer Chromatography

Supernatant liquid is applied in 150-μl portions as a spot (diameter approximately 2 cm) to 2.5 × 20 cm channels of silica gel G
that had been activated by heating just prior to use. The resolution of the substrate and product is such that there are no
adverse effects on the separation following application of 150 μl
of aqueous sample. Chromatograms were developed over approximately 60 minutes in sandwich chambers in benzene:acetone (1:1 [v/v])*
The air-dried sheet was sprayed lightly with water** and the region
containing mevalonolactone (R_f 0.5-0.9) was removed using a razor
blade and transferred to a 6 ml glass vial to which 4.0 ml of
0.7% butyl-PBD, 8% naphthalene in 3-methoxyethanol:toluene
(2:3 [v/v]) was added as a scintillation fluid. DPM from ^{3}H and
^{14}C were determined using a liquid scintillation spectrometer
using the external standard channel ratio method. The quantity
of [^{14}C] mevalonate produced was calculated from [^{14}C] DPM using

---

* William Bensch has recently shown that toluene may be substituted for benzene. This appears valid for rat liver reductase,
  although the mobility of mevalonolactone is significantly
  less than in the standard solvent.
** This facilitates removal of the solvent which contains the
  desired product (mevalonolactone) in a thin roll. Application
  of the correct quantity of water requires practice. We hold
  the TLC sheet at arm's length and spray from a compressed gas
  powered sprayer.

[^{3}H]DPM to correct for variation in recovery of product. Use of
this internal standard (21) thus makes it unnecessary to measure
the volume of sample taken for TLC with precision. Although not
done intentionally, variations of $\pm$ 50% in the volume of sample
applied can readily be tolerated.

Reductase activity is expressed in standard International
Enzyme Units (picamoles mevalonate formed per minute at 37° C)
and the specific activity of HMG-CoA reductase in terms of picamoles mevalonate formed per minute per milligram protein.

Analysis of Protein

Protein was assayed by a modified florescamine procedure (19).

RESULTS

Verification of the Presence in Rat Leukocyte Microsomes of
HMG-CoA Reductase

Among those who study HMG-CoA reductase, there is a widespread
and regrettable tendency when reaction mixtures are resolved in
the standard (benzene:acetone) solvent to equate counts in the
mevalonolactone region with mevalonolactone without seeking independent verification of this assumption. Because this failing has
analogies in other areas of quantitative TLC, the limitations of
the assumption will be considered.

Although it has been repeatedly shown by independent analyses
that the above assumption is valid for microsomal preparations of
liver and other tissues from the rat and certain other animals,
instances do arise where this assumption is invalid. A case in
point is human lymphocyte preparations (see below). It is therefore essential that, whenever a different animal or tissue is
studied, the investigator verify that the product counted is
solely mevalonolactone. Determination of the ^{14}C/^{3}H ratio in the
mevalonolactone region, even if performed for several mobile
phases, is not by itself adequate. The standard and presumptively
identical product must be present in a 1:1 ratio throughout the
region counted. The chromatograms should therefore be divided
into 0.5-cm sections in order to establish product identity and
the product-to-standard ratio should remain constant throughout
the entire region of the chromatogram that contains product. Figure 4 illustrates application of these criteria to the product
formed by rat leukocyte HMG-CoA reductase.

Microsomes from rat leukocytes convert [^{14}C]HMG-CoA to a
product that, after lactonization, comigrates with [^{3}H] mevalonic

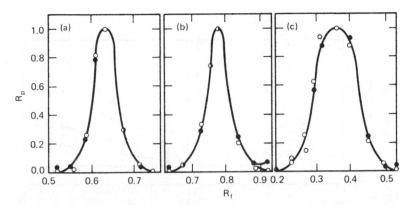

Figure 4. Chromatography of the product formed from HMG-CoA by leukocyte microsomes. Leukocyte microsomes (600 µl) from female rats were incubated with substrate-cofactor solution for 90 minutes at 37° C, then with acid for 15 minutes at 37° C. After removal of precipitated material by centrifugation, the mixture was applied to six channels (each 2.5 × 17.5 cm) of a silica gel TLC sheet and chromatographed in a sandwich chamber with the standard solvent, benzene-acetone 1:1 (v/v). One channel was sectioned at 0.5-cm intervals and the ^{14}C and ^{3}H in each section were counted (A). The region corresponding to R_f's of 0.5-0.75 was scraped from the remaining five channels, pooled, and eluted with acetone and methanol. Portions of the eluate were applied to 2.5 × 17.5 ml channels of silica gel TLC sheets and chromatographed in a chamber equilibrated with an acidic mobile phase, butanol-1-propionic acid-water 10:4:1 (v/v) (B). In addition, a portion was treated with 1.0 N NH₄OH (15 minutes at 37° C) and then applied to a 2.5 × 17.5 cm channel of a cellulose TLC sheet and chromatographed in a chamber with solvent-saturated vapor with an alkaline solvent, butanol-1-ammonia-water 20:1:1 (v/v) (C). Each chromatogram was sectioned and counted as above. Data for the [^{3}H]-mevalonate internal standard (o) and the ^{14}C-labeled product (•) are expressed as the ratio of dpm in each fraction to dpm in the peak fraction (R_f) versus R_f. The peak fraction contained approximately 800 ^{14}C dpm and 6000 ^{3}H dpm in each case.

lactone during TLC in an alkaline system or with [^{3}H] mevalonolactone after treatment with acid and TLC in neutral or acidic solvents (Figure 4). The formation of this product requires NADPH. It follows that the product is mevalonic acid, and that HMG-CoA reductase is present in microsomal preparations from rat leukocytes.

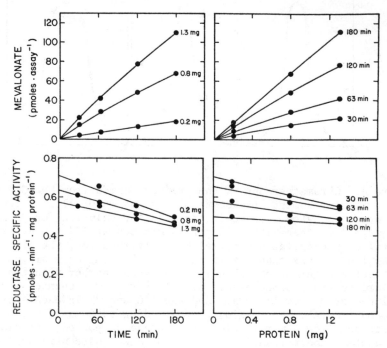

Figure 5. Kinetics of mevalonate production as a function of
incubation time and of microsomal protein concentration. Portions of a suspension of leukocyte microsomes containing the
amount of protein indicated were incubated under standard conditions except that time was varied as shown.

Kinetics of Mevalonate Formation

Mevalonate production increased with incubation time (T) and with
amount of microsomal protein added (P) (Figure 5). These
responses appear to be, but are not exactly, linear (19). The
kinetics of mevalonate formation as a function of HMG-CoA concentration are shown in Figure 6. K_m for (R,S) HMG-CoA is 7 μM.

Regulation of Rat Leukocyte HMG-CoA Reductase by Fasting and by Cholestyramine

In rats fasted for 24 hours, mean HMG-CoA reductase activity in
leukocyte microsomes fell to 45% that of fed controls. In rats

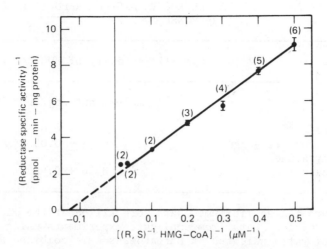

Figure 6. Double-reciprocal plot of the substrate-dose response of HMG-CoA reductase from rat leukocytes. Microsomal suspensions were incubated for 90 minutes under standard assay conditions except that the concentration of (R,S)-HMG-CoA was varied as shown. The utilization of substrate varied from 1.5% at μM HMG-CoA to 0.2% at 60 μM HMG-CoA. Data are mean values ± SEM for the number of replicate assays shown in parentheses.

fed cholestyramine (a drug known to elevate hepatic HMG-CoA reductase activity) for 3 days, mean reductase activity rose to 155% that of controls (Table I).

Regulation of Rat Leukocyte HMG-CoA Reductase by Cholesterol

When rats were fed a chow diet containing 1% commercial grade cholesterol for 3 days, mean HMG-CoA reductase activity in leukocytes declined exponentially with a half-time of about 1.8 days (Figure 7).

Absence of a Diurnal Rhythm

HMG-CoA reductase activity was measured in leukocyte microsomes and, as a positive control, in rat liver at 4-hour intervals for 24 hours. Although hepatic HMG-CoA reductase activity varied over sixfold from a minimum of 65 ± 12 (1600 hours) to a maximum of $413 + 95$ pmole min^{-1} mg protein^{-1} (2400 hours), the reductase activity of leukocyte microsomes did not appear to vary

TABLE I.   VARIATION OF RAT LEUKOCYTE HMG-CoA REDUCTASE ACTIVITY
IN RESPONSE TO DIET[a]

Diet	Specific Activity of HMG-CoA Reductase
	pmole/minute/mg Protein
Normal (10)	0.51 ± 0.04
Fasted 24 hours (11)	0.24 ± 0.03
2% Cholestyramine for 3 days (13)	0.80 ± 0.05

[a] Diets were fed ad libitum. Rats were killed at the midpoint of
the dark cycle. Data from duplicate experiments using the total
number of rats indicated in parentheses were combined and are
present as mean values $\pm$ SEM.

significantly (mean value was 0.50 ± 0.07 pmole $min^{-1}mg$ $protein^{-1}$)
(Figure 8).

Preliminary Application of the Rat Leukocyte Reductase Technology to Human Leukocytes

This method has been modified for application to human lymphocytes. Modifications were required because of the presence in
human lymphocyte preparations of highly active enzymes that
cleave HMG-CoA. This results in diminished HMG-CoA concentration
in the assay and in radioactive metabolites that comigrate with
mevalonolactone in the standard TLC system. To deal with these
problems, (1) cleavage enzyme activity was decreased by preincu-
bation of whole cell homogenates in phosphate buffer, (2) the HMG-
CoA concentration in the assay was increased to 300 µM, (3) an
additional TLC step was employed to physically remove a large portion of the contaminating material, and (4) an NADP minus control
was used to allow correction for residual contaminating material.
The additional TLC step is as follows: after the usual chromatography the remaining mevalonolactone is hydrolyzed to mevalonic
acid by spraying the sheet with 0.1 N NaOH and incubating at 80°
C for 10 minutes. The sheet is dried and rechromatographed in
the reverse direction with ethanol:acetone (1:4). The mevalonic
acid is stationary and about 70% of the contaminating material is
removed (Figure 9).

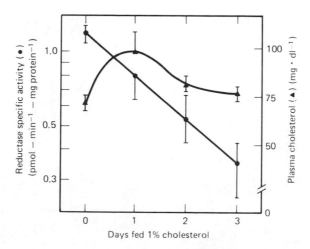

Figure 7. Regulation of leukocyte HMG-CoA reductase by dietary cholesterol. Twelve female rats weighing 325 $\pm$ 5 g were fed a normal diet supplemented with 1% cholesterol for 0-3 days. ▲ Plasma cholesterol (A) and leukocyte microsomal reductase activity (●) were assayed in all rats on the same day at the midpoint of the light phase. Data (mean $\pm$ SEM for three rats) for cholesterol are plotted on a linear scale and for reductase activity on a logarithmic scale.

DISCUSSION

Methods for isolation of the microsomal fraction and for assaying for HMG-CoA reductase activity previously used for rat liver (20) were modified for application to rat leukocytes. Because the specific activity of HMG-CoA reductase in rat leukocytes is extremely low (10^{-2} to 10^{-3} that of rat liver), the assay was modified in order to increase the sensitivity approximately five hundredfold. The principal change was the use of higher specific activity substrate. More microsomal protein was used, time was extended, and a larger fraction of the incubation mixture was subjected to separation by TLC.

Despite its low activity, HMG-CoA reductase in rat leukocyte microsomes can be assayed reliably. The product comigrates with authentic mevalonic acid/mevalonolactone in three TLC systems. The response of leukocyte reductase in the assay to variation in substrate concentration, protein concentration, and incubation time mimics that of rat liver reductase. The coefficient of

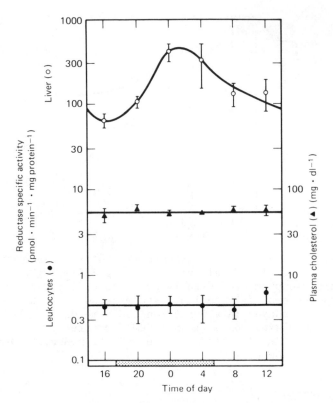

Figure 8. Activity of liver and of leukocyte HMG-CoA reductase
at various times of day. Eighteen female rats weighing 334 ± 10
g were kept on a light-dark schedule as indicated by the horizontal bar on the abscissa. At the time of day indicated, microsomal reductase activity in liver (o), leukocytes (●), and plasma
cholesterol (▲) were assayed in each of three rats. Data for the
mean ± SEM are plotted on a logarithmic scale.

variation in activity between rats subjected to similar manipulations was the same (53%) in leukocytes and liver. Inter-rat variability thus is no worse for leukocyte than for liver HMG-CoA
reductase.

HMG-CoA reductase activity in leukocytes, like that in brain,
testis, lung, and spleen (22), but unlike that of liver and intestine (23), does not appear to exhibit a diurnal variation. From
the standpoint of utilizing leukocyte reductase activity as a

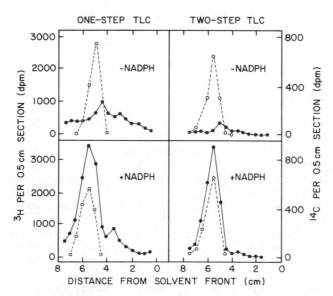

Figure 9.   Illustration of reversed TLC.   Symbols: (●) = [14]C;
(o) = [3]H mevalonolactone (authentic sample included as internal
standard).   Even in the absence of NADPH (top left), significant
nonmevalonate-^{14}C tends to migrate with mevalonolactone.   This
spurious material is removed by a second, reverse development
(top and bottom right).

diagnostic tool in humans, if a rhythm proves also to be absent
from human leukocytes, there are distinct advantages. Effects on
reductase would not be complicated by the time of day at which
blood samples are drawn.
     The assay of HMG-CoA reductase activity in cell-free prepara-
tions from human lymphocytes is complicated (1) by the presence
of high levels of enzymes other than HMG-CoA reductase that metabolize HMG-Coa, and (2) the comigration of certain radioactive by-
products with mevalonolactone in the standard TLC system.   It
appears that these difficulties may be circumvented by modifications that include in situ conversion of mevalonolactone back to
the free acid immediately following the standard TLC separation.
This is followed by reverse TLC prior to removing and counting
the mevalonate.

ACKNOWLEDGMENTS

This work was supported by a postdoctoral fellowship to N. Young
from the National Institutes of Health and by grants from the
Indiana Heart Association and the National Institutes of Health
(HL 19223).

REFERENCES

1. J. L. Goldstein, H. G. Schrott, W. R. Hazzard, E. L. Bierman,
   and A. G. Motulsky, J. Clin. Invest. 52, 1544 (1973).
2. J. L. Goldstein and M. S. Brown, Proc. Nat. Acad. Sci. USA 70,
   280 (1973).
3. J. L. Goldstein, M. J. E. Harrod, and M. S. Brown, Am. J. Hum.
   Genet. 26, 199 (1974).
4. R. Schoenheimer and F. Breusch, J. Biol. Chem. 103, 439
   (1933).
5. R. G. Gould and C. B. Taylor, Fed. Proc. 9, 179 (1950).
6. R. G. Gould and G. Popják, Biochem. J. 66, 51p (1957).
7. N. L. R. Bucher, P. Overath, and F. Lynen, Biochim. Biophys.
   Acta 40, 491 (1960).
8. M. D. Siperstein and M. J. Guest, J. Clin. Invest. 39, 642
   (1960).
9. V. W. Rodwell, J. L. Nordstrom, and J. J. Mitschelen, Adv.
   Lipid Res. 14, 1 (1976).
10. D. F. Pawlinger and J. C. Shipp, J. Clin. Invest. 44, 1084
    (1965).
11. A. K. Khachadurian, Lancet 2, 778 (1969).
12. G. C. K. Liu, E. H. Ahrens, Jr., P. H. Schreibman, P. Samuel,
    D. J. McNamara, and J. R. Crouse, Proc. Nat. Acad. Sci. USA
    72, 4612 (1975).
13. G. Nicolau, S. Shefer, G. Salen, and E. H. Mosbach, J. Lipid
    Res. 15, 94 (1974).
14. J. M. Dietschy and M. S. Brown, J. Lipid Res. 15, 508 (1974).
15. A. M. Fogelman, J. Edmon, J. Seager, and G. Popják, J. Biol.
    Chem. 250, 2045 (1975).
16. M. J. P. Higgins and D. J. Galton, Eur. J. Clin. Invest. 7,
    309 (1977).
17. F. Lynan and M. Grassl, Hoppe-Seyler's Z. Physiol. Chem. 313,
    291 (1958).
18. J. L. Nordstrom, V. W. Rodwell, and J. J. Mitschelen, J. Biol.
    Chem. 252, 8924 (1977).
19. N. L. Young and V. W. Rodwell, J. Lipid Res. 18, 572 (1977).
20. D. J. Shapiro, J. L. Nordstrom, J. J. Mitschelen, V. W. Rod-
    well, and R. T. Schimke, Biochim. Biophysica Acta 370, 369
    (1974).

21. S. Goldfarb and H. C. Pitot, J. Lipid Res. $\underline{12}$, 512 (1971).
22. G. C. Ness, Fed. Proc. $\underline{35}$, 1404 (1976).
23. S. Shafer, S. Hauser, V. Lapar, and E. H. Mosbach, J. Lipid Res. $\underline{13}$, 571 (1972).

CHAPTER 30

Separation of Metabolites
of Biogenic Amines

Henry Weiner

It is often necessary to quantitate the level of biogenic amines
in samples from urine, blood, or tissue homogenates. When inves-
tigating drugs or chemicals that can alter the metabolism of the
amines, it may be necessary to measure the level of transient
intermediates produced during the enzymatic degradation of the
amine, in addition to the amine itself.

The metabolism of three major biogenic amines, serotonin,
noradrenaline, and dopamine, are similar to each other and are
similar in every tissue. The basic reactions involved are illustrated in equation (1):

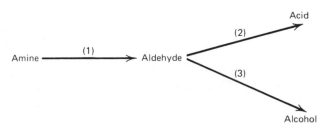

Here (1) is the enzyme monoamine oxidase, (2) is the aldehyde
dehydrogenase, and (3) is the aldehyde reductase. Thus, four
major products can be expected from the metabolism of any one
amine. Unfortunately, the problem can become more complicated.
In the case of noradrenaline, for example, 3,4-dihydroxyphenyl-
ethanolamine, the 3-hydroxyl position can be methylated to form
normetanephrine. This, in turn, can be metabolized as illustrated
in equation (1). Hence, from noradrenaline, a total of eight
metabolites can be found: two amines, two acids, two alcohols,
and two aldehydes.

442

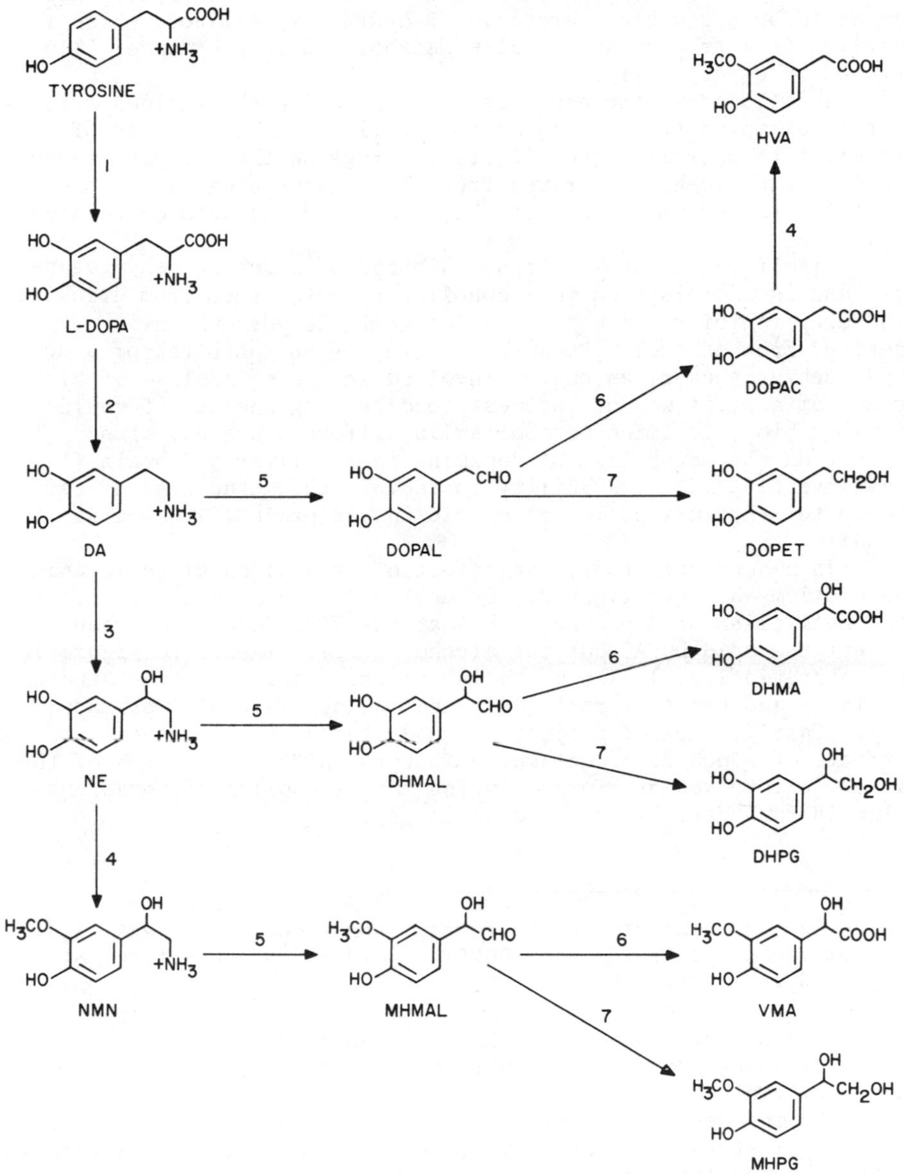

Figure 1.   Metabolites of dopamine and noradrenaline.

443

In Figure 1 the relationship between the catechol amines, noradrenaline and dopamine, is illustrated. The other biogenic amine to be discussed, serotonin (5-hydroxytryptamine), which is derived from tryptophan, is also metabolized in the manner illustrated by equation (1).

To illustrate the necessity of separating the various metabolites, examples from our laboratory will be given. It was of interest to determine the effects of drugs on the ultimate metabolism of the aldehydes derived from the various biogenic amines, that is, the enzymatic conversion of aldehyde to acid or to alcohol.

Disulfiram is an in vitro inhibitor of aldehyde dehydrogenase and is administered to alcoholics to deter them from drinking. The drug inhibits the metabolism of acetaldehyde, which is derived from ethanol. Because the drug is an inhibitor of aldehyde dehydrogenase, an enzyme involved in the metabolism of biogenic amines, it was of interest to determing whether it could inhibit biogenic amine metabolism in different organs. Its effect on the metabolism of dopamine in rat liver and brain (1,2) was investigated. Its ability to inhibit the metabolism of serotonin to 5-hydroxyindoleacetic acid in rat brain (3) was also studied.

In a separate study, the effect of ethanol on biogenic amine metabolism was investigated. Ethanol has been reported to alter the metabolism of dopamine such that the final excreted product is not the acid DOPAC but the alcohol DOPET. (Refer to Figure 1 for abbreviations.) The mechanism of this conversion was studied in liver and brain of rats (4,5) and in the brain of a primate (6). Lastly, a pilot project was undertaken to study the effects of aldehyde reductase inhibitors on the conversion of the aldehyde intermediate formed during the metabolism of noradrenaline in rat heart, liver, and brain (7).

Abbreviations used are the following: DA--dopamine; DOPAL--3,4-dihydroxyphenylacetaldehyde; DOPET--3,4-dihydroxyphenylethanol; DOPAC--3,4-dihydroxyphenylacetic acid; NE--noradrenaline; DHMAL--3,4-dihydroxymandelaldehyde; DHGP--3,4-dihydroxyphenylglycol; DHMA--3,4-dihydroxymandelic acid; NMN--normetanephrine; MHMAL--4-hydroxy-3-methoxyphenylglycol; VMA--vanillylmandelic acid; and HVA--homovanillylic acid.

The enzymes responsible for the transformation are the following: 1--tyrosine hydroxylase; 2--dopa-decarboxylase; 3--dopamine β-hydroxylase; 4--catechol-O-methyltransferase; 5--monoamine oxidase; 6--aldehyde dehydrogenase; and 7--aldehyde reductase.

The aldehydes formed from biogenic amines are not normally found in biological fluids. Their steady-state level must be vanishingly small because of high chemical and enzymatic activity. It was necessary to measure exogenous levels of the aldehyde by performing various in vivo and in vitro incubations with radioactive substrates. Thus, quantitation was by scintillation counting. No attempts were made to measure endogenous levels of the amine metabolites. The details of the incubation procedure are not the subject of this discussion and are described in the various publications mentioned. In general, the labeled amine was incubated either with isolated tissue or perfused through organs. The incubation media or perfusate was acidified to pH 4 and subjected to centrifugation to remove precipitated protein. The supernatant was subjected to the chromatographic separations to be outlined.

PROCEDURES

Dopamine

Various thin layer chromatographic systems were employed to separate the metabolites derived from dopamine. In all cases the metabolites did not separate from each other. It was found, however, that a combination chromatographic-electrophoretic procedure would allow for the complete separation of the various expected metabolites of dopamine. In addition to the metabolites shown in Figure 1, an additional intermediate of interest was found. This was the isoquinoline alkaloid tetrahydropapaveroline (THP) derived from the condensation of dopamine with the biogenic aldehyde (DOPAL).

Paper was employed for the separations, although thin layer chromatography on cellulose is equally effective. Ten to fifty microliters of an incubation solution was spotted in the lower left hand corner of a strip of Whatman 3 MM paper (9 × 2.5 in.). Chromatography was performed first in order to separate the alcohol from the aldehyde. Electrophoresis was then performed in borate buffer, as originally described by Schweitzer and Friedhoff (8). The electrophoresis allowed for all of the charged particles to separate from each other. Electrophoresis was performed in borate buffer for two reasons. First, the catechol ring is unstable and is subject to air oxidation. Borate can complex with the catechol ring, thus stabilizing the hydroxyl groups. Further, after borate condenses, it acts as a negatively charged species (pK ca. 10). Thus, the neutral materials, alcohol and aldehyde, will have a negative charge and will migrate during electrophoresis. A representative chromatograph is shown in Figure 2. The details of the separation can be found else-where (4) and are outlined briefly in the figure legend.

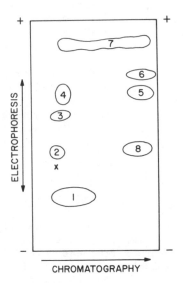

Figure 2. The separation of dopamine and its metabolites by the 2-dimensional paper chromatographic-electrophoretic technique. The x denotes the site at which the sample is applied and the arrows indicate the direction of chromatography and electrophore-sis. The symbol (+) represents the anode. Chromatography was performed with ethyl acetate-methanol-acetic acid (19:6:0.3). Electrophoresis was then performed for 3 hours at 150 V at 4° C. The spots represent the following compounds: 1, 3-0-methyltyra-mine; 2, dopamine (DA); 3, tetrahydropapaveroline (THP); 4, 3,4-dihydroxyphenylacetaldehyde (DOPAL); 5, 3,4-dihydroxyphenylethanol (DOPET); 6, homovanillic acid (HVA); 7, 3,4-dihydroxyphenylacetic acid (DOPAC); 8, 3-methoxy-4-hydroxyphenylethanol (MOPET).

    The two-dimensional separation procedure, unlike chromatog-raphy, allows many samples to be run simultaneously. Routinely, one investigator can perform as many as 16 incubations simultane-ously. After spotting the various samples on separate sheets of paper, it takes less than a day to analyze various incubations to determine product formation.

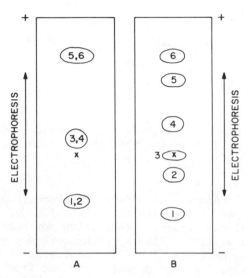

Figure 3. Separation of the metabolites of noradrenaline. Electrophoresis at pH 6.5 in 50 mM sodium phosphate (A) without and
(B) with 7 mM sodium borate was for 3 hours at 200 V at 4° C.
Components are the following: 1, normetanephrine (NMN); 2, noradrenaline (NE); 3, 4-hydroxy-3-methoxyphenylglycol (MHPG); 4,
3,4-dihydroxyphenylglycol (DHPG); 5, vanillylmandelic acid (VMA);
6, 3,4-dihydroxymandelic acid (DHMA); X, origin.

Noradrenaline

It was found that simple electrophoresis in the presence of borate allowed for the total separation of the major metabolites
derived from noradrenaline (7). The importance of borate in the
media is illustrated by comparing the two electrophoretograms
presented in Figure 3. In the absence of borate, both acid products migrate with the same mobility as do the two amines.
Neither the alcohol nor aldehyde leaves the origin. Borate, as
mentioned previously, interacts with the catechol ring, imparting
a negative charge, and causes a differential charge to be formed
on each of the classes of compounds. Thus, the methylated and
nonmethylated acids as well as amines separate from the nonmethylated compounds. It was interesting to find that the two neutral
components would separate from each other. Conceivably borate
complexes with an enediol form of the aldehyde as illustrated
below:

One could postulate that aldehyde then binds two moles of borate
whereas alcohol binds only one through the catechol ring.

Serotonin

A two-dimensional chromatographic-electrophoretic procedure was
originally employed to separate the metabolites of serotonin (9).
It was found that two separate steps allow for better quantitation of the metabolites. Thin layer chromatography on silica gel
separated the two neutrals, 5-hydroxyindoleethanol and 5-hydroxy-
indoleacetaldehyde, from the parent amine and from the acid product, 5-hydroxyindoleacetic acid. Electrophoresis was performed
on a separate sample to separate the acid and amine from each
other as well as from the neutrals.
     The chromatographic step was similar to that originally
employed by Lahti and Majchrowicz (13). The details of the
chromatographic separation are indicated in the legend of Table
I. Electrophoresis was performed on a 9-in. paper strip at pH
6.5 in 50 mM phosphate for 3 hours at 200 V. The acid migrated
2 in. toward the anode while the amine migrated 2 in. toward the
cathode. The alcohol and aldehyde remained at the origin. Thus,
the two separate procedures or the original two-dimensional procedure resulted in an accurate determination of the percent of
metabolites found as acid or alcohol and those that remained as
the aldehyde. Because serotonin does not contain the catechol
ring, unlike the other two biogenic amines discussed, it was not
necessary to use borate.

Quantitation

If [14][C]-amines were employed, it was possible either to cut out
the paper in the region corresponding to the metabolite or scrape
the silica directly into 3 ml of scintillation fluid. However,
if [3][H]-amines were employed, it was necessary to elute the activity from the paper prior to counting (7). The paper in the region
to be counted was cut into small segments and placed in a

TABLE I.   SEPARATION OF SEROTONIN METABOLITES
BY THIN LAYER CHROMATOGRAPHY[a]

Compound	R_f
5-Hydroxyindoleacetaldehyde	0.58
5-Hydroxyindoleethanol	0.28
5-Hydroxyindoleacetic acid	0.0
Serotonin	0.0

[a] Chromatography was performed on 0.2-mm
silica gel 60 TLC plates on an aluminum
backing obtained from EM Laboratory. The
solvent was ethyl acetate-chloroform (3:2).
The plate was sprayed with Ehrlich reagent
(11) in order to visualize the components.

scintillation counting vial. Water (0.5 ml) was added and
allowed to stand for 1 hour. It was found that essentially all
of the counts were eluted by this procedure. At this stage, 3 ml
of cocktail was added and solutions counted. The scintillation
cocktail was composed of toluene (2800 ml), Triton X-100 (1200
ml), diphenyl oxazold (16 g), and 1,4-bis[2(4-methyl-5-phenyl-
oxazolyl)]benzene (400 mg).

Routinely, standards were added directly to the media spotted
immediately before separation. After separation, the chromatograms were sprayed with Erlich reagent, followed by sodium carbonate in order to visualize the components. These were found not to
interfere with the efficiency of the radioactive counting.

REPRESENTATIVE RESULTS

Dopamine

The Metabolism of Dopamine in Various Tissues. Dopamine is primarily oxidized to DOPAC after the amine is converted to the
intermediate aldehyde. The metabolism of dopamine in liver
slices, isolated heart cells, and brain tissue was investigated.
The use of the chromatography allowed for separation of the metabolites. Inasmuch as a different amount of amine was employed in

different incubations and a different level of monoamine oxidase activity exists in each tissue, the data were reported as the percent of deaminated products isolated as a particular component. Some representative data are presented in Table II. It can be noted that in all tissue the bulk of the metabolites was the acid and little was isolated as the alcohol.

In Vivo Inhibition of Aldehyde Dehydrogenase by Disulfiram. Disulfiram, sold in the United States as the drug Antabuse, is given to alcoholics to deter them from drinking. The drug was administered to rats and the ability of the brain and liver to convert the biogenic aldehyde into the acid was measured. In the liver the oxidation of dopamine to DOPAC was inhibited by 16%, whereas in brain slice incubations, ca. 50% of the reaction was inhibited. By surgically implanting push-pull perfusion canula into specific regions of the rat brain, it was possible to determine both the metabolism of dopamine and its inhibition by disulfiram in different regions of the brain (2).

Effect of Ethanol on Dopamine Metabolism. One of the major effects of ethanol on metabolism in liver is that compounds that are normally oxidized are reduced. This is because a large quantity of NADH is produced during the metabolism of ethanol and acetaldehyde. This form of the coenzyme can cause the reduction of other compounds. It was of interest to determine whether the metabolism of ethanol in liver would alter the metabolism of dopamine such that alcohol and not acid was the product. Experiments were also performed in the brain because ethanol is not metabolized to any measurable degree there. It was found that the metabolism of ethanol did indeed cause an alteration of the metabolism of dopamine in liver slices but not in brain tissue (4,5, 12). In the liver the ratio of DOPET/DOPAC increased from 0.2 in the absence of ethanol to 4 in the presence of 40 mM ethanol.

Serotonin

Thus far, only limited experiments have been performed on the metabolism of serotonin. One study examined whether disulfiram inhibited the ultimate oxidation of serotonin in brain, as it did dopamine (3). The brain was coarsely dissected to produce fore and hind quarters after being cut laterally to produce the left and right hemisphere. Tissue was obtained from control and disulfiram fed animals. It was found that the degree of inhibition of 5-hydroxyindoleacetic acid formation was only 10 and 20% in the fore and hind brain, respectively. This is in contrast to the approximately 50% inhibition found with dopamine as a substrate (1).

TABLE II. METABOLISM OF EXOGENOUS DOPAMINE IN RAT TISSUE

Tissue	THP	DOPAL	Percent Metabolism DOPET	HVA	DOPAC
Liver slice[a]	11.1 ± 2.2	29.7 ± 3.5	9.8 ± 2.7	2.2 ± 0.9	41.1 ± 3.2
Brain slice[b]	10.4 ± 4.4	21.2 ± 4.1	30.2 ± 4.5[c]		38.2 ± 9.3[d]
Heart cells[e]	12.2	20.6	2.8	45.0	21.6

a From Reference 12--standard deviation for five rat livers.

b From Reference 9--standard deviation for eight rat brains.

c Subsequent studies revealed that if the concentration of dopamine in the incubation was lowered from 2 to 0.2 mM, the DOPET level decreased to 2-4% (2).

d HVA and DOPAC were not counted separately.

e From Reference 7--duplicate runs for cells isolated from a heart.

Currently under investigation is the reason for the differential inhibition of acid formation in brain. It is not yet known whether the cause is related to the fact that different forms of aldehyde dehydrogenase are actually involved in the metabolism of each of the aldehydes or if in one region of the brain the drug is more rapidly destroyed than in another.

Noradrenaline

Limited experimentation was performed with the metabolism of nor-adrenaline in rat tissue. The metabolism in liver, brain, and heart was studied. With heart, both isolated myocytes (individual heart cells) and perfusion of an intact isolated beating rat heart were employed. The distribution of metabolites obtained from these incubations is presented in Table III for each tissue employed. In addition, experiments were performed in the presence of phenobarbital and probenecid, two in vitro aldehyde reductase inhibitors. It can be noted from the data that in each tissue the bulk of the deaminated product was glycol (alcohol) and not acid, as was the case with dopamine as inhibitor. Thus, noradrenaline is preferentially metabolized to a reduced product whereas dopamine, which differs only by a missing β-hydroxyl group, is primarily oxidized to an acid. The aldehyde reductase inhibitors were effective in the liver and brain and not in the heart. Thus far, experimentation in isolated heart tissue has not yielded concrete data. This is not due to any limitations in the separation technique but more a result of our lack of involvement in pursuing the answer.

CONCLUSIONS

Though there was no attempt to optimize the separation techniques employed, the techniques presented do offer a convenient and rapid method for separating the products of biogenic amines. Endogenous levels were not sought by quantitative techniques. It is very likely that, with proper reagents, good densitometry methods can be developed, enabling one to look for the endogenous levels of these components. The use of electrophoresis with borate combined with chromatography offers a rapid method for separating components with greatly different charges. It should be possible to use borate to separate many other dihydroxy or polyhydroxy compounds.

The total separation of all the catechol products from one another is an area for future research, as dopamine is the precursor of noradrenaline. An investigation of two-dimensional systems for this separation purpose is under way. The main difficulty so far has been separating the aldehydes.

TABLE III.   METABOLISM OF EXOGENOUS NORADRENALINE IN RAT TISSUE[a]

| Tissue | Percent Metabolites | | | | |
	NMN	DHPG	DHMAL	VMH	DHMA
Heart cells	6.1	44.6	21.5	18.5	9.9
Brain mince	10.6	51.8	14.4	10.7	12.2
Liver slice	3.2	48.0	7.9	9.2	31.6

[a] Data from Reference 7.

     The main advantage of the techniques presented here is that
the compounds do not have to be quantitatively extracted from
tissue or biological fluids. This means that not only will there
be less loss in recovery, but a more rapid work-up of many samples
can be accomplished in the minimum amount of time.

     ACKNOWLEDGMENTS

The work in the author's laboratory was sponsored in part by a
grant from the Public Health Service AA01395 and National Science
Foundation BMS 03926. Mr. Krug was supported in part by a grant
from the Indiana Heart Association and Dr. Tank by a David Ross
Fellowship from Purdue University.

     REFERENCES

1. D. Berger and H. Weiner, Biochem. Pharmacol. 26, 742 (1977).
2. H. Weiner, C. W. Simpson, J. A. Thurman, and R. D. Myers,
   Brain Res. Bull. 3, 541 (1978).
3. H. Weiner and J. A. Thurman, manuscript in preparation.
4. A. W. Tank, H. Weiner, and J. A. Thurman, Ann. N. Y. Acad.
   Sci. 273, 219 (1976).
5. A. W. Tank and H. Weiner, in Effect of Ethanol on Dopamine
   Metabolism in the Brain in Alcohol Intoxication and With-
   drawal, edited by M. Gross (Plenum, New York, 1977).
6. H. Weiner, R. D. Myers, C. W. Simpson, and J. A. Thurman,
   The Effects of Ethanol on the Metabolism of Exogenously Added
   Dopamine in the Brain of a Monkey, in press.

7. E. C. Krug, The Preferential Reduction of Norepinephrine in Rat Heart, Liver and Brain. M.S. Thesis, Purdue University, 1977.
8. J. W. Schweitzer and A. J. Friedhoff, Life Sci. 8, 173 (1969).
9. D. Berger and H. Weiner, Biochem. Pharmacol. 26, 841 (1977).
10. C. W. Easley, Biochim. Biophys. Acta 107, 386 (1965).
11. A. W. Tank, Ph.D. Thesis, Purdue University, 1976.
12. H. Weiner and J. A. Thurman, in preparation.
13. R. D. Lahti and E. Majchrowicz, Quart. J. Stud. Alcohol 35, 1 (1974).

CHAPTER 31

Quantitation of Free and Esterified Cholesterol by Densitometry of High Performance Thin Layer Chromatograms

David Kritchevsky, Charlene Z. Hirsch, and Joseph C. Touchstone

The need for sensitive methodology is apparent in enzyme studies that require isolation and quantitation of a specific component. These are limited by the smallest amount of material that can be measured. The use of radioactive substrates permits investigators to scale down various assays, but the use of radioactive materials limits the study to available labeled substances. Studies of aortic cholesteryl esterase (E.C. 3.1.1.13) activity in acetone powder preparations of aorta were dependent upon the assay method. The amount of cholesteryl esterase activity per aorta, both synthetic and hydrolytic, is small and thus dictates the number of specimens required for metabolic studies of this enzyme system. Furthermore, the assays are limited by the availability of suitable substrates of high specific activity. This enzymatic system (1,2) has been used for the study of the differences in cholesteryl ester metabolism between various species (3), for the assessment of the effects of cholesterol-lowering drugs on aortic metabolism (4), and for the study of esterification of sterols other than cholesterol (5).

The use of direct analysis of lipids was applied by Blank et al. (6) for lipids but these earlier methods were relatively insensitive. For this reason densitometry had given way to use of radioactive substrates. However, later Touchstone et al. (7) showed that densitometry can be a sensitive and reproducible technique. As indicated more recently there is a general acceptance of in situ quantitation by densitometry (8). In later work Touchstone et al. (9) showed that the use of improved sorbents (high-performance thin layer chromatography) and adaptation of these layers to spectrodensitometry led to microtechniques suitable for assay of as little as 10 ng of lipids. The use of the cupric acetate/phosphoric acid charring reagent (10) led to greater reproducibility and sensitivity. The present report

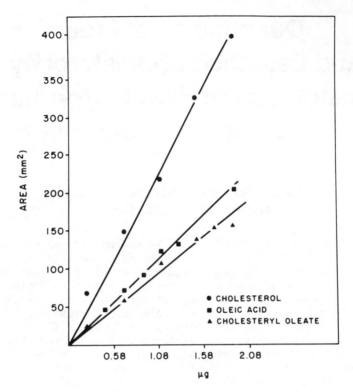

Figure 1.   Standard curves for cholesterol, cholesterol oleate, and oleic acid.   Detecting agent: Cupric acetate-H_3PO_4.

describes the use of HPTLC in the microdetermination in cholesterol and lipids in aortic tissues and incubations.

MATERIALS AND METHODS

A Model 3000 spectrodensitometer equipped with a Model 300D density computer (both from Schoeffel Instrument Corp., Westwood, N. J.) was used for scanning.   Plates were scanned at 600 nm in the dual-beam transmission mode of the spectrodensitometer (range up to 1.0 A).   The beam length used to scan 20 × 20 cm plates was 9 mm; that used for the 10 × 10 cm plates was 4.5 mm.   Peak areas were calculated by the multiplication of peak height by peak width at half height.   With baseline taken at half peak height,

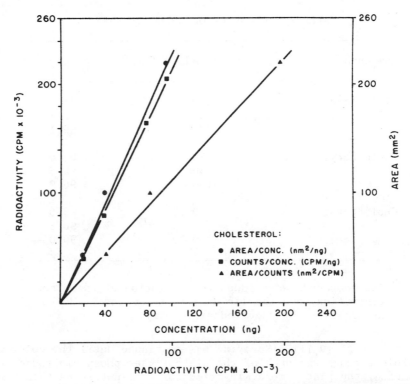

Figure 2. Results of analysis of known concentrations of [4-^{14}C]-cholesterol and spectrodensitometric analysis of comparable concentrations.

separation of the cholesterol peak from the origin was sufficient to give reproducible calculations.

For TLC, either 20 × 20 cm Silica Gel G plates (250 microns thick) or 10 × 10 cm Silica Gel HPTLC-F plates (100 microns thick) were used. They were purchased from Quantum Industries, Fairlawn, N. J., or from Analtech, Newark, Del. The plates were scored into lanes 10 mm (20 × 20 plates) or 5 mm (10 × 10 plates) wide with a Schoeffel scoring device or a dissecting scalpel and glass straight-edge. All plates were prewashed in the developing mobile phase of hexane-diethyl ether-acetic acid (83:16:1 by vol) before chromatography. After development, the chromatograms were dried for 5 minutes at 120° C before being sprayed with reagent.

The spray reagents used were either a solution of phospho-molybdic acid in absolute ethanol (50 g/liter) (10,11) or a 30-g/liter solution of cupric acetate in phosphoric acid solution

TABLE I.   RECOVERY OF ADDED LIPIDS[a]

Compound	Amount Added (μg)	Recovery (%)
Oleic acid	200	93.6
	500	98.3
	1200	99.1
Cholesteryl oleate	20	95.0
	50	95.3
	100	96.9
Cholesterol	0.12	98.0
	0.24	90.3
	0.30	90.0

[a] The compounds were added to a solution of known concentration, and the lipids were separated by HPTLC and analyzed by spectrodensitometry.

(80 ml/liter) (9,11). Spraying was continued until the color was visible through the back of the plate. When phosphomolybdic acid was used, the plates were air-dried for 15 minutes and then heated at 120° C for 5 minutes; when the reagent was cupric acetate-H_3PO_4, the plates were kept at 130° C for 30 minutes (12).

Cholesterol, cholesteryl oleate, and oleic acid were purchased from Sigma Chemical Co., St. Louis, Mo., and assayed for purity before use by thin layer chromatography. The radioactive substrates, [4-^{14}C]-cholesterol and [4-^{14}C]cholesterol oleate, were purchased from New England Nuclear Corp., Boston, Mass. The specific activities of the two radioactive substrates, the purity of which was ascertained by TLC, were 53.5 and 58.2 μCi/mole, respectively. Sodium taurocholate (Sigma Chemical Co.) was used without prior purification.

For the enzymic studies, normal aortas of male rabbits were obtained from Rockland, Gilbertsville, Pa. The acetone powder was prepared according to the method of Kothari et al. (1) and stored at -20° C until use. The enzyme was extracted by suspension of the requisite amount of powder in 20 times its weight of NH_4Cl-NH_4OH buffer (pH 8.0, 20 nmole NH_4/liter). The suspension was stirred for 30 minutes and centrifuged at 10,000 x g for 15 minutes. The supernate contained 0.5-0.9 mg protein/ml, as determined by a microscale biuret procedure (13).

TABLE II.   RESULTS OF PANCREATIC CHOLESTEROL ESTERASE REACTION:
COMPARISON OF RADIOASSAY AND SPECTRODENSITOMETRY
(AVERAGE OF 10 EXPERIMENTS)

	Hydrolysis (% Reaction)	
	Spectrodensitometry	Radioassay
Blank + S.D.	12.50 + 0.98	14.69 + 7.77
Variance[a]	7.84	52.90
Reaction + S.D.	70.73 + 4.00	58.86 + 4.00
Variance	5.69	6.80
True % reaction[b]	57.73 + 2.49	44.17 + 5.89
	Synthesis (% Reaction)	
Blank + S.D.	13.34 + 1.18	2.08 + 0.79
Variance	8.85	37.98
Reaction	39.87 + 1.27	30.85 + 1.34
Variance	3.19	4.35
True % reaction	26.47 + 1.23	28.77 + 1.07

[a] (S.D. ÷ average % reaction) × 100.

[b] % Reaction -blank.

For hydrolysis, 0.5 µmole cholesteryl ester was added to 0.5
ml phosphate buffer (pH 6.6, 150 nmole/liter) containing 150
nmole sodium taurocholate per liter. The mixture was homogenized
for 1 minute in a Potter-Elvehjem homogenizer and then sonicated
for 15 minutes at 30 kilocycles/second with a WL40 sonicator
(Ultrasonics, Inc., Plainview, N. Y.). For synthesis, 1.0 µmole
cholesterol and 3.0 µmole oleic acid were added to 0.5 ml acetate
buffer (pH 5.2, 4 nmole/liter) containing 50 nmole sodium taurocholate per liter.
    After adding 0.5 ml enzyme preparation to the appropriate
substrate, the mixture was incubated at 37° C with shaking for 4
hours in a Dubnoff metabolic shaker. For aortic enzyme, the
reaction was linear from the beginning; with porcine pancreatin,
enzyme activity increased exponentially after a lag time of 3.5
hours. The reaction was terminated by the addition of a 1 ml cold
NaCl solution (0.9 nmole/liter) and 4 ml benzene and vigorous
agitation of the mixture for 1 minute with a vortex-type mixer.
The mixture was centrifuged at 3000 rpm, and the organic layer

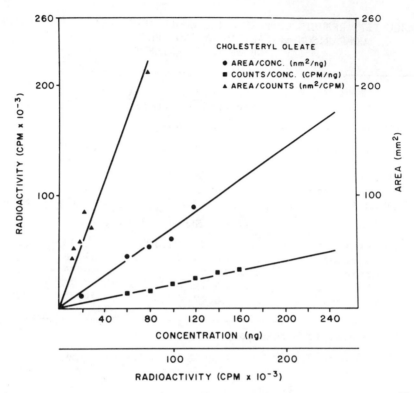

Figure 3. Results of analysis of known concentrations of [4-^{14}C]-cholesteryl oleate and spectrodensitometric analysis of comparable concentrations.

was aspirated and evaporated to a small volume under a stream of nitrogen.

Cholesteryl ester hydrolysis of porcine pancreatin (E.C. 3.1.1.3; Sigma, St. Louis, Mo.) was also studied. The enzyme was dissolved in NH_4Cl-NH_4OH buffer as described above and the reaction mixture treated in a manner similar to that described for the aortic enzyme.

For spectrodensitometry, the volume of extracted reaction mixture was adjusted to 0.1 ml and aliquots were applied to TLC plates for development. Quantities of 8-16 and 2-5 μl were spotted on 20 × 20 and 10 × 10 high-performance thin layer chromatography (HPTLC) plates, respectively. On HPTLC plates that had been divided into 20 lanes, samples were applied to lanes 5, 6, 9, 10, 13, 14, 15, and 16 to permit scanning in the dual-beam mode

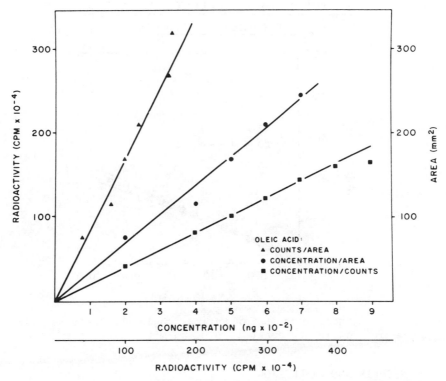

Figure 4. Results of analysis of known concentrations of $[1-^{14}C]$-cholesterol and spectrodensitometric analysis of comparable concentrations.

after the spots had been made visible with phosphomolybdic acid or cupric acetate-H_3PO_4. For assay of radioactivity, the developed thin layer plates were exposed to iodine vapor (10), the iodine was permitted to sublime, the appropriate areas made visible by the iodine were scraped into scintillation vials, 5 ml of Aquasol was added (14), and the radioactivity was determined in a Beckman LS 100 C liquid scintillation spectrometer.

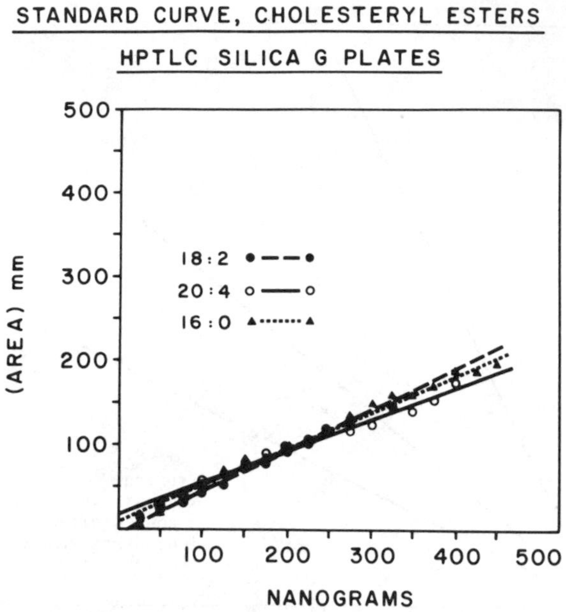

STANDARD CURVE, CHOLESTERYL ESTERS

HPTLC SILICA G PLATES

Figure 5.   Standard curves for various cholesteryl esters.   18:2,
linoleate; 20:4, arachidonate; 16:0, palmitate.

RESULTS AND DISCUSSION

Comparison of the detection methods revealed that phosphomolybdic
acid could be used for quantitation of cholesterol in the range
of 0.1-1.00 µg, but that oleic acid and cholesterol oleate were
not detectable in these amounts.   Cupric H_3PO_4, on the other hand,
was useful for measuring all three substances in the range 0.01-
1.00 µg.   Figure 1 shows standard curves for cholesterol, cholesterol oleate, and oleic acid when HPTLC was used in conjunction
with the cupric acetate-H_3PO_4 spray and spectrodensitometry.   To
evaluate this method further, analytical recovery of known concentrations of carbon-14-labeled cholesterol, cholesteryl oleate, and
oleic acid was compared.   The results (Figures 2-4) make it evident that spectrodensitometry is quite comparable to isotopic
assay.   Moreover, this method has the advantage of being very
sensitive and obviating use of labeled substrates, thus broaden-
ing the range of substrates that may be used.

STANDARD CURVE, CHOLESTERYL ESTER MIX

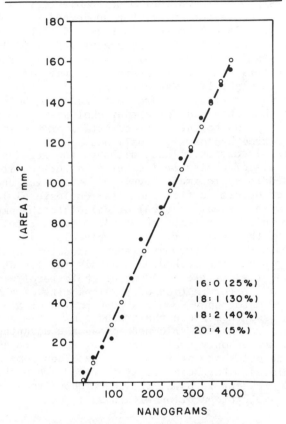

Figure 6. Standard curves for mixed cholesteryl esters. 16:0, palmitate; 18:1, oleate; 18:2, linoleate; 20:4, arachidonate. Closed circles, analytical data; open circles, best-fit line.

The analytical recovery of added cholesterol, cholesteryl oleate, and oleic acid is shown in Table I.

After an experiment involving pancreatic cholesterol ester-ase (Table II) the analytical recovery of material by spectroden-sitometry was compared to radioassay. Results by the two methods agreed well for assay of the product of synthesis, cholesteryl oleate, but were less similar with regard to the measurement of the products of the hydrolysis reaction. The very large variance

in the radioisotope assay could be due to radiation-induced formation of extraneous reaction products.

Since only cholesteryl oleate was used in these studies, it
was of interest to learn whether other cholesteryl esters would
exhibit similar characteristics when subjected to spectrodensitometric analysis. As can be seen in Figure 5, the esters of palmitic (16:0), linoleic (18:2), and arachidonic (20:4) acids gave
very similar standard curves. Also tested was a mixture of
cholesteryl esters combined to approximate the proportion found
in serum. A straight line was obtained (Figure 6).

As seen in Figure 1, the area obtained for cholesteryl oleate
is less than that obtained for either cholesterol or oleic acid.
This is probably due to the characteristics of the spray reagent
(affinity for free hydroxyl groups?) but speculation is beyond the
scope of this discussion. Thus, although the reading obtained
through analysis of cholesteryl ester is accurate, it cannot
simply be used to approximate levels of esterified cholesterol.
One can either determine free and total cholesterol in a sample
(before and after saponification) or calculate the amount of
cholesterol, which represents about 64.5% (myristate) to 59.4%
(linoleate) of the cholesteryl ester molecule.

The combination of chromatography and spectrodensitometry
offers a very sensitive method of analysis for lipids, permitting
determination of as little as 10 ng of cholesterol without derivative formation. Determination of this quantity of cholesterol by
gas-liquid chromatography requires preparation of a halogenated
ether. When used to measure the products of enzymic reactions,
the HPTLC-spectrodensitometric method has the advantage that it
requires the use of only small amounts of substrate and that it
eliminates the necessity of recovering reaction components from
thin layer plates. Furthermore, it is rapid; when the cupric
acetate-H_3PO_4 reagent is used, the spectrodensitometry yields
data from a sample preparation within 90 minutes.

REFERENCES

1. H. V. Kothari, B. F. Miller, and D. Kritchevsky, Biochim.
   Biophys. Acta 296, 446 (1973).
2. H. V. Kothari and D. Kritchevsky, Lipids 10, 322 (1975).
3. D. Kritchevsky and H. V. Kothari, Steroids Lipids Res. 5, 23
   (1974).
4. D. Kritchevsky and H. V. Kothari, in Atherosclerosis III,
   edited by G. Schettler and A. Weizel (Springer Verlag, Berlin,
   1974), pp. 39-45.
5. D. Kritchevsky, S. A. Tepper, and H. V. Kothari, Artery 1,
   437 (1975).

6. M. L. Blank, J. A. Schmit, and O. S. Privett, J. Am. Oil Chem. Soc. 41, 371 (1964).
7. J. C. Touchstone, S. S. Levin, and T. Murawec, Anal. Chem. 43, 858 (1971).
8. J. C. Touchstone and J. Sherma, eds., Densitometry in Thin Layer Chromatography (Wiley-Interscience, New York, 1979).
9. J. C. Touchstone, M. F. Dobbins, C. Z. Hirsch, A. R. Baldino, and D. Kritchevsky, Clin. Chem. 24, 1496 (1978).
10. M. E. Fewster, B. J. Burns, and J. E. Mead, J. Chromatogr. 43, 120 (1969).
11. D. Kritchevsky and M. R. Kirk, Arch. Biochem. Biophys. 35, 346 (1952).
12. E. A. MacMullen and J. E. Heveran, in Quantitative Thin-Layer Chromatography, edited by J. C. Touchstone (John Wiley, New York, 1973), p. 224.
13. Technical Bulletin 201, Charring Techniques (Quantum Industries, Fairfield, N. J., 1975).
14. R. F. Itzhaki and D. M. Gill, Anal. Biochem. 9, 401 (1964).
15. D. Krithevsky and S. Malhotra, J. Chromatogr. 52, 498 (1970).

CHAPTER 32

Application of TLC to the Study of Phenobarbital Metabolism

**Sidney S. Levin, Joseph C. Touchstone,
and David Y. Cooper**

The barbiturates are all derivatives of barbituric acid, which was first produced by the condensation reaction of urea and malonic acid by Adolph von Baeyer in the year 1864. Barbituric acid is not a central depressant--it does not have hypnotic or sedative effect. It is used in the manufacture of plastics as well as sedatives and hypnotics. The two hydrogens on the carbon atom in position 5 must be substituted by an alkyl or aryl group to induce a sedative effect. ⬧

Urea + Malonic Acid = Barbituric Acid

	R_1	R_2
Barbiturate		
Barbitual	C_2H_5	C_2H_5
Phenobarbital	C_2H_5	C_6H_6

The first hypnotic derivative of barbituric acid, barbital, was introduced by Fischer and von Mering in 1903 and was called "Veronal." It is still considered an excellent hypnotic but has been supplanted to some extent by some of the newer more controllable compounds.

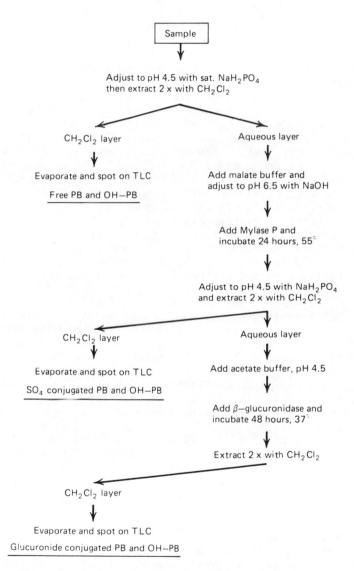

Figure 1. Flow sheet for extraction of phenobarbital and its metabolites.

467

In 1912 the second of the barbiturates, phenobarbital (PB), was introduced and marketed under the trade name, "Luminal." The substitution of the phenyl group for an alkyl group gives the compound selective anticonvulsant activity.

The barbiturates as a group depress the activity of all excitable tissues; however, not all tissues are affected to the same degree. The CNS is most sensitive; therefore a hypnotic or sedative dose will have little or no effect on the skeletal, cardiac, or smooth muscle. It is only in acute barbital depression, that is, intoxication, that serious defects in cardiovascular and other peripheral functions occur.

In the adult human the hypnotic dose of phenobarbital is about 2 mg/kg body weight. The sedative dose, which is one-third to one-quarter the hypnotic dose, will induce sleep for 6 to 7 hours. It is the best of the anticonvulsants but slower acting than some of the newer drugs. PB is used as a preanesthetic drug to reduce the excitatory state of the general anesthetics. The drug is the most effective compound used to stimulate the hepatic enzyme activity (1,2). In neonatal jaundice PB is used to stimulate the liver to conjugate bilirubin (3). Remmer demonstrated that PB stimulated the microsomal enzyme systems (4). Maximum induction, that is, stimulation of the hepatic microsomal P-450 cytochrome oxidase system in the rat, occurs after the administration of PB for a period of four days. PB was administered to Sprague Dawley rats weighing 200-300 g at a daily dose level of 80 mg/kg body weight. Although the dose level is 40 times the average human adult hypnotic dose, there was little effect on activity, food intake, or weight gain. These results indicated that, in the rat, there is either rapid metabolism to an inactive form, rapid excretion, or both. Butler (5) has shown that in man and the dog the major metabolite is parahydroxyphenobarbital [5-ethyl-5-(p-hydroxyphenyl)barbituric acid (pOHPB), which may then be conjugated with either glucuronic acid or sulfate and eventually excreted. Preliminary studies on the rat indicate similar results (7,8).

The distribution of the PB and its metabolites was determined through the use of thin layer chromatography of extracts of homogenates of feces and sections of intestine, muscle, liver, and kidney as well as urine and bile. PB and (2-^{14}C) PB were administered intraperitoneally to the rat or used as substrates for in vitro studies. Initially analysis was attempted by gas chromatography (9); however, although PB was readily detected, it was very difficult to determine the PB metabolites without derivitization. Fluorescent thin layer plates were used because the products as well as the substrate absorbed somewhat at 254 nm and could therefore be visualized and quantitated on the plate by nondestructive methods (10).

MATERIALS AND METHODS

Preparation for extraction depended on the type of sample. Bile
and urine were extracted directly with dichloromethane. The
feces was weighed and then homogenized with a Polytron homogenizer using a minimal amount of water. The homogenizer was
operated in a series of 30-second pulses, until the homogenate
appeared smooth. The homogenizer was rinsed two times with minimal amount of water that was then combined with the initial homogenate; the volume was noted. An aliquot of the combined homogenate was taken for extraction. Tissues were homogenized in a similar fashion. When $2\text{-}^{14}C$ PB was used, a 50-μl aliquot was added
to counting vials containing 10 ml Hydromix scintillation cocktail (Yorktown). These samples were counted directly with an
Intertechnique scintillation counter to determine the total activity of the tissue.

The general extraction procedure for all samples was carried
out in conical heavy walled tubes with screw caps following the
scheme in Figure 1 (11,12). The samples were adjusted to a pH
4.5 with a saturated solution of monobasic sodium phosphate and
then extracted with 5 volumes of distilled methylene chloride by
shaking for 1 minute. The tubes were then centrifuged and the
methylene chloride layer aspirated. The extraction was repeated;
the extracts were combined and then evaporated to dryness under
nitrogen at 37° C. The dry residue was dissolved in 200 ml of a
methylene chloride-acetone-methanol (1:1:1) mixture and an aliquot
was streaked on a previously scored thin layer plate. This fraction represents the "free" or unconjugated fraction of the PB
metabolites.

Maleic buffer (0.5 M) was added to the aqueous phase following the extractions, in a ratio of 1 maleate : 4 sample; then the
pH was adjusted to 6.5 with 5 M NaOH, Mylase P (Aspergillus oryzae)
(3) (5 mg/ml total volume) which was then added to each sample.
The samples were hydrolyzed at 55° C for 24 hours. At this pH
and temperature the Mylase is specific for sulfate hydrolysis
(13). After incubation the mixture was readjusted to pH 4.5 with
saturated monobasic sodium phosphate, extracted with methylene
chloride, and evaporated to dryness as described in the previous
paragraph. This extract represented the "sulfated" metabolites
of PB.

The aqueous phase from this extraction was treated with 0.5
M sodium acetate buffer, pH 4.5 (1 ml acetate : 3 ml sample).

For each milliliter of final sample volume 35 mg β-glucuroni-
dase (78000 Sigma) was then added. The samples were then hydrolyzed at 37° C for 48 hours, followed by extraction in the same
manner as the previous fractions. This is the "glucuronide"
fraction.

Analtech 20 × 20 cm silica gel GF 254 precoated plates were scored into 10-mm lanes with a Schoeffel scoring instrument and then activated, that is, heated at 110° C for 30 minutes before use. Depending on the concentration, the samples were streaked using 10-μl aliquots on each of three to five adjacent lanes. A blank lane was left between each of the sample series to act as reference for quantitation on the densitometer in the double beam mode. Authentic PB and pOHPB standards were applied on each chromatogram as an external standard as well as an overspot on one lane of each sample series.

The plates were developed with an ethyl acetate-methanol-ammonium hydroxide mobile phase (82:14:4). The ammonium hydroxide was Baker Analyzed with an assay of 27.6% NH. The chromatograms were scanned with a Schoeffel densitometer (Model 3000) at 255 nm in transmission, double beam mode. The absorption bands were marked and the lanes were sectioned. The chromatograms were then sprayed with water to deactivate the sorbent and to reduce the loss of powder into the air. The deactivation increased the recovery rate from the TLC plate from 60-70% on the unsprayed plate to 80-90% on the wetted plate. The lanes were scraped in a serial manner starting from the solvent front. The sorbent was collected in counting vials, allowed to dry completely; then 10 ml of toluene solution of Omnifluor was added to each vial and radioactivity was assayed. The combination of radioactive assessment and the densitometer scans permitted differentiation between the metabolites of the radioactive and nonradioactive PB and PB substrates.

RESULTS

The total activity was determined on aliquots of tissue homogenate obtained from PB preinduced rats 18 hours after a single intraperitoneal injection of 2-^{14}C PB. The drug was distributed to all tissues as in the process of being excreted in the feces and urine. There was higher activity in the gastrointestinal tract (per gram of tissue) than in either the liver or kidney (Table I). In the gut the radioactivity was lowest in the duodenum and increased through the jejunum and ileum to the colon. At the time of sacrifice 37% of the injected radioactivity had been excreted in the urine and 6.1% in the feces. The urine and feces were extracted and hydrolyzed for metabolite analysis as described previously (Table II). In the feces 84% of the activity was excreted as pOHPB metabolites, of which 63% was unconjugated, 18% was sulfate, and 19% was glucuronide. In the urine, however, only 36% of the radioactivity was excreted as pOHPB metabolites,

TABLE I.   DISTRIBUTION OF ^{14}C IN RAT TISSUES
18 HOURS AFTER INTRAPERITONEAL INJECTION OF 2-^{14}C-PB

Tissue	DPM/g	Total Counts $(\times 10^3)$
Spleen	6,641	9.6
Liver	7,835	142.9
Kidney	13,246	39.1
Duodenum	34,035	95.0
Jejunum	86,585	143.7
Ilium	161,618	221.4
Colon	364,353	477.3
Brain	6,372	12.2
Muscle	12,306	1,477.0
Fat	15,155	99.1
Packed RBC	29,825	268.4
Plasma	23,710	213.4

Sample	Total Collected	Total Counts $(\times 10^3)$	% of ^{14}C Injected
Urine	32 ml	37,990	37
Feces	1.91 g	5,761	6

of which 16% was unconjugated, 59% was sulfate, and 24% was glucuronide. Of the unmetabolized PB in the urine and feces 93% and 97%, respectively, was excreted unconjugated, that is, methylene chloride extractable.

In another study, where 1 μCi per day 2-^{14}C PB was injected for four consecutive days, urine and feces were collected on the second, third, and fourth days of injection as well as on the sixth and eighth days after the first injection (Tables IIIA and IIIB). On the second day 36% of the activity excreted in the feces was excreted as pOHPB; this increased to a maximum level of 82 and 84% on the sixth and eighth days, respectively. Forty-seven percent of the pOHPB was unconjugated on the second day; however, this increased to 69% by the eighth day. The sulfate conjugates of the pOHPB that were at 30% of the pOHPB on the second day leveled off and remained at a 24% level for the balance

TABLE II. EXCRETION OF PHENOBARBITAL AND ITS METABOLITES 18 HOURS AFTER 2-^{14}C-PB INJECTION

Sample	Phenobarbital (%)	p-OH-Phenobarbital (%)
Urine	64	36
Feces	16	84

Sample	Phenobarbital (%)			p-OH-Phenobarbital (%)		
	Unconjugated	Sulfate (%)	Glucuronide	Unconjugated	Sulfate (%)	Glucuronide
Urine	93	3	4	16	59	24
Feces	91	7	2	63	18	19

TABLE IIIA.   PHENOBARBITAL AND p-OH-PHENOBARBITAL (pOPHB)
             EXCRETED IN FECES

Experimental Day	1	2	3	4	6	8
Injection	[1 µCi 2-^{14}C-PB]					
	↓	↓	↓	↓		
Total PB (%)		64	58	32	18	16
Total p-OHPB		36	68	82	84	

TABLE IIIB.   PHENOBARBITAL AND ITS METABOLITES IN FECES

Experimental Day	1	2	3	4	6	8
Injection	[1 µCi 2-^{14}C-PB]					
	↓	↓	↓	↓		
PB						
Unconjugated (%)		72	74	82	69	58
Sulfate (%)		7	14	13	22	32
Glucuronide		20	12	5	9	10
p-OHPB						
Unconjugated (%)		47	57	57	57	69
Sulfate (%)		30	30	24	24	1
Glucuronide		23	13	19	19	37

473

of the study. The glucuronides that made up 25% of the pOHPB
initially also decreased and remained constant at 19%. The
unmetabolized PB that was initially 72% of the PB excreted in the
feces increased to 82% by the fourth day of injection and then
decreased during the post-injection period.

In another series of studies PB was mixed into the standard
diet and added to the drinking water of rats for a three-day
period. The rats were then killed and the gastrointestinal tract
from the duodenum to the caecum was excised and sectioned into
segments approximating the duodenum, jejunum, ileum, and caecum.
The ends of each segment were tied so that the lumen became a sac
in which was placed 1 µCi 2^{14}C PB in 50 mM Tris buffer, pH 7.5.
The intestinal segments were then incubated for 3 hours in small
beakers containing Tris buffer adjusted to pH 7.8 at 37° C. The
reaction was stopped with cold methylene chloride (MC&B), and the
incubates were extracted, spotted, and developed as previously
described. The chromatograms were scanned at 254 nm by densitometry for quantitation against authentic standards. The results
(Table IV) indicated that the total pOHPB metabolites increased
progressively from 5% of the extracted PB-pOHPB in the duodenum
to 34% in the caecum. The sulfates increased successively in the
descending segments. However, the glucuronides increased in the
jejunum and ileum and then dropped in the caecum at the same time
that the proportion of free pOHPB increased. Perhaps the glucuro-
nidase of the flora of the ileum hydrolyzed the glucuronides.
The plates were then sectioned and scraped for scintillation
counting. The results (Table V) indicate that more than 99% of
the CPM was unmetabolized PB. However, the distributions of the
PB and pOHPB metabolites as determined with the densitometer were
not radioactive and were the result of the ingested PB. The incubation of the 2-^{14}C PB did not metabolize the PB.

These results indicate that in the rat at least there is an
active oxidative metabolism of phenobarbital in the gastrointestinal tract. Further studies of the role of sulfation and glucu-
ronidation in the liver as well as the part taken by the gastrointestinal tract are necessary in order to assess the true status
of drug metabolism in the enterohepatic system.

TABLE IV. DISTRIBUTION OF INGESTED PHENOBARBITAL (PB)
AND p-HYDROXYPHENOBARBITAL (OHPB) IN GASTROINTESTINAL TRACT

	Duodenum	Jejunum	Ileum	Caecum
		Intestinal Segment		
PB	µg	µg	µg	µg
Free form	2088	2592	2289	1346
OHPB				
Free form	85 (78%)	111 (50%)	268 (37%)	415 (60%)
Conjugates				
Sulfate	13 (12%)	43 (19%)	151 (21%)	185 (26%)
Glucuronide	11 (10%)	68 (31%)	306 (42%)	97 (14%)
Total	109 (100%)	222 (100%)	725 (100%)	697 (100%)
Total PB + OHPB	2197	2814	3013	2043
PB/OHPB	19	12	3	2

TABLE V. DISTRIBUTION OF PHENO[2-^{14}C]BARBITAL AND ITS METABOLITES FOLLOWING $\underline{IN\ VITRO}$ INCUBATION

| | Intestinal Segment DPM | | | |
	Duodenum	Jejunum	Ileum	Caecum
PB				
Free form	1,225,580	1,321,227	1,701,255	1,373,046
OHPB				
Free form	2,825 (53%)	2,748 (47%)	4,730 (69%)	2,669 (66%)
Conjugates				
Sulfates	2,115 (39%)	1,944 (33%)	1,457 (21%)	948 (22%)
Glucuronides	444 (8%)	1,141 (20%)	673 (10%)	487 (12%)
Total	5,384 (100%)	5,833 (100%)	6,860 (100%)	4,104 (100%)
Total PB + OHPB	1,230,964	1,327,060	1,708,116	1,377,150
PB/OHPB	99.6	99.6	99.6	99.7

REFERENCES

1. D. Y. Cooper, H. Schleyer, J. H. Thomas, H. M. Vars, and O. Rosenthal, in Cytochrome P-450 and b-5: Structure, Function and Interaction, edited by D. Y. Cooper et al. (Plenum Press, New York, 1975), pp. 81-102).
2. L. Rosenthal, H. M. Vars, H. Schleyer, D. Y. Cooper, S. S. Levin, and J. C. Touchstone, in Microsomes and Drug Oxidations, edited by V. Ullrich et al. (1977), p. 598.
3. J. F. Crigler and N. J. Gold, J. Clin. Invest. 45, 998 (1966).
4. H. Remmer and H. J. Merker, Ann. N. Y. Acad. Sci. 125, 79 (1965).
5. T. C. Butler, J. Pharmacol. 113, 326 (1955).
6. T. C. Butler, J. Pharmacol. Exp. Ther. 116, 326 (1956).
7. J. C. Touchstone, S. S. Levin, B. Novack, and D. Y. Cooper, Fed. Proc. 36, 843 (1977).
8. S. S. Levin, D. Y. Cooper, J. C. Touchstone, and O. Rosenthal, Fed. Proc. 37, 1472 (1978).
9. J. C. Touchstone, M. Schwartz, and S. S. Levin, J. Chromatogr. Sci. 15, 528 (1977).
10. J. C. Touchstone, S. S. Levin, and T. Murawec, Quantitative Thin Layer Chromatography, edited by J. C. Touchstone (John Wiley, New York, 1973), p. 1.
11. S. K. Levy and T. Schwartz, Clin. Chem. Acta 54, 19 (1974).
12. R. C. Baselt and L. J. Casarett, J. Chromatog. 57, 139 (1971).
13. J. C. Touchstone, J. W. Greene, R. C. McElroy, and T. Murawec, Biochemistry 2, 653 (1963).

CHAPTER 33

Thin Layer Radiochromatographic Technique for Quantitative Determination of Prostaglandins, Thromboxanes, and Prostacyclin

James G. Hamilton and Lawrence D. Tobias

The original thin layer chromatographic (TLC) mobile phases of Green and Samuelsson (1) were developed at a time (1964) when prostaglandins E and F were the only prostaglandins known to occur naturally. The mobile phase* used to separate E and F prostaglandins was benzene-dioxane-acetic acid (20:20:1) (A I); Benzene-dioxane (5:4) (M I) was used for separation of the methyl esters. The separation of PGE_1, E_2, and E_3, or $PGF_{1\alpha}$, $F_{2\alpha}$, and $F_{3\alpha}$ required silica gel containing silver nitrate. The organic layer from ethyl acetate-acetic acid-methanol-isooctane-water (110:30:35:10:100) (A II) was used to separate free acids and the organic phase from ethyl acetate-methanol-water; either 8:2:5 (M II) or 32:5:20 (M III) was used to separate methyl esters. Good separation of all of these compounds was obtained.

A synthetic program for diasteromeric prostaglandins led Andersen (2) in 1969 to devise a number of other TLC mobile phases for prostaglandins. Many of these were variations of ones devised by Green and Samuelsson. System P-11 consists of ethyl acetate-hexane-water-methanol-acetic acid and is similar to Green and Samuelsson's A II. The D-series systems, which consist of benzene-dioxane-acetic acid, are similar to A I. The H series consists of hexane-methylene dichloride-THF-acetic acid in varying proportions, and the F series is based on ethyl acetate with more polar additives.

In 1966, Hamberg and Samuelsson (3) added TLC mobile phases A VII, A IX, and A X to the ones developed by Green and Samuelsson.

---

* All mobile phases covered by this report can be found in Table I. Designations given each mobile phase in Table I appear in parentheses after ratio.

A VII and A IX are the organic layer from ethyl acetate-acetic acid-isooctane-water, 110:20:30:100 for A VII and 90:20:50:100 for A IX. A X is the organic layer from ethyl acetate-isooctane-water (50:100:100).

Radiochromatography is a technique that is extremely useful in all types of chromatography. It has been used extensively in both paper and thin layer chromatography almost from their inception. (For a review of the early application to thin layer chromatography see Mangold [4]).

The first radiochromatography of prostaglandins played an important role in elucidating the pathway of prostaglandin biosynthesis. Van Dorp et al. (5) and Bergstrom et al. (6) in 1964 incubated [^{3}H]arachidonic acid with sheep seminal vesicular glands and showed that [^{3}H]arachidonic was converted into [^{3}H]PGE$_2$. Van Dorp et al. used a mobile phase of chloroform-methanol-acetic acid-water (90:6:1:0.75), whereas Bergstrom et al. used solvent systems A I and A II. This work established the biosynthetic relationship of polyunsaturated fatty acids and prostaglandins and was the basis for all subsequent radiochromatography.

Pace-Asciak et al. in 1970 (7) using radiochromatography showed that rat stomach homogenate converted PGF$_{1\alpha}$ into metabolites. The time course of conversion and the effect of substrate concentration on the conversion to the two metabolites were elucidated. A mobile phase of chloroform-methanol-acetic acid-water (90:9:1:0.65) and a Packard Model 7201 radiochromatogram scanner were used. The metabolites were isolated by preparative thin layer chromatography and their structure determined by gas chromatography-mass spectrometry.

In 1973, Crowshaw (8) examined the conversion of [^{3}H]arachidonic acid into products by rabbit renal cortex and renal medulla. The mobile phase used consisted of chloroform-methanol-acetic acid (90:5:5); products were monitored with a Packard Model 7201 radiochromatogram scanner. Tissue slices were incubated in phosphate buffer with 2 mM glutathione and [1-^{14}C]arachidonic acid. Slices of both medulla and cortex incorporated the arachidonic acid into triglycerides and phospholipids. No prostaglandins were found in either tissue. Both PGE$_2$ and PGF$_{2\alpha}$ were released into the medium from the slices of medulla but not from the cortex. Arachidonic acid was found in both. Previous work showed that PGE$_2$ and PGF$_{2\alpha}$ were present in medulla but not in cortex.

Stone et al. (9) examined the effect of various substances on prostaglandin synthesis from [1-^{14}C]arachidonate using sheep seminal vesicles as a source of enzyme. The mobile phase used was 1% formic acid in ethyl acetate; products were analyzed with a Panax radiochromatogram scanner. Gold salts and phenylbutazone selectively inhibited the synthesis of PGF$_{2\alpha}$ and PGE$_2$,

TABLE I. MOBILE PHASES FOR THIN LAYER CHROMATOGRAPHY

Composition	Compounds Separated	Designation	Reference(s)
Benzene-dioxane			
5:4	ME esters of PG	M I	1
3:2	PG	D I	2
Ethyl acetate-methanol-water[a]			
8:2:5	Me esters of PG	M II	1
16:2.5:10	Me esters of PG	M III	1
Ethyl acetate-acetic acid-methanol-isooctane-water[a]			
110:30:35:10:100	PG	A II	1,6
110:10:15:10:100	PG	A II	3
Ethyl acetate-acetic acid-methanol-hexane-water[a]			
4:1:1:2:2	PG	P II	2
Ethylacetate-isooctane-water[a]			
50:100:100	PG	A X	3
50:100:100	Endoperoxides		11
75:75:100	Hydroxy FA		12
Benzene-dioxane-acetic acid			
20:20:1	PG	A I	1
20:20:1	PG	D IV	2
40:10:1	PG	D II	2
20:10:1	PG	D III	2,13,14
20:10:1	PG and TXB_2		17,31,32

Solvent system		Code	Ref.
Chloroform–THF–acetic acid 10:2:1	PG	C I	2
Hexane–THF–methylene dichloride 1:1:1	PG	N I	2
Hexane–THF–methylene dichloride–acetic acid 6:2:2:1	PG	H I	2
30:3:10:3	PG	H II	2
10:10:10:1	PG	H IV	2
Ethylacetate–formic acid 100:1	PG	F I	2,9
400:5	PG	F IV	2
Ethylacetate–acetic acid 99:1	PG and TXB$_2$		26
Ethylacetate–ethanol–acetic acid 100:1:1	PG	F V	2
Ethylacetate–acetone–acetic acid 90:10:1	PG	F VI	2,29
Cyclohexane–ethylacetate–acetic acid 60:40:2	PG	F VII	2
Chloroform–methanol–acetic acid–water 90:9:1:0.65	PG and 6-keto PGF$_{1\alpha}$		7,21,24,27

Table I (continued)

Composition	Compounds Separated	Designation	Reference(s)
Chloroform-methanol-acetic acid-water			
90:8:1:0.8	PG and TXB_2		30
90:6:1:0.75	PG		5
Chloroform-methanol-acetic acid			
90:5:5	PG		8
90:10:2	PG and 6-keto $PGF_{1\alpha}$		19
Ethylacetate-isooctane			
55:45	Me esters of PG		10
Ethylacetate-isooctane-acetic acid			
50:50:0.5	Endoperoxides		10
Ethylacetate-acetic acid-isooctane-water[a]			
90:20:50:100	PG	A IX	3
90:20:50:100	Hydroxy FA		13
90:20:50:100	PG and 6-keto $PGF_{1\alpha}$		20
11:2:5:10	PG and 6-keto $PGF_{1\alpha}$		24,26,29,32,34
97.5:17.5:44.5:89.5	PG and 6-keto $PGF_{1\alpha}$		25
110:20:30:100	PG		3
80:20:50:100	Hydroxy FA	A VII	16
Ethylacetate-isooctane-water[a]			
50:100:00	PG	A X	3
75:75:100	PG and thromboxanes		12

Composition	Compounds Separated	Designation	Reference(s)
Isopropyl ether-methyl ethyl ketone-acetic acid 50:50:1	PG and TXB_2		
Hexane-ether-acetic acid			
90:15:1	Hydroxy acids		13
15:85:0.1	Neutral lipid		28
25:75:1	Endoperoxides		33
	PG and hydroxy FA		
Ether-methanol-acetic acid			
90:1:2	PG and TXB_2		14,16
90:2:0.1	Endoperoxides		28
90:1:2	PG endoperoxides		15

a Less polar phase.

483

respectively. Most nonsteroidal antiinflammatory compounds were
found to inhibit PGE_2 and $PGF_{2\alpha}$ synthesis equally.

In 1974 Hamberg et al. (10) isolated two prostaglandin endoperoxides by incubating $[1-^{14}C]$arachidonic for short periods of
time with sheep seminal vesicular microsomes. The mobile phase
used for chromatography was ethyl acetate-isooctane-acetic acid
(50:50:0.5) (for free acids). A Berthold Dunnschicht scanner was
used for determination of radioactive products. The less polar
endoperoxide (15-hydroperoxy-9α, 11α-peroxidoprosta-5,13-dienoic
acid) was given the trivial name PGG_2 and the more polar endoperoxide (15-hydroxy-9α,11α-peroxidoprosta-5,13-dienoic acid)
was given the trivial name PGH_2. PGG_2 and PGH_2 were found to be
50-450 times more active than PGE_2 in causing contraction of a
rabbit aorta strip. The half-life in aqueous media of 5 minutes
was significantly longer than that of "rabbit-aorta-contracting
substance" released from guinea pig lung, which indicated that
they were not identical. Addition of the endoperoxide to washed
human platelets caused rapid aggregation. Hamberg and Samuelsson
(11) investigated the transformation of arachidonic acid and endoperoxides in platelets. Arachidonic acid incubated with human
platelets was converted into three compounds: 12L-hydroxy-5,8,10,
14-eicosatetraenoic acid (HETE), 12L-hydroxy-5,8,10-hepadecatrie-
noic acid (HHT), and the hemiacetal derivative of 8(1-hydroxy-3-
oxopropyl)9,12L-dihydroxy-5,10-hepatadecadienoic acid (TBX_2).
Aspirin and indomethacin inhibited the formation of the latter
two compounds, indicating that their formation proceeds by pathways involving fatty acid cyclooxygenase. HETE is formed by the
action of a lipoxygenase. 5,8,11,14-Eicosatetraynoic acid inhibited both the lipoxygenase and cyclooxygenase. The endoperoxides
were transformed almost exclusively into HHT and HETE with minimal formation of PGE_2 and $PGF_{2\alpha}$. The TLC mobile phase was the
organic layer of ethyl acetate-isooctane-water (50:100:100), with
a Berthold scanner used for product analysis.

Hamberg (12) et al. detected an intermediate (TXA_2) in the
formation of thromboxane B_2 from arachidonic acid and endoperoxides. The structure was deduced from trapping experiments with
appropriate nucleophiles. Incubation of arachidonic acid or endoperoxide with washed platelets led to formation of an unstable
factor that induced irreversible platelet aggregation. The properties and mode of formation indicated that this factor was identical to thromboxane A_2. Evidence suggested that the more unstable and major component of rabbit-aorta-contracting substance
formed in platelets and guinea pig lung was also TXA_2. The
mobile phases consisted of the organic layer of ethyl acetate-
isooctane-water (75:75:100); a Berthold scanner was used to monitor products.

Bills et al. (13) made a thorough study of the metabolism of [^{14}C]arachidonic acid in human platelets. There was a time-dependent incorporation of [1-^{14}C]arachidonate into platelet phosphatidicylcholine, phosphatidylinositol, phosphatidylethanolamine, and phosphatidylserine when incubated with platelet-rich plasma. When the labeled platelets were washed and treated with thrombin, there was a major decrease in the radioactivity of phosphatidy-choline and phosphatidylinositol. This decrease accounted for the corresponding increase in the formation of HETE, HHT, and TXB$_2$. The type of scanner used to detect radioactivity was not specified. Neutral lipids were separated on silica gel F-254 using hexane-ether-acetic acid (90:15:1). Hydroxy fatty acids (HHT and HETE) were separated on Ag-impregnated silica gel F-254 using the organic layer of ethyl acetate-acetic acid-isooctane-water. Phospholipids were separated on Whatman Q5W silica gel using chloroform-methanol-13.5 M ammonia-water (70:30:4:1). The phosphatidylserine-phosphatidylinositol zone was eluted from the thin layer chromatogram and separated on Whatman SG81 paper at 4° C using 2,6-dimethyl-4-heptanone-acetic acid-water (40:20:3).

Bailey et al. (14) examined a number of mobile phases for their ability to separate thromboxqnes from prostaglandins. A Vanguard two-dimensional radioscanner, Model 940, equipped with a recorder and dot printer, provided both qualitative and quantitative information. Of nine mobile phases tested, only diethyl ether-methanol-acetic acid (90:1:2) gave reasonable separation of TXB$_2$ from prostaglandins. Diethyl ether-methanol-acetic acid (90:1:2) was originally developed by Nugteren and Hazelhof (15) for the separation of the products of the enzymatic conversion of polyunsaturated fatty acids by sheep seminal vesicles.

Lagarde et al. (16) worked out a radiochemical assay for HETE, HHT, and 11-mono-O-methylthromboxane B$_2$ (the product formed when thromboxane is trapped with methanol). Hydroxy fatty acids were separated using the organic phase of ethyl acetate-acetic acid-isooctane-water (80:20:50:100). O-Methyl TXB$_2$ was separated from prostaglandins with ether-methanol-acetic acid (90:1:2).

Ali et al. (17) found that bovine gastric mucosal microsomes synthesized prostaglandins from arachidonic acid but thromboxanes were the principal product. The mobile phase used was benzene-dioxane-acetic acid (20:10:1), one of the mobile phases that Bailey et al. (12) had considered inadequate for separation of prostaglandins from thromboxanes. A Packard Model 7201 radio-chromatogram scanner was used. Gas chromatography-mass spectrometry of the TXB$_2$ fraction from the TLC plate confirmed the identification of TXB$_2$.

Anderson et al. (18) examined the kinetics of the conversion of the endoperoxide PGH$_2$ by thromboxane synthase. Both HHT and

TXA_2 were formed from PGH_2 by the same enzyme, thromboxane synthase, and TXA_2 broke down exclusively to TXB_2. A reaction
mechanism was postulated in which one molecule each of TXA_2 and
HHT are produced simultaneously from two molecules of PHG_2.
Ethyl acetate-isooctane-water-acetic acid (75:75:100:4) was used
for separation of O-methyl-TXB_2 and TXB_2, and light petroleum-
ether-acetic acid (50:50:1) for separation of HHT, HETE, and
arachidonic acid. Plates were scraped in 0.4-cm sections from
the origin to the solvent front, and the radioactivity was determined by liquid scintillation counting.

    Researchers at Wellcome Laboratories in 1976 (19) found that
arterial walls contain an enzyme that converts endoperoxides
into a substance that inhibits platelet aggregation. The substance, PGX, relaxed mesenteric and coeliac muscle strips but did
not contract rabbit aorta, pulmonary artery, or vena cava strips.
At pH 7.6 it lost its antiaggregatory activity within 20 minutes
at 22° C. Coinciding with the loss of antiaggregatory activity
was the appearance of a chemically stable substance that differed
in R_f from any prostaglandin known to be derived directly from
PGH_2, when monitored with a TLC mobile phase consisting of chloroform-methanol-acetic acid (90:10:2) coupled with a Packard radiochromatogram scanner. This group combined with the Upjohn group
(20) and identified PGX as 9-deoxy-6,9α-eopoxy-Δ^5-$PGF_{1\alpha}$, which
was given the trivial name, prostacyclin. The A IX mobile phase
of Hamberg and Samuelsson (3), ethyl acetate-acetic acid-isooctane-
water (90:20:50:100) was used to confirm the structure. The exact
radiochromatographic scanner was not specified but the scans
revealed a clear separation of 6-keto $PGF_{1\alpha}$ from PGE_2 and $PGF_{1\alpha}$.

    Pace-Asciak and Wolfe (21) had previously shown that rat
stomach homogenates produced 6-keto $PGF_{1\alpha}$, which they isolated
and identified.(22). The significance of this finding was not
realized at that time. Chloroform methanol-acetic acid-water
(90:9:1:0.65) was the mobile phase of choice. Incubation of
stomach homogenate with [^{3}H]arachidonate resulted in the majority
of the product-related radioactivity becoming associated with the
PGE region. This distribution of radioactivity was determined by
radiochromatography. Only about 13% of the radioactivity in the
PGE region was PGE. The remainder of the material was subjected
to argentation TLC using the A II solvent system of Green and
Samuelsson (11). Two compounds were isolated that were structurally related to 6-keto $PGF_{1\alpha}$.

    Pace-Asciak (23) found that 6-keto $PGF_{1\alpha}$ and its derivatives
were also produced by acetone powders of sheep seminal vesicles.
This was in contrast to the majority of other workers, who found
that PGE and $PGF_{2\alpha}$ were produced from SSV.

    In 1977 Cottee et al. (24) studied the products formed from
arachidonic acid and sheep seminal vesicular cyclooxygenase under

various conditions. At low substrate concentrations and in the absence of reduced glutathione the major product was determined to be 6-keto $PGF_{1\alpha}$. It was noted that 6-keto $PGF_{1\alpha}$ cochromatographed with PGE_2 in several common systems. Seven different mobile phases were tested, with only the organic phase of ethyl acetate-acetic acid-isooctane-water producing adequate separations. A Panax radiochromatogram scanner was used to monitor products.

Wallach (25) showed that preparations of hog aortas, ram seminal vesicles, and bovine corpora lutea are all active in converting labeled PGH_2 to prostacyclin. The reaction was followed by quantitation of the 6-keto $PGF_{1\alpha}$ spontaneously formed from the prostacyclin. The A IX system, which consists of the organic phase of ethyl acetate-isooctane-acetic acid-water (97.5:44.5:17.5:89:5), was the mobile phase of choice. A Vanguard radiochromatogram scanner was used for product analysis.

Since the identification of thromboxanes and prostacyclin as important metabolites of arachidonic, and in a number of tissues as the dominant metabolites, TLC and radiochromatogram scanning have played an important role in defining the exact nature and distribution of the metabolites. Sun et al. (26) examined a large number of tissues using $[1-^{14}C]PGH_\alpha$, which was prepared from $[1-^{14}C]$arachidonic acid by incubation for short periods of time with sheep seminal vesicles. The mobile phase used for separation of $PGF_{2\alpha}$ from PGE_2, and PGD_2 from TXB_2 was 1% acetic acid in ethyl acetate. The mobile phase for the separation of PGE_2, PGD_2, and 6-keto $PGF_{1\alpha}$ was the organic phase of ethyl acetate-acetic acid-isooctane-water (100:20:50:100). Double development was necessary to get adequate separations. Radioactivity was located with a Vanguard Model 930 TLC scanner. Large amounts of TXB_2 were formed by guinea pig and rabbit lung, guinea pig and monkey spleen, and human and monkey platelet. 6-Keto $PGF_{1\alpha}$ was formed by rabbit, monkey, and human uterus, cow and horse corpus luteum, pig and sheep aorta, and rat stomach and small intestine. In some tissues, such as rabbit lung, the distribution of metabolites depended on substrate and enzyme concentrations; low substrate-to-enzyme ratios favored prostacyclin formation, whereas high ratios favored thromboxane formation.

Pace-Asciak and Rangaraj (27) examined the distribution of products synthesized from arachidonic acid in various tissues of the rat. The technique involved isolating fractions by TLC, and quantitative determination by gas chromatography-mass spectrometry. The spleen formed primarily TXB; the lung formed 6-keto $PGF_{1\alpha}$ with significant amounts of TXB_2. The stomach fundus almost exclusively formed 6-keto $PGF_{1\alpha}$. Liver and kidney formed primarily PGE_2.

Miyamoto et al. (28) studied the conversion of $[1-^{14}C]8,11,14$ eicosatrienoic acid to prostaglandin E_1. An unstable intermediate, which was identified as the endoperoxide, was formed. In the presence of glutathione the endoperoxide formed PGE_1. Ethyl ether-petroleum ether-acetic acid (85:15:0.1) or ethyl ether-methanol-acetic acid (90:2:0.1) were the TLC mobile phases of choice, coupled with a Packard 7201 radiochromatogram scanner. A similar study was made by Flower et al. (29) using $[^3H]$arachidonic acid. By a combination of radiochemical and spectrophotometric methods it was determined that bovine seminal vesicles produced PGE_2, $PGF_{2\alpha}$, PGD_2, and malondialdehyde from arachidonic acid. The organic phase from ethyl acetate-water-isooctane-acetic acid (11:10:5:2) or ethyl acetate-acetone-acetic acid (90:10:1) was used for development. Radioactivity was obtained from 0.5-cm strips of the chromatogram.

In 1977 Moncada et al. (30) found that imidazole was an inhibitor of thromboxane synthase from PGH_2. PGH_2 was formed from arachidonate and ram seminal vesicular microsomes. Incubation with platelet microsomes produced TXA_2, which was assayed with rabbit thoracic aorta strips. The methodology involved synthesizing $[1-^{14}C]PGH_2$ from $[1-^{14}C]$arachidonic acid, incubating the product with platelet microsomes, and TLC of the extracts in chloroform-methanol-acetic acid-water (90:8:1:0.8). The IC_{50} for imidazole inhibition was determined to be 22 μg/ml (approximately 320 μM). Of a number of compounds tested as inhibitors of TXA_2 synthase only 1-methyl imidazole was more active than imidazole. Imidazole affected cyclooxygenase only at much higher concentrations.

Needleman et al. (31) studied the effects of imidazole, N-0164, and U-51605, reported thromboxane synthase inhibitors, on platelet thromboxane synthase. All three compounds blocked platelet microsomal thromboxane synthesis without affecting platelet adenyl cyclase. N-0164 also blocked cyclooxygenase. U-51605 inhibited thromboxane synthesis and platelet aggregation. Imidazole blocked synthesis without affecting aggregation. The TLC mobile phase was benzene-dioxane-acetic acid (60:30:3). This gave a good separation of TXB_2 from PGE_2 and $PGF_{2\alpha}$. Radioactivity was determined with a Vanguard strip scanner.

Hsueh and Needleman (32) labeled isolated, perfused rat hearts and hydronephrotic kidneys with $[1-^{14}C]$arachidonic acid by two different techniques. Fatty acid and prostaglandins were released with a bradykinin or transient ischemia stimulus. The venous effluent from the perfused heart or hydronephrotic kidney was examined for $[^{14}C]$-PG and $[^{14}C]$arachidonic acid by TLC and radiochromatographic scanning. For separation of 6-keto $PGF_{1\alpha}$, PGE_2, PGD_2, and arachidonic acid, the organic phase from ethyl

acetate-water-isooctane-acetic acid (110:100:50:20) was used.
For separation of $PGF_{2\alpha}$, PGE_2, PGD_2, TXB_2, and arachidonic acid,
benzene-dioxane-acetic acid (60:30:3) was the mobile phase.

In 1970, Ahern and Downing (33) found that 5,8,11,14-eicosa-
traynoic acid irreversibly inhibited the conversion of [^{3}H]arachi-
donic acid to PGE_2 by acetone powders of sheep seminal vesicles
(SSV). It also inhibited the hydroxylation of linoleic and lino-
lenic acid by SSV. Hexane-ether-acetic acid (25:75:1) was used
as mobile phase. Autoradiograms were prepared by spraying the
chromatograms with a scintillation reagent and exposing x-ray
film to the plates overnight. The developed films were scanned
with a photodensitometer, Photovolt Model 530, attached to a
strip chart recorder.

Tomlinson et al. (34) tested indomethacin, naproxen, and
aspirin, all of which were nonsteroidal antiinflammatory agents,
as inhibitors of the conversion of arachidonic acid into PGE_2 by
bovine seminal vesicle microsomes. Their relative potency was
determined to be 2140 to 150 to 1, respectively, which correlated
with their potency as antiinflammatory agents. Ethyl acetate-
acetic acid-isooctane-water (110:20:50:100) was used for separation. The radioactive zones were located with a Packard radiochromatogram scanner.

Human platelets contain the enzymes lipoxygenase, cyclooxygenase, and thromboxane synthetase. The mixture of products from
incubating platelets with arachidonic consists of the hydroxy
fatty acids, HETE and HHT, and TXB_2. Some PGE_2 and PGD_2 are also
formed. This system can be simplified by using platelet microsomes and PGH_2. The enzymatic products formed are TXB_2 and HHT,
which are the products formed by thromboxane synthase. Some PGE_2
and PGD_2 are formed from the spontaneous breakdown of PGH_2.

With the mobile phase isopropyl ether-methyl ethyl ketone-
acetic acid (50:50:1), TXB_2 has a higher R_f value than $PGF_{2\alpha}$ and
PGE_2 (35). HHT is also clearly separated from HETE. Bovine
aorta microsomes enzymatically convert PGH_2 to prostacyclin,
which, in acid solution, spontaneously decomposes to 6-keto-$PGF_{1\alpha}$.
Nonenzymatic products are PGE_2 and PGD_2. The IPE-MEK-acetic acid
mobile phase does not resolve 6-keto $PGF_{1\alpha}$ and $PGF_{2\alpha}$, but since
no PGF   is formed by bovine aorta microsomes from PGH_2, the system is adequate for measuring the conversion of PGH_2 to 6-keto
$PGF_{1\alpha}$.

The method was used in this laboratory to test the effectiveness of a number of compounds as inhibitors of thromboxane synthase and prostacyclin synthase. Radioactive PGH_2 was prepared
from arachidonic acid, either tritium or carbon-fourteen labeled,
by a short incubation with sheep seminal vesicular microsomes,
followed by extraction and purification as previously described

(26). The scanner was a Berthold LB-2760 TLC Scanner. This
scanner was particularly convenient because a number of lanes
could be automatically scanned. Figure 1 shows a tracing from a
chromatogram with a number of standard compounds. Lane 1 (lanes
are numbered from top to bottom, which corresponds to left to
right on the TLC plate) is a mixture of $[^{14}C]PGE_2$ and PGA_2, lane
2 is $[^{14}C]PGO_2$, and lane 3 is $[^{14}C]TXB_2$. Lane 4 is $[^{14}C]PGH_2$
incubated with bovine aortic microsomes; this reaction was stopped
before complete conversion of PGH_2 to 6-keto $PGF_{1\alpha}$. Lane 5 is
$[^{14}C]PGH_2$ (note some spontaneous decomposition to PGE_2 and PGD_2),
and lane 6 is $[^{14}C]PGF_{2\alpha}$.
        Figure 2 is a TLC tracing demonstrating the technique of
testing inhibitors of thromboxane synthase. Lane 1 is $[^{14}C]PGH_2$
carried through the reaction sequence without enzyme. Lane 2 is
the reaction mixture without inhibitor. The conditions were
adjusted so that a small amount of PGH_2 remained at the end of the
reaction. The reaction of PGH_2 with thromboxane synthase, whether
from platelets or guinea pig lung, produced approximately equal
proportions of HHT and TXB_2. Lanes 3 and 4 demonstrate attempts
at inhibition of the enzymatic reaction.
        Any inhibitor of thromboxane synthase is tested in this
laboratory for inhibition of prostacyclin synthase and cyclooxygenase. L-8027, 2-isopropyl-3-nicotinyl indole, was determined to
be a good inhibitor of the enzyme from human platelets (IC_{50} =
2.0 µM) but a weak inhibitor of the enzyme from guinea pig lung
(IC_{50} = 630 µM). Imidazole inhibited both enzymes to the same
extent (IC_{50} = 600 µM). L-8027 also inhibited cyclooxygenase
(IC_{50} - 2 µM) and prostacyclin synthase (IC_{50} between 0.1 and
1 mM). Imidazole at 1 mM had no effect on either prostacyclin
synthase or cyclooxygenase (36).

        CONCLUSION

This report is not intended to be an exhaustive review of the
literature on the use of radiochromatography and/or prostaglandins.
It is meant, however, to illustrate the important role that thin
layer radiochromatography played in the discovery and isolation of
endoperoxides, thromboxanes, and prostacyclin. The first two
papers (5,6), which showed the relationship between polyunsaturated fatty acids and prostaglandins, were the logical beginning
of a complete understanding of arachidonic acid metabolism. If
arachidonic acid formed prostaglandins, it was logical to ask
whether there were other potent biologically active compounds
formed from arachidonic acid.

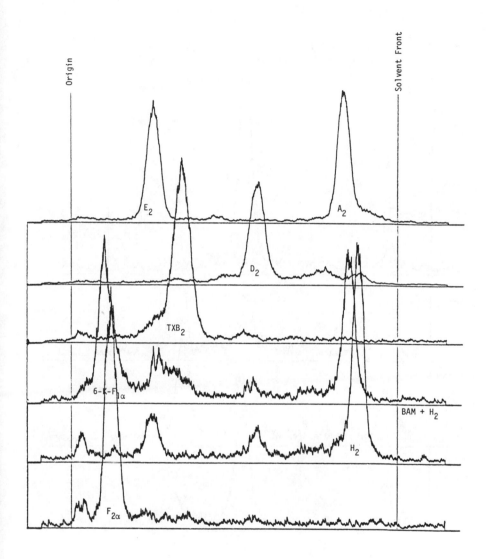

Figure 1. Representative separations of radiolabeled ([^{3}H] or [^{14}C]) TXB$_2$ and PG on silica gel O (New England Nuclear), using isopropyl ether-methyl ethyl ketone-acetic acid, 50:50:1. 6-Keto PGF$_{1\alpha}$ was formed by the incubation of bovine aorta microsomes with PGH$_2$.

491

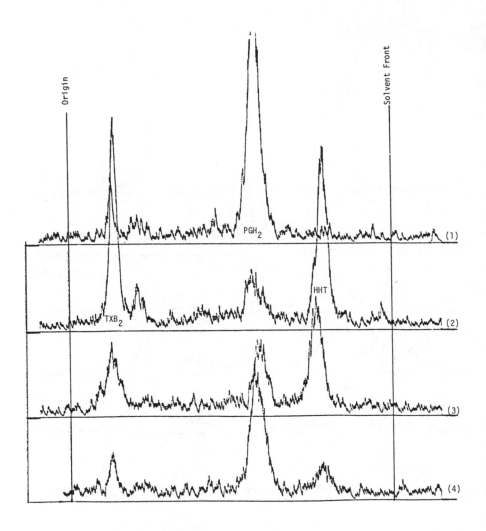

Figure 2. Results of extraction and subsequent chromatography on
ITLC-SG (Gelman Instrument Co.) using isooctane-methyl ethyl
ketone-acetic acid (100:19:1). (1)-heat-denatured human platelet
microsomes incubated with PGH_2. (2)-viable microsomes incubated
with PGH_2. (3),(4)-reactions with two compounds evaluated as
inhibitors.

Thus, beginning in 1964 the stage was set for the story to unfold. This work was ahead of its time and only retrospectively were these compounds shown to be a part of prostacyclin biosynthesis. What is missing in a review such as this is the relationship of the separation methodology to biological activity. After isolation and identification of PGH_2, the biological activity of PGH_2 indicated that PGH_2 and rabbit-aorta-contracting substance were different, which led to the isolation and identification of the thromboxanes. The biological activity of prostacyclin led to the recognition that this was a unique metabolite of arachidonic acid.

Radiochromatography also played a role in the inactivation and degradation of prostaglandins. The inactivation of prostaglandins by 15-keto dehydrogenases is an important biological event. This part of prostaglandin metabolism was not reviewed.

The Packard radiochromatogram scanner 7201 first became available in 1964. Its design was primarily as a strip scanner for paper chromatograms. It efficiently measures one spot on a 5 × 20 cm plate. Other scanners are much more convenient but have about the same efficiency as the Packard scanner (37).

REFERENCES

1. K. Green and B. Samuelsson, J. Lipid Res. 5, 117 (1964).
2. N. H. Anderson, J. Lipid Res. 10, 316 (1969).
3. M. Hamberg and B. Samuelsson, J. Biol. Chem. 241, 257 (1966).
4. H. Mangold, in Thin-Layer Chromatography, A Laboratory Handbook, edited by E. Stahl (Academic Press, New York, 1965), p. 58.
5. D. A. Van Dorp, R. K. Beerthuis, D. H. Nugteren, and H. Vonkeman, Biochim. Biophys. Acta 90, 204 (1964).
6. S. Bergström, H. Danielsson, and B. Samuelsson, Biochim. Biophys. Acta 90, 207 (1964).
7. C. Pace-Asciak, K. Morawska, and L. S. Wolfe, Biochim. Biophys. Acta 218, 288 (1970)
8. K. Crowshaw, Prostaglandins 3, 607 (1973).
9. K. J. Stone, S. J. Mather, and P. P. Gibson, Prostaglandins 10, 241 (1975).
10. M. Hamberg, J. Svensson, T. Wakabayashi, and B. Samuelsson, Proc. Nat. Acad. Sci. USA 71, 345 (1974).
11. M. Hamberg and B. Samuelsson, Proc. Nat. Acad. Sci. USA 71, 3400 (1974).
12. M. Hamberg, J. Svensson, and B. Samuelsson, Proc. Nat. Acad. Sci. USA 72, 2994 (1975).
13. T. K. Bills, J. B. Smith, and M. J. Silver, Biochim. Biophys. Acta 424, 303 (1976).

14. J. M. Bailey, R. W. Bryant, S. J. Feinmark, and A. N. Makjeha, Prostaglandins 13, 479 (1977).
15. D. H. Nugteren and E. Hazelhof, Biochim. Biophys. Acta 326, 448 (1973).
16. M. Lagarde, A. Gharib, and M. Dechavanne, Clin. Chim. Acta 79, 255 (1977).
17. M. Ali, J. Zamecnik, A. L. Cerskus, A. J. Stoessl, W. H. Barnett, and J. W. D. McDonald, Prostaglandins 14, 819 (1977).
18. M. W. Anderson, D. J. Crutchley, B. E. Tainer, and T. E. Eling, Prostaglandins 16, 563 (1978).
19. R. J. Gryglewski, S. Bunting, S. Moncada, R. J. Flower, and J. R. Vane, Prostaglandins 12, 685 (1976).
20. R. A. Johnson, D. R. Morton, J. H. Kinner, R. R. Gorman, J. C. McGuire, F. F. Sun, N. Whittaker, S. Bunting, J. Salmon, S. Moncada, and J. R. Vane, Prostaglandins 12, 915 (1976).
21. C. Pace-Asciak and L. S. Wolfe, Biochemistry 10, 3657 (1971).
22. C. Pace-Asciak, J. Am. Chem. Soc. 98, 2348 (1976).
23. C. Pace-Asciak, Biochemistry 10, 3664 (1971).
24. F. Cottee, R. J. Flower, S. Moncada, J. A. Salmon, and J. R. Vane, Prostaglandins 14, 413 (1977).
25. D. P. Wallach, Prostaglandins 15, 671 (1978).
26. F. F. Sun, J. P. Chapman, and J. C. McGuire, Prostaglandins 14, 1055 (1977).
27. C. R. Pace-Asciak and G. Rangaraj, Biochim. Biophys. Acta 486, 579 (1977).
28. T. Miyamoto, S. Yamamoto, and O. Hayaishi, Proc. Nat. Acad. Sci. USA 71, 3645 (1974).
29. R. J. Flower, H. S. Cheung, and D. W. Cushman, Prostaglandins 4, 325 (1973).
30. S. Moncada, S. Bunting, K. Mullane, P. Thorogood, J. R. Vane, A. Raz, and P. Needleman, Prostaglandins 13, 611 (1977).
31. P. Needleman, B. Bryan, A. Wyche, S. D. Bronson, K. Eakins, J. A. Ferrendelli, and M. Minkes, Prostaglandins 14, 897 (1977).
32. W. Hsueh and P. Needleman, Prostaglandins 16, 661 (1978).
33. D. G. Ahern and D. T. Downing, Biochim. Biophys. Acta 210, 456 (1970).
34. R. V. Tomlinson, H. J. Ringold, M. C. Qureshi, and E. Forchielli, Biochem. Biophys. Res. Commun. 46, 552 (1972).
35. J. G. Hamilton and L. D. Tobias, Prostaglandins 13, 1019 (1977).
36. L. D. Tobias and J. G. Hamilton, Int. Prostagl. Conf., in press.
37. S. Prydz, Anal. Chem. 45, 2317 (1973).

CHAPTER 34

Secretions of the Sebaceous Gland

Donald T. Downing

The skin surface and hair of mammals are coated with mixtures of lipids that arise principally from the sebaceous glands that are attached to hair follicles. A small proportion of the surface lipid is provided by the keratinized cells of the surface of the skin. The sebaceous lipids have been of interest. Each mammalian species produces a completely different mixture of neutral lipid classes (Table I), among which there are unusual branched chains, odd-carbon chains, and abnormal positions of unsaturation (1-3). The size and activity of the sebaceous glands are controlled and modified by the hormones from a variety of endocrine organs, principally the gonads, adrenals, and pituitary (4-7). Since sebum output (and sometimes even the physical size) of the glands can be measured by noninvasive means, sebaceous gland activity has frequently been used to monitor changes in the hormonal status of patients (7) or experimental animals (4,6). Acne vulgaris, the prevalent, disfiguring skin disease characteristic of adolescence, does not occur when the sebaceous glands are not active and can be alleviated by reducing sebum production (8-10). For this reason alone there has been extensive investigation of the composition of human sebum and of endogenous and exogenous factors that can affected sebum production and composition. Such studies have required almost 50,000 sebum analyses in our laboratories, a volume of work that has only been possible because of the development of an efficient procedure for the quantitative analysis of lipids by thin layer chromatography (TLC) (11). This technique has been applied to many investigations of the physiology and biochemistry of sebaceous lipids as well as to lipids of blood and other tissues of animals and plants, and to problems in lipid biochemistry and biophysics. Because of the wide applicability of this method and because some of its essential principles can be extended to

495

TABLE I

Constituent	Human	Sheep	Rat	Mouse	G.pig	Dog	Horse
Squalene	12	--	0.5	--	--	--	--
Sterols	1.5	12	6	13	9		
Sterol esters	2.5	46	27	10	33	42	38
Wax esters	25	10	17	5	--	--	--
Wax diesters	--	21	21	65	24	32	--
Glyceryl ethers	--	--	8	--	28	--	--
Triglycerides	41	--	--	6	--	--	--
Free fatty acids	16.5	--	1	--	--	--	--
Free fatty alcohols	--	11	--	--	6	--	--
Giant-ring lactones	--	--	--	--	--	--	48

the analysis of other classes of compounds, the procedure will be described in some detail. None of the steps in this method is unprecedented but each makes an important contribution and the combination of steps appears to be unique in regard to the efficiency, accuracy, and linearity of the overall procedure for the analysis of mixtures of neutral lipids.

PROCEDURE

Preparation and Storage of Chromatographic Plates

The essential requirement is a clean, uniform layer of silica gel 0.2 to 0.3 mm thick, which contains no organic binder and is supported on 20 × 20 cm glass plates 3-4 mm thick. The adsorbent layer must be soft enough to be scored down to the glass without chipping or cracking the layer. These requirements are most easily and economically met by preparing the chromatographic plates in the laboratory.

Best results are obtained with a plate leveler (e.g., Quickfit-Reeve Angel), which keeps the upper surfaces of the glass plates in register with each other. After the plates are mounted, the upper surfaces should be thoroughly cleaned with an ethanol-moistened paper towel. When dry, any lint or dust should be removed with a clean paint brush. After application of the usual 2:1 mixture of distilled water and silica gel, small irregularities may be removed by pounding on the sides of the plate leveler.

   As soon as the coating has set, the plates are heated in a
drying oven at 110° C for 2 hours. After cooling, the plates
should be developed in chloroform-methanol to move any lipid
contamination to the top edge. The upper 1 cm of the plate may
then be isolated by a horizontal line scored in the sorbent.
   Lanes 6 mm apart are then scored in the layer. This may be
done accurately and efficiently with a simple device that can be
easily constructed from Lucite, 6D box nails, and #12 rubber
bands (Figure 1).

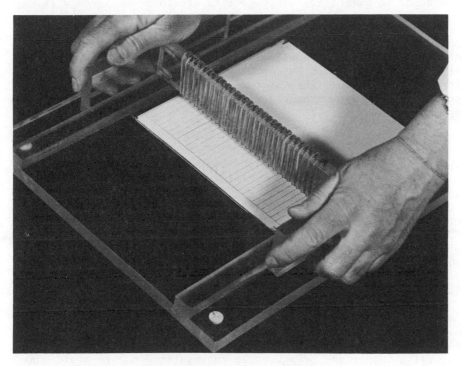

Figure 1.   Device for ruling chromatographic lanes on TLC plates.
Accurate spacing is required when the densitometer has provision
for automatically positioning each lane for scanning.

   These plates may be kept in usable condition for months if
they are stored in an all-glass container. The essential requirement is prevention of the vapors of organic compounds in the
atmosphere from becoming adsorbed on the layer. The vapors from
wood and cardboard are a common source of such contamination.

Sample Application

Several studies have appeared regarding the most accurate means
of applying precise volumes of solutions to chromatograms (12-14)
and they have suggested that inaccuracies in application can be
the largest source of error in quantitative TLC. In many
instances accuracy is not affected by some variation in the volume of solution applied. This is true when the relative amounts
of components in a mixture are to be determined, and it is true
when absolute quantities of components are to be measured with
the aid of an internal standard. Therefore, whenever possible,
quantitative TLC should be conducted so that it is unnecessary to
apply precise volumes of solution. When it is inescapable, the
most accurate means of application is the capillary pipet, and in
the technique described here the most suitable volume for application is 3 µl, because this is the largest volume that can be
applied at one time without some chance of crossover to adjacent
lanes.

Solvent Development

Although many mixtures of neutral lipids may be adequately
resolved on TLC by a single development with a solvent or solvent
mixture, human skin surface lipids require multiple development
because of the number of constituents and their range of polarity.
Successive development with hexane (to 19 cm), toluene (to 19 cm),
and twice with hexane-ether-acetic acid (70:30:1, to 10 cm), has
been adopted as a standard. This provides adequate resolution of
all of the lipid classes present in human surface lipids (Figure
2) as well as the other mammalian surface lipids that have been
studied (Table I).
      For quantitative analysis it is important that solvent devel-
opment be carried out under ideal conditions in order to avoid
uneven migration and formation of distorted spots. For this
reason it is mandatory that the chromatography tank be fully lined
with filter paper and be housed in an insulated box to avoid con-
vection currents in the vapor phase. After placing the tank in
the insulated container, sufficient time should elapse to permit
temperature equilibration. When development is to proceed to the
termination line drawn across the top of the chromatographic plate,
sufficient time should be allowed for each chromatographic lane to
become fully saturated with solvent. After each development, the
chromatographic plate must be allowed to dry for at least 10 minutes before redevelopment or final visualization.

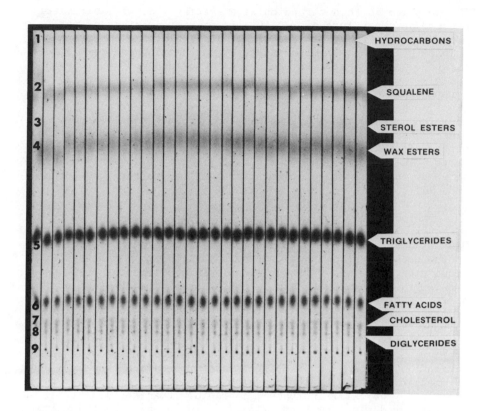

Figure 2.   Chromatograms of human skin surface lipid after charring.

Visualization

Charring is the method of choice for the visualization of lipids for quantitation, because of its sensitivity and universality. When followed by adequate heating, 50% sulfuric acid is an entirely satisfactory spray reagent, giving equal yields of carbon from a wide range of lipids (11).   The chromatograms must be evenly sprayed with a fine mist of acid until the first signs of wetness begin to appear.   Suitable heating conditions are then achieved by placing the chromatographic plate in a cold aluminum slab, 20 cm × 20 cm × 9 mm thick, lying on an electric hotplate.

The hotplate is then turned on at a setting that will ultimately produce a temperature of 220° C. It is at this stage of the procedure that the thickness of the glass plate is important-- thinner glass becomes dish-shaped during heating and as a result is inadequately heated at the edges. Provided both the back of the glass plate and the surface of the aluminum slab are kept scrupulously clean, the glass will not shatter during heating.

The heating is continued for 10-15 minutes beyond the time when fumes cease to be driven off the chromatograms. Recent studies have shown that, under these conditions, even the methyl esters of stearic, oleic, and linoleic acids give equally intense spots. The use of stronger oxidizing agents is not only unneces- sary but is likely to produce erroneous results through overoxida- tion (11).

Photodensitometry

Many studies of the in situ photodensitometric quantitation of spots on chromatograms have revealed that the most important requirement is for a constant relationship between spot size and the length of the light beam. It is the primary function of the narrow lanes scored in the adsorbent to impose a constant spot size so that each chromatogram can be scanned with a single, unin- terrupted passage of the light beam. Quantitation is performed on the basis of the areas of the peaks produced on the strip- chart record. However, several reports have discussed the propo- sition that, because of the light-scattering properties of chromatographic media, the measured optical density (and there- fore peak area) will increase less rapidly than the quantities of material producing the spots (15-17). In an early photodensito- metric system (Photovolt Corp., New York) the Varicord Model 42B recorder made it possible to overcome this nonlinearity by selec- tion of a quasilogarithmic response to the photometer output. As a result, linear calibration lines (Figure 3) were obtained for the peak areas produced by the varying quantities of lipids (11). Other photodensitometers are calibrated to give a linear response to optical density standards, and as a result give curvilinear correlations with varying quantities of standards chromatographed (14-17). Theoretical treatments based on the Kubelka-Munk expres- sion for light-scattering media can improve these standard curves (15-17) but the calculations are so complex that computer assist- ance is essential. We have found that simple triangulation fol- lowed by calculation of the peak area according to the formula:

$$\text{Area} = (H^{1.4} \times W)/2$$

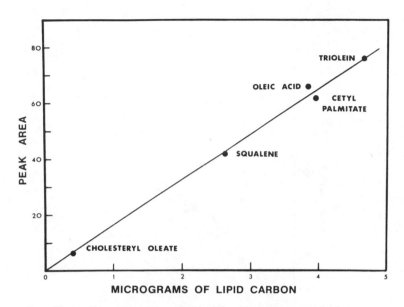

Figure 3. When the amounts of lipids chromatographed are
expressed in terms of their carbon content, a single calibration
line is obtained, passing through the origin. These data were
obtained with a Photovolt densitometer and Varicord Model 42B
recorder that has a variable adjustment for deviations from the
Beer-Lambert Law.

provides adequate correction, so that the separate curves for a
variety of lipids (Figure 4) become a single straight line correlation representing all of the lipids (Figure 5). Under these
circumstances the relative amounts of the components present in a
mixture can be determined without the use of reference compounds
and absolute quantitation can be achieved simply by addition of
an internal standard. In this way, none of the lanes on a chromatographic plate need be used for standards, so that up to 30
samples can be analyzed on each plate.

    Evaluation of the Technique

As originally described (11), this procedure for quantitative TLC
was evaluated using synthetic mixtures of lipids similar in composition to human sebum. These mixtures were analyzed, without the
use of reference compounds, and the accuracy and precision of the

502 Secretions of the Sebaceous Gland

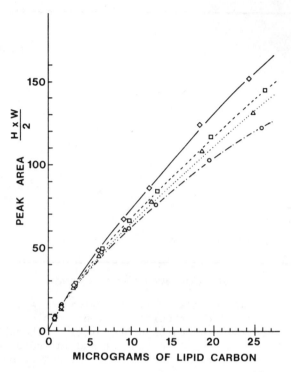

Figure 4. On densitometers calibrated for linear response to optical density, the deviation from the Beer-Lambert Law inherent in light-scattering media results in a series of nonlinear standard curves.

method were both about ± 5% (Table II). When the peak areas obtained with the Photovolt densitometer were plotted against the carbon content of all the lipids applied to the chromatograms, the values for all of the lipids fell on a single straight line that passed through the origin (Figure 3). This was possible because the peak areas obtained were independent of the distance of migration, which in turn resulted from the confinement of spot size by the narrow chromatographic lanes. When wider lanes (1.4 cm) were used, peak areas were no longer independent of the distance of migration (11). More recently we have employed a Clifford photodensitometer, Model 445 (Corning Medical Products, Medfield, Mass.). This instrument has the advantages of a linear response to optical density over a dynamic range of 3 O.D. units, high scanning speed (1 cm/second), and provision for scanning all of the chromatographic lanes on a 20 × 20 cm plate automatically. With the simple mathematical correction for deviation from the Lambert-Beer Law described above, the linearity and accuracy of the TLC analysis have been maintained (Figure 5).

ANALYSIS OF HUMAN SKIN LIPIDS

Collection and Preparation of the Lipid Samples

For many studies an adequate sample of skin surface lipid can be obtained by wiping the forehead with a small cylindrical (1 × 3/8 in.) hexane-moistened polyurethane sponge mounted on a stainless-steel wire handle. This collection device can be transported before and after use in a standard culture tube having a Teflon-lined screw cap and containing 5 ml of hexane (Figure 6). For analysis, the solvent is evaporated under a stream of nitrogen and the residue is redissolved in 0.2 ml of hexane. Usually, 3 μl of this solution is sufficient for application to a thin layer chromatogram for analysis of the relative amounts of the lipid constituents.

When necessary, the measurement of the absolute amount of surface lipids per unit area of skin can be performed by pressing an open-ended glass cylinder onto the skin and collecting the lipid from the delineated area in a small volume of hexane (Figure 7). Quantitation is most readily achieved if the collection solvent contains a known quantity of an internal standard, such as fatty acid methyl ester. The collected solution is then evaporated to dryness as before and the residue redissolved in 0.2 ml hexane for analysis.

The sebum content of skin biopsies can also be estimated by TLC, although this must be done indirectly because subcutaneous fat and epidermal lipids are likely to be major components of the extracts of such specimens. The lipids can be efficiently extracted from skin biopsies simply by immersing the tissue in chloroform-methanol (2:1) overnight. The tissue is then removed and the solvent and extracted water are evaporated under a stream of nitrogen. The lipid residue is then redissolved in 100 μl of toluene, the low volatility of which is an advantage when measurements of absolute amounts of lipids must be made without the use of an internal standard.

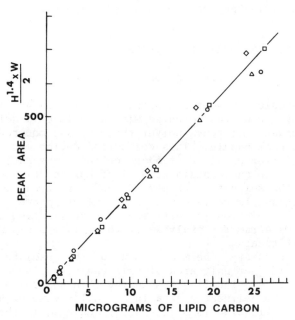

Figure 5. Recalculation of the data for Figure 4, after applying the exponential correction factor to the peak heights, results in a single calibration line for all of the lipids chromatographed.

TABLE II. ACCURACY AND PRECISION OF THE TLC ANALYSIS OF SEBUM

| Constituents | Composition (wt %) | | |
	Known	Found	SD
Cholesterol	2.5	2.9	0.37
Cholesterol oleate	2.5	2.6	0.28
Triolein	30.0	28.8	1.58
Oleic acid	25.0	24.3	1.20
Cetyl palmitate	25.0	26.0	0.79
Squalene	15.0	15.4	0.62

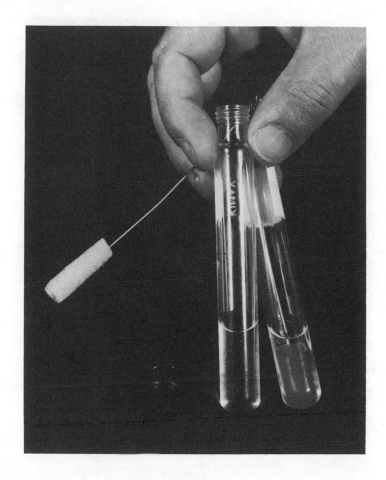

Figure 6.   Collections of human skin surface lipids can conveniently be made away from the laboratory using the solvent-soaked
sponge carried in the culture tube having a Teflon-lined screw
cap.

TLC Analysis

As described above, TLC has provided an accurate and efficient
analysis of human skin lipids. Each of the significant components
is adequately resolved (Figures 2 and 8) and quantitated when the
sample to be analyzed is from the skin surface. With the lipids

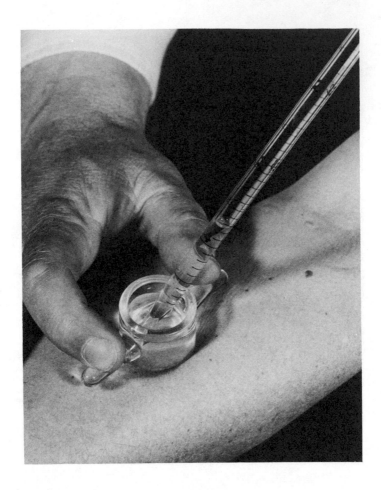

Figure 7. When the weights of skin surface lipids per unit area
must be measured, collections can be made from an area defined by
an open-ended cylinder pressed to the skin. The collecting solvent contains methyl oleate as internal standard.

from biopsy specimens the triacylglycerol fraction from the subcutaneous fat is so large that only the squalene and wax ester components can be quantitated. Nevertheless, this can be of value
for several kinds of investigations, including the study of biosynthesis when coupled with photodensitometry of autoradiograms
of ^{14}C-labeled lipids recovered from skin specimens (Figure 9).

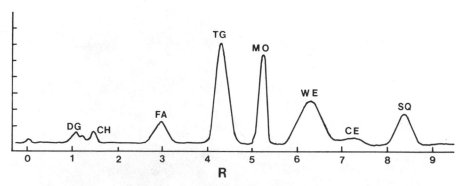

Figure 8. Densitometer scan of human skin surface lipids
resolved on TLC and then charred with H_2SO_4: DG--diglycerides;
CH--cholesterol; FA--free fatty acids; TG--triglycerides; MO--
methyl oleate (internal standard); WE--wax esters; CE--cholesterol
esters; and SQ--squalene. The resolution on silica gel G was
carried out by successive development with hexane (to 19 cm),
toluene (to 19 cm), and hexane-ether-acetic acid (to 10 cm twice).

   Results

Variability in Composition. There are many indications that under
normal circumstances the composition of the lipid mixture from
human sebaceous glands does not vary greatly between individuals.
This can be seen in the results of analyses, by several workers
using different chromatographic techniques, of surface lipid samples from different subjects and anatomical sites (18-22), as
shown in Table III. This comparison serves to substantiate the
values obtained by TLC. In the analysis of surface lipids from
17 adult male subjects (Table IV) the proportions of constituents
varied over a relatively narrow range, except for differences in
the relative proportions of triacylglycerols and free fatty acids.
The sum of these two, however, was relatively constant, resulting
from individual differences in the degree to which sebaceous
tricylgylcerols were hydrolyzed to free acids by cutaneous bacteria (23-25). Analysis of surface lipid samples collected twice
weekly for 4-14 months showed no significant changes with time in
the surface lipid of individuals (21). It would appear, therefore, that skin surface lipid composition is a stable parameter in
which imposed changes should be readily discernible.

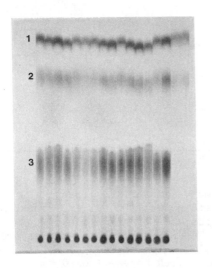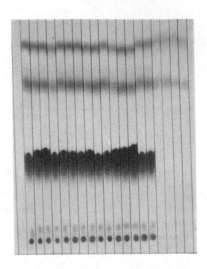

Figure 9. An autoradiogram (left) and charred chromatogram of
the lipids from full thickness human skin resolved by development
with (a) hexane and (b) toluene. The squalene (1) and wax
esters (2) are entirely of sebaceous gland origin, whereas the
triglycerides (3) come both from the glands and the subcutaneous
fat.

TABLE III. ANALYSES OF HUMAN SURFACE LIPID
FROM SEBUM-RICH AREAS (WT %)

Constituent			Anatomical Origin			
	Back	Back	Scalp	Scalp	Brow	Face
Squalene	11.4	15.6	12.8	11.7	12.0	11.6
Wax esters	21.5 ⎤	24.2	20.2	21.8	25.0	22.6
Sterol esters	2.9 ⎦		3.3	2.6	2.1	2.5
Free fatty acids	16.0	16.2	29.6	33.1	16.4	27.2
Glycerides	46.4	42.6	31.7	29.3	43.2	35.4
Sterols	1.8	1.4	2.4	1.5	1.4	0.7
Paraffins	1.3	1.8	0.8	0.6	---	1.3
Reference	18	19	18	20	21	22

Anatomical Variation. On areas of the skin where sebaceous glands
are numerous and active, as on the head, chest, and back, between
90 and 99% of the surface lipid is produced by the sebaceous
glands, with the remainder coming from the keratinized cells of
the epidermis (26). At the higher levels of surface lipid
(500 μg/cm^2), an 80% reduction of sebum output would be required
before the proportion of epidermal lipid (5 μg/cm^2) in the samples rose from 1% to 5%. Therefore, large variations in sebum
output can occur in these areas without significantly affecting
the composition of the surface lipid mixture. However, on other
anatomical sites, such as the limbs, where sebum production is
low, epidermal lipid can be a major component of the surface
lipid, which changes in composition as sebum output varies (Figure
10). Extrapolation of data for samples collected from various
anatomical sites allowed estimation of the amount of epidermal
lipid recoverable from the skin surface and calculation of its
composition (26).

Variation During Development. In children the sebaceous glands
are active at birth; they then decline in activity during the
first year and continue to produce a minimal amount of sebum until
reawakened by androgen stimulation at puberty. The resulting
changes in skin surface lipid composition were observed by TLC
analysis of samples collected from 50 children ranging in age from
birth to puberty (27), which showed the reciprocal changes in the
concentration of cholesterol (coming mainly from the epidermis)
and the wax esters (which are produced solely by the sebaceous
glands). However, it was apparent from the changes in surface
lipid composition that the glands began reactivation around the
age of 8 years, well before the generally recognized age of
puberty. Although this phenomenon remains to be explained, it
seems possible that it is due to early stimulation of the glands
by androgens from the adrenals rather than from the gonads (6).

Effect of Drugs. Although some drugs, such as estrogens (4,7,10)
and the acetylenic analog of arachidonic acid (28), can reduce
sebum output, the changes are rarely sufficient to affect the composition of the surface lipid. Recently, however, the surface
lipids of patients who had been treated with the synthetic retin-
oid 13-cis-retinoic acid have been studied (29). Here the reduction in sebaceous gland activity is so profound that the surface
lipid contains similar quantities of epidermal lipid and sebum,
with the result that the composition is changed dramatically.

Effect of starvation. All of the changes in skin surface lipid
composition described above resulted from changes in the ratio of

TABLE IV.   VARIABILITY OF HUMAN SURFACE LIPIDS

Constituents	Mean	Range
Triglycerides	41.0	19.5 - 49.0
Diglycerides	2.2	2.3 - 4.3
Free fatty acids	16.4	7.9 - 39.0
Wax esters	25.0	22.6 - 29.4
Squalene	12.0	10.1 - 13.9
Cholesterol	1.4	1.2 - 2.3
Cholesterol esters	2.1	1.5 - 2.6

sebum to epidermal lipid rather than from any change in the composition of sebum itself. Only in subjects undergoing long-term
fast has any imposed change in sebum composition been detected.
Both in grossly obese patients who fasted for weeks (30) and in
normal subjects who fasted for 10 days (31), the surface lipid
showed a dramatic increase in the concentration of squalene,
which rose from an average of about 12% to an average value of
25%. This change seems to result from the suppression of the
synthesis of all sebum constituents except squalene, presumably
as the result of suppression of fatty acid synthesis. The change
in surface lipid composition began after about 5 days of fasting,
not, apparently, because of such a delay in the effect of starvation, but because of the time that elapses between the synthesis
of sebum and its appearance on skin surface (32). Changes in the
skin surface lipids of subjects suffering from malnutrition (pel-
lagra) have also been observed, as well as the return towards
normal when the patient received treatment (33,34).

Effect of Bacteria. Sebum is subject to the action of the normal
cutaneous microflora as it passes up the hair follicle to the
skin surface. However, only the triacylglycerol is hydrolyzed by
the bacteria, and individuals differ widely in the degree to
which these are converted to free fatty acids. Several antibiotics, notably the tetracyclines and erythromycin (35), can reduce
this hydrolysis. TLC analysis has shown that the hydrolysis is
mainly carried out by Propionibacterium acnes residing in the
follicular canal, and that reduction in the populations of other
organisms by specific antibiotics did not significantly affect
the composition of the surface lipid.

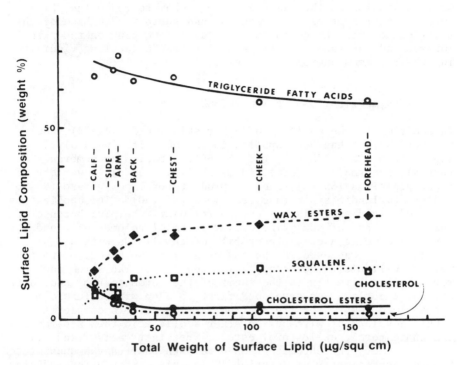

Figure 10. Anatomical variation in the amount of human skin surface lipid accumulating in 3 hours after washing with soap and
water on various parts of the anatomy. At low levels of surface
lipid the composition begins to change from that of sebum to that
of epidermal lipid.

Analysis of Sebaceous Lipids from Other Mammals

A variety of methods have been used for the collection of skin
surface lipids from other species, the most common being immersion
of the animal in a suitable solvent or extraction of clipped fur
(1). The recovered lipids can be adequately resolved in most
cases by the same system of solvent development described above
for the resolution of human lipids.

Quantitation with TLC was used to obtain the composition of
the lipids from sheep (1,26), guinea pig (37), dog (38), and
horse (39) as well as from man and from the only species of

reptile that appears to have been studied (40). TLC was also
used in studies of the time taken for sebum to reach the skin
surface after synthesis in both man and sheep (32). Each of the
major components showed the same delay in emerging on the skin
surface and the same specific activity of ^{14}C labeling, indicating their common anatomical origin.

OTHER APPLICATIONS OF THE METHOD

In addition to the analysis of other skin lipids (41-54), the
basic technique has been applied to the quantitative analysis of
lipids from blood (55-65), gut (66-68), subcellular membranes
(69-74), abnormalities of lipid metabolism (75-84), a variety of
lower organisms (85-93), and the products of lipoxygenase (94-96)
and prostaglandin synthetase reactions (97-100). The application
to blood lipid analysis has been particularly helpful because of
the capability of analyzing extremely small volumes of blood
serum, producing a complete profile of the lipids with a 2-μl
sample. This was used in the analysis of mouse blood, where the
volume available is necessarily limited (57). The serum samples
were applied directly to the lanes of the chromatograms and, after
drying, the plate was developed first to 5 cm with chloroform-
methanol to remove the lipids from any lipoprotein complexes. A
single development with hexane-ether-acetic acid then separated
the cholesterol, cholesterol esters, free fatty acids, and tria-
cylglycerides; the chromatograms were then charred and quantitated
by photodensitometry as described. In this instance the addition
of an internal standard is not feasible, so that volumes of serum
and of a standard (docosanol) had to be accurately measured with
micropipets.
     The general procedure for quantitation described here can
also be applied to phospholipids. The principal problems in this
application are the difficulty of obtaining phospholipid standards
that can be accurately weighed and the fact that one-directional
development does not resolve all of the naturally occurring phospholipids. Nevertheless, much information can be obtained with
this method and with much less effort than any other technique;
it seems likely that it will continue to find applications in a
wide variety of studies.

ACKNOWLEDGMENT

This research was supported in part by USPHS Grant AM 22083.

REFERENCES

1. D. T. Downing, in Chemistry and Biochemistry of Natural Waxes, edited by P. E. Kolattukudy (Elsevier, Amsterdam, 1976).
2. D. T. Downing and J. S. Strauss, J. Invest. Dermatol. 62, 228 (1974).
3. T. Nikkari, J. Invest. Dermatol. 62, 257 (1974).
4. F. J. Ebling, J. Invest. Dermatol. 62, 161 (1974).
5. M. E. Stewart and P. E. Pochi, Internat. J. Dermatol. 17, 167 (1978).
6. P. E. Pochi, J. S. Strauss, and D. T. Downing, J. Invest. Dermatol. 69, 485 (1977).
7. P. E. Pochi and J. S. Strauss, J. Invest. Dermatol. 62, 191 (1974).
8. J. S. Strauss, P. E. Pochi, and D. T. Downing, Ann. Rev. Med. 26, 27 (1975).
9. J. S. Strauss, P. E. Pochi, and D. T. Downing, J. Invest. Dermatol. 62, 321 (1974).
10. J. S. Strauss, P. E. Pochi, and D. T. Downing, J. Invest. Dermatol. 67, 90 (1976).
11. D. T. Downing, J. Chromatog. 38, 91 (1968).
12. J. W. Fairbairn and S. J. Relph, J. Chromatog. 33, 494 (1968).
13. M. S. J. Dallas, J. Chromatog. 33, 337 (1968).
14. E. A. MacMullan and J. E. Heveran, in Quantitative Thin Layer Chromatography, edited by J. C. Touchstone (Wiley, New York, 1973), p. 217.
15. J. Goldman and R. R. Goodall, J. Chromatog. 32, 24 (1968).
16. J. Goldman and R. R. Goodall, J. Chromatog. 40, 345 (1969).
17. L. R. Treiber, J. Chromatog. 100, 123 (1974).
18. C. A. Lewis and B. Hayward, in Modern Trends in Dermatology, edited by P. Borrie (Butterworths, London, 1971), vol. 4, pp. 89-121.
19. E. Haahti, Scand. J. Clin. Lab. Invest. 13, Suppl. 59.
20. N. Nicolaides and R. C. Foster, J. Am. Oil Chem. Soc. 33, 404 (1956).
21. D. T. Downing, J. S. Strauss, and P. E. Pochi, J. Invest. Dermatol. 53, 322 (1969).
22. C. B. Felger, J. Soc. Cosmet. Chem. 20, 565 (1969).
23. R. R. Marples, A. M. Kligman, L. R. Lantis, and D. T. Downing, J. Invest. Dermatol. 55, 173 (1970).
24. R. R. Marples, D. T. Downing, and A. M. Kligman, J. Invest. Dermatol. 56, 127 (1971).
25. R. R. Marples, D. T. Downing, and A. M. Kligman, J. Invest. Dermatol. 58, 155 (1972).
26. R. S. Greene, D. T. Downing, P. E. Pochi, and J. S. Strauss, J. Invest. Dermatol. 54, 240 (1970).

27. R. S. Greene, D. T. Downing, P. E. Pochi, J. S. Strauss, J. Invest. Dermatol. 54, 139 (1970).
28. J. S. Strauss, P. E. Pochi, and E. N. Whitman, J. Invest. Dermatol. 48, 492 (1967).
29. G. L. Peck, T. G. Olsen, F. W. Yoder, J. S. Strauss, D. T. Downing, M. Pandya, D. Butkus, and J. Arnaud-Battandier, N. Eng. J. Med., in press.
30. P. E. Pochi, D. T. Downing, and J. S. Strauss, J. Invest. Dermatol. 55, 303 (1970).
31. D. T. Downing, J. S. Strauss, and P. E. Pochi, J. Clin. Nutr. 25, 365 (1972).
32. D. T. Downing, J. S. Strauss, P. Ramasastry, M. Abel, C. E. Lees, and P. E. Pochi, J. Invest. Dermatol. 64, 215 (1975).
33. M. Dogliotti, M. Liebowitz, D. T. Downing, and J. S. Strauss, Brit. J. Derm. 97, 25 (1977).
34. J. S. Strauss, J. J. Vitale, D. T. Downing, and D. Franco, Am. J. Clin. Nutr. 31, 237 (1978).
35. J. S. Strauss and P. E. Pochi, J. Invest. Dermatol. 47, 577 (1966).
36. P. Ramasastry, Ph.D. Thesis (Boston University, 1970).
37. D. T. Downing and D. M. Sharaf, Biochim. Biophys. Acta 431, 378 (1976).
38. D. M. Sharaf, S. J. Clark, and D. T. Downing, Lipids 12, 786 (1977).
39. S. W. Colton VI and D. T. Downing, unpublished.
40. D. G. Ahern and D. T. Downing, Lipids 9, 8 (1974).
41. R. R. Marples, J. J. Leyden, R. N. Stewart, O. H. Mills, and A. M. Kligman, J. Invest. Dermatol. 62, 37 (1974).
42. G. Rajka, Arch. Dermatol. Forsch. 251, 43 (1974).
43. S. M. Puhvel, R. M. Reisner, and M. Sakamoto, J. Invest. Dermatol. 64, 406 (1975).
44. S. M. Puhvel, J. Invest. Dermatol. 64, 397 (1975).
45. R. L. Baer, S. M. Leshaw, and A. R. Shalita, Arch. Dermatol. 112, 479 (1976).
46. S. M. Puhvel and M. Sakamoto, J. Invest. Dermatol. 69, 401 (1977).
47. S. M. Puhvel and M. Sakamoto, J. Invest. Dermatol. 68, 93 (1977).
48. R. M. Lavker, J. Invest. Dermatol. 65, 93 (1975).
49. C. J. Skerrow and A. G. Matoltsy, J. Cell Biol. 63, 524 (1974).
50. V. R. Wheatley, L. T. Hodgins, W. M. Coon, M. Kumarasiri, H. Barenzweig, and J. M. Feinstein, J. Lipid Res. 12, 347 (1971).
51. A. G. Matoltsy and M. N. Matoltsy, J. Ultrastr. Res. 41, 550 (1972).

52. O. H. Mills, R. R. Marples, and A. M. Kligman, Arch. Dermatol.
    106, 200 (1972).
53. J. E. Fulton, G. Plewig, and A. M. Kligman, J. Am. Med.
    Assoc. 210, 2071 (1969).
54. S. J. Lewis, A. R. Shalita, and W.-L. Lee, J. Invest.
    Dermatol. 71, 370 (1978).
55. R. M. Loria, S. Kibrick, and G. E. Madge, J. Invest. Dis. 133,
    655 (1976).
56. B. Sears, R. J. Deckelbaum, M. J. Janiak, G. G. Shipley, and
    D. M. Small, Biochemistry 15, 4151 (1976).
57. R. M. Loria, S. Kibrick, D. T. Downing, G. Madge, and L. C.
    Fillios, Nutr. Rep. Int. 13, 509 (1976).
58. R. J. Nicolosi, M. G. Herrera, M. Ellozy, and K. C. Hayes,
    J. Nutr. 106, 1279 (1976).
59. A. R. Tall, R. J. Deckelbaum, D. M. Small, and G. G. Shipley,
    Biochim. Biophys. Acta 487, 145 (1977).
60. R. J. Deckelbaum, G. G. Shipley, and D. M. Small, J. Biol.
    Chem. 252, 744 (1977).
61. R. J. Deckelbaum, A. R. Tall, and D. M. Small, J. Lipid Res.
    18, 164 (1977).
62. R. J. Deckelbaum, G. G. Shipley, D. M. Small, R. S. Lees, and
    P. K. George, Science 190, 392 (1975).
63. A. R. Tall, V. Hogan, L. Askinazi, and D. M. Small, Biochem-
    istry 17, 322 (1978).
64. P. H. R. Green, A. R. Tall, and R. M. Glickman, J. Clin.
    Invest. 61, 528 (1978).
65. C. T. Chan, H. Wells, and D. M. Kramsch, Circ. Res. 43, 115
    (1978).
66. S. J. Robins, D. M. Small, J. S. Trier, and R. M. Donaldson,
    Biochim. Biophys. Acta 233, 350 (1971).
67. R. N. Redinger, A. H. Herman, and D. M. Small, Gastroenterol-
    ogy 64, 610 (1973).
68. J. B. Watkins, C. M. Bliss, R. M. Donaldson, and R. Lester,
    Pediatrics 53, 511 (1974).
69. T. W. Keenan and D. J. Moore, Biochemistry 9, 19 (1970).
70. T. W. Keenan, R. Berezney, L. K. Funk, and F. L. Crane,
    Biochim. Biophys. Acta 203, 547 (1970).
71. J. Comte, D. Gautheron, F. Peypoux, and G. Michel, Lipids 6,
    882 (1971).
72. J. E. Kinsella, Lipids 7, 165 (1972).
73. T. W. Keenan, R. Berezney, and F. L. Crane, Lipids 7, 212
    (1972).
74. C. R. Loomis, M. J. Janiak, D. M. Small, and G. G. Shipley,
    J. Mol. Biol. 86, 309 (1974).
75. A. V. Chobanian, G. C. Gerritse, P. I. Brecher, and M. Kessler,
    Diabetology 10, 595 (1974).

76. T. Castellani, P. Kjelgaardnielsen, and J. Wolffjensen, J. Chromatog. 104, 123 (1975).
77. J. L. Hojnacki, R. J. Nicolosi, and K. C. Hayes, J. Chromatog. 128, 133 (1976).
78. W. V. Allen, Am. Zool. 16, 631 (1976).
79. S. S. Katz, G. G. Shipley, and D. M. Small, J. Clin. Invest. 58, 200 (1976).
80. G. H. Rothblat, J. M. Rosen, W. Insull, A. O. Yau, and D. M. Small, Exp. Mol. Path. 26, 318 (1977).
81. S. S. Katz, D. M. Small, J. G. Brook, and R. S. Lees, J. Clin. Invest. 59, 1045 (1977).
82. L. Prutkin, T. G. Rossman, and V. R. Wheatley, Cancer Res. 37, 551 (1977).
83. P. Polgar, L. Taylor, and L. Brown, Mech. Age Dev. 7, 151 (1978).
84. M. R. Anver, R. D. Hunt, and L. V. Chalifoux, J. Med. Primatol. 1, 241 (1972).
85. J. Barrett, C. W. Ward, and D. Fairbairn, Comp. Biochem. Physiol. 38, 279 (1971).
86. J. Barrett and P. E. Butterworth, Lipids 6, 763 (1971).
87. W. M. Walter, A. P. Hansen, and A. E. Purcell, J. Food Sci. 36, 795 (1971).
88. J. K. Testerman, Biol. Bull. 142, 160 (1972).
89. W. V. Allen, Comp. Biochem. Physiol. 47, 1297 (1974).
90. J. D. Saide and W. C. Ullrick, J. Mol. Biol. 87, 671 (1974).
91. N. N. Stepanichenko, S. D. Gusakova, A. A. Tyshchenko, S. Z. Mukhamedzhanov, A. U. Umarov, and O. S. Otroschchenko, Khimiya Prirodnuikh Soedinenii 1976, 431 (1976).
92. L. N. Ten, A. A. Tyshishenko, N. N. Stepanichenko, S. D. Gusakova, S. Z. Muhamedjanov, O. S. Otroshtshenko, and A. G. Kasyanenko, Khimiya Prirodnyikh Soedinenii 1977, 632 (1977).
93. W. V. Allen, Comp. Biochem. Physiol. 57, 41 (1977).
94. H. W. Gardner and D. Weisleder, Lipids 5, 678 (1970).
95. H. W. Gardner, R. Kleiman, and D. Weisleder, Lipids 9, 696 (1974).
96. H. W. Gardner, Lipids 10, 248 (1975).
97. D. G. Ahern and D. T. Downing, Biochim. Biophys. Acta 210, 456 (1970).
98. D. T. Downing, D. G. Ahern, and M. Bachta, Biochem. Biophys. Res. Commun. 40, 218 (1970).
99. D. T. Downing, Prostaglandins 1, 437 (1972).
100. D. T. Downing, J. S. Barve, F. D. Gunstone, F. R. Jacobsberg, and M. Lie Ken Jie, Biochim. Biophys. Acta 280, 343 (1972).

CHAPTER 35

Surveillance of Sulfonamide Residues in Food-Producing Animals

J.L. Wooley, Jr., M.E. Grace and C.W. Sigel

Before a drug is approved for use in food-producing animals, analytical methods must be devised that can be used for surveillance and enforcement to insure that unacceptable concentrations of the drug do not occur in edible tissues intended for human consumption. The tissue in which the total residues require the longest time to reach safe levels is called the target tissue and the compound that is selected as an indicator of the total residues is called the marker residue. During development of a method for surveillance, the relationship between the depletion of the marker residue and the total residue in the target tissue must be established. The surveillance method must have the sensitivity for measurement of the marker residue down to a concentration that indicates the total residues have reached an acceptable level (1).

Current FDA guidelines recommend using radiolabeled compounds to establish the marker residue and target tissue (1). Many of the antibacterial sulfonamides were approved for use long before these guidelines were introduced and over the years the USDA has relied on a nonspecific procedure based on the Bratton Marshall (B-M) reaction (2) for estimation of the "total sulfonamide residues" and for monitoring the marker residue. The extent to which the B-M procedure provides an indication of total sulfonamide residues, especially when total residues approach the tolerance level, has never been established. Because some other drugs and agricultural chemicals as well as the various sulfonamides currently in use couple with the Bratton-Marshall reagent, a positive Bratton-Marshall reaction only indicates that a sulfonamide might be present; additional methods are needed to identify the particular sulfonamide or mixture responsible for the positive reaction.

Recognizing the limitations of the B-M procedure, several laboratories have investigated more sensitive and specific

approaches to characterize sulfonamide residues. Recently Goodspeed et al. of the USDA described a GLC method they have developed for qualitative analysis of B-M positive residues. The
method was validated for five sulfonamides and retention time
data were presented for the fourteen most widely used sulfonamides (3).
     In our laboratory, interest in methodology useful for sulfon-
amides arose with the development of the antibacterial combinations of trimethoprim, with various sulfonamides that are used
for both human* and veterinary** medicine. Initially a quantitative TLC method for assaying sulfonamides at therapeutic concentrations (μg/ml) was developed (4). This technique employed in
situ absorbance measurements using a scanning spectrodensitometer,
and was sufficiently selective and sensitive for routine sulfonamide determinations in plasma and other fluids. However, it
lacked the sensitivity needed for measurement of residues in tissue. This deficiency was overcome by forming fluorescamine
derivatives directly on silica gel plates, the resulting zones
being quantitated by using a scanning spectrodensitometer (5,6).
The quantitative technique was further modified by employing two-
dimensional thin-layer chromatography in conjunction with fluorescamine derivatization to provide a highly sensitive and selective
screen for veterinary sulfonamides (7). In this chapter, methods
for measuring tissue residues of sulfadiazine (SDZ) and sulfaquinoxaline (SQX) are described. The results of a SDZ residue depletion study are reported, and the relevance to sulfonamide surveillance is discussed.

MATERIALS AND METHODS

Quantitative Procedure for SQX in Chicken Tissues

Minced tissue samples (1 g) were placed into 20-ml plastic centrifuge tubes with 2 ml of 0.1 M citric acid-0.2 M sodium phosphate
buffer, pH 7.0, and homogenized for 15 seconds with a Polytron
PCU-2 homogenizer equipped with a 10-ST generator. An additional
milliliter of buffer was added to the homogenate and the sample
was homogenized for another 15 seconds. Two milliliters of buffer
were used to rinse the generator. For recovery experiments, 10 or

---

 * Trimethoprim/sulfamethoxazole (1:5, w/w), called Septra[R],
   Burroughs Wellcome Co.
** Trimethoprim/sulfadiazine (1:5), w/w) called Tribrissen[R],
   Burroughs Wellcome Co.

100 ng SQX were added to the homogenate, and the sample was mixed
with the homogenizer for 15 seconds. Homogenates of muscle were
extracted at this point, but fat, gizzard, heart, kidney, liver,
or skin homogenates were capped and centrifuged at 15,000 rpm for
20 minutes in a Sorvall RC-5 centrifuge equipped with a SM-24
head. The supernatants were poured carefully into 15-ml glass
centrifuge tubes. The homogenate pellet was resuspended with the
homogenizer in 1 ml of buffer and centrifuged as above.

Each sample (muscle homogenate or supernatant from other tissues) was adjusted to pH 5.6 with 8.5% H_3PO_4 and extracted twice
with 5 ml of ethyl acetate. The extracts were combined, 5 µl
each of Dowtherm A (26.5% biphenyl in phenyl ether) and N,N-
diethylaniline added, and the solvent evaporated with a stream of
dry nitrogen gas at room temperature.

The residue was dissolved in 1 ml of acetone (0.5 ml for samples in the 10-ng/g range) and 100 µl was applied to a prescored
TLC plate (E. Merck, Silica Gel 60, 0.25 mm, 20 × 20 cm without
fluorescent indicator). A series of standards in the range 1-10
ng/spot were applied to the same plate, which was then developed
twice in diethyl ether to 15 cm. The tank for TLC development
was a rectangular Desaga multiplate developing chamber without a
saturation pad. Fresh solvent was placed daily into the tank at
least 1 hour before development of the plate. Grease affected
compound migration, so it was not applied to the chamber lid.

After drying with a stream of air at room temperature, the
plate was dipped (~2 seconds) into a stainless-steel tank containing 25 mg of fluorescamine in 250 ml of acetone to form the fluorescent derivative. When fluorescing spots (as observed by long-
wavelength uv light) appeared, the plate was dipped (~2 seconds)
into a second tank containing a 0.5% triethanolamine solution to
stabilize the derivative (5). Quantitation of the drug was accomplished by scanning the TLC plate for fluorescence at an excitation wavelength of 290 nm using a Schoeffel Model SD3000 spectrodensitometer. Total emission above 400 nm, transmitted by a
Corning 3-74 cutoff filter, was measured and recorded on a Honey-
well Electronik 194 Recorder. Peak areas were calculated by a
Spectra-Physics Autolab System I computing integrator. The sample
preparation scheme for SQX is summarized in Figure 1.

Quantitative Procedure for Sulfadiazine in Calf Tissues

The previous reported method (5) was used with the following
modifications: (i) Liver, kidney, and muscle homogenates and
plasma were extracted with ethyl acetate-acetone-methanol (90:9:1,
v/v) instead of ethyl acetate. (ii) For samples in which the
concentration of SDZ was expected to be above 0.5 ppm, 20-100 µl

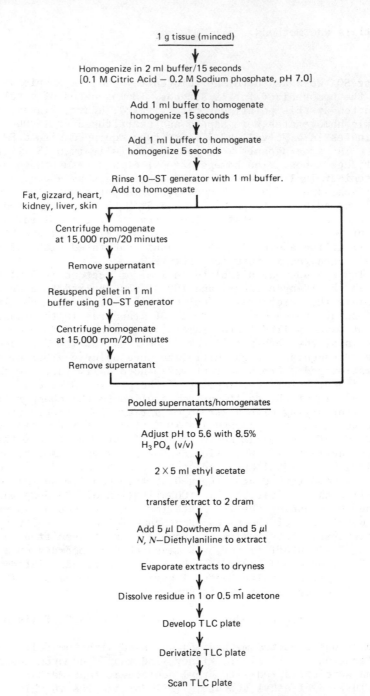

Figure 1. Sample preparation scheme for measurement of SQX in chicken tissues.

of a 4-ml extract of 0.4-1.0 ml of plasma or 2-3 ml of homogenate (equivalent to 1 g of tissue) was spotted. (iii) For plasma and muscle samples with SDZ concentrations below 0.5 ppm, 1 ml of plasma or 2-3 ml of homogenate (equivalent to 1 g of tissue) were extracted twice with 4.0 ml of solvent. The extracts were passed through a sodium sulfate column, evaporated under a stream of nitrogen to 1.0 ml, and 100 μl was spotted. (iv) For kidney and liver samples with concentrations below 0.5 ppm, homogenates were extracted twice, and the combined extracts were back-extracted twice with 1.0 ml of 1 N NaOH. The alkaline phase was adjusted to pH 5.5 and reextracted twice with ethyl acetate-acetone-methanol (90:9:1, v/v), which was treated as described in (ii) above. (v) MN silica gel G plates developed twice in diethyl ether were used for the analysis.

Study of Depletion of Sulfadiazine Residues in Calves

Nine male calves aged 1-2 weeks (40-50 kg) received orally one Tribrissen$^{(R)}$ bolus/day (200 mg trimethoprim and 800 mg sulfadiazine) for five consecutive days. Three calves were killed on each of days 1, 3, and 7. Plasma, liver, kidney, psoas, and quadriceps muscle were obtained from each animal. All samples were stored at -20° C until analysis.

Procedure for Surveillance of Sulfadiazine in Calf Tissue

A two-dimensional thin layer chromatographic method that can be used to distinguish sulfadiazine from the other 12 sulfonamides possibly present in calf tissue has been described previously (7). As an extension to that work, it has been shown that all thirteen sulfonamides can be reproducibly separated using a 20 × 20 cm Whatman plate that has a 3 × 20 cm strip of KC_{18} (reversed phase silica), with the remainder of the plate being the K5 silica formulation. The sample is applied to the reverse phase part of the plate, which is developed with toluene-acetonitrile (80:20, v/v) to 16 cm, oven dried at 110° C for 15 minutes, and developed in the second dimension with ethyl acetate-methanol-NH_4OH (85:15: 0.6, v/v) to 14 cm.

Sulfonamide Standards

Sulfamethazine (SMT), sulfamerazine (SMR), sulfadiazine (SDZ), sulfathiazole (STZ), sulfaquinoxaline (SQX), sulfabromethazine (SBR), sulfadimethoxine (SDX), sulfachlorpyridazine (SCP), sulfaethoxypyridazine (SEP), sulfanilamide (SNL), sulfapyridine (SPD), sulfaguanidine (SGD), and sulfisoxazole (SIX) were obtained from commercial sources. All compounds were checked for purity by TLC in two mobile phases.

RESULTS

The methodology for measuring tissue residues of SDZ and SQX was validated by adding known quantities of each compound to tissue homogenates that were assayed as described above. For SQX, chicken tissues were "spiked" at a concentration of 0.1 ppm (Table I), except the skin, which was spiked at the 0.01 ppm level. This latter concentration was chosen in order to illustrate the sensitivity of the method and its applicability to monitoring the residue depletion of SQX in chicken tissues. For all tissues, extracts of control samples were assayed by spotting the quantity of extract that would be analyzed if SQX concentrations were at the 0.01 ppm level. None of the tissues showed significant blanks. Scans of a typical standard curve (1 to 10 ng/spot) and the extracts from spiked samples and blanks are shown in Figures 2 and 3. Figure 4 shows a standard curve.

TABLE I.   RECOVERY OF SQX FROM SPIKED CHICKEN TISSUE HOMOGENATES[a]
(MEAN ± SD)

Tissue	Concentration	Percent Recovery
Fat	0.1 ppm	73.5 ± 5.2 (4)
Gizzard	0.1 ppm	58.8 ± 7.6 (5)
Heart	0.1 ppm	57.0 ± 2.6 (5)
Kidney	0.1 ppm	58.4 ± 7.6 (5)
Liver	0.1 ppm	52.6 ± 5.9 (5)
Muscle	0.1 ppm	75.9 ± 6.6 (5)
Skin	0.1 ppm	61.4 ± 5.3 (5)
Skin	0.01 ppm	77.0 ± 13.7 (5)

[a] Number of samples in parentheses.

    The use of an internal standard was not investigated. It was shown, however, that by spiking and extracting samples in the range of concentrations included in the standard curve, a

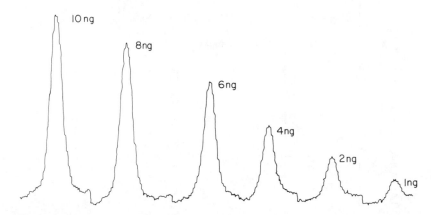

Figure 2.   Trace of typical peaks observed when scanning the
fluorescamine derivative of SQX standards, 1 to 10 ng.

"corrected standard curve" was generated that could be used for
the assay of unknowns. The corrected recovery for spiked (0.01
ppm level) skin samples was 103%.
      Recovery data for extraction of SDZ from calf tissue are
shown in Table II. Mean recoveries over the concentration range
of 0.01 to 25 ppm were greater than 60%. Using the method as
described, the concentrations of SDZ in the plasma and edible
tissues from three calves killed on each of days 1, 3, and 7 following five daily doses of Tribrissen[R] were measured and are
shown in Table III. By day 7 the mean concentrations of SDZ were
at or below the tolerance level of 0.1 ppm for the plasma and
tissues. Of the edible tissues, kidney had the highest residues
of intact drug. The depletion of SDZ from tissues paralleled
that from plasma (Figure 5), and in general the SDZ concentrations
were higher in plasma than in the tissues.
      To distinguish SDZ from the 12 other sulfonamides that might
be present in calf tissue a qualitative TLC method was used (7).
A portion of an extract was spotted on a 20 × 20 cm silica gel 60
plate that was developed in two dimensions. Shown in Figure 6 is
a schematic drawing of a two-dimensional chromatogram showing the
identity of the thirteen sulfonamides that might be present in
calf tissue. A chromatogram of an extract from the control calf
kidney is shown in Figure 7; the R_f of SDZ is indicated by an
arrow. Chromatograms showing blank calf kidney spiked with and

TABLE II. RECOVERY OF SULFADIAZINE FROM SPIKED CALF PLASMA AND TISSUE HOMOGENATES[a]
(MEAN ± SD)

Concentration	Percent Recovery from			
	Plasma	Kidney	Liver	Muscle
25.0 ppm	95.0 ± 2.5 (5)	---	---	---
10.0 ppm	---	82.1 ± 4.8 (10)	76.4 ± 2.8 (10)	84.3 ± 2.5 (10)
2.5 ppm	90.6 ± 3.7 (5)	---	---	---
1.0 ppm	92.8 ± 4.6 (5)	78.2 ± 5.2 (10)	76.4 ± 4.2 (10)	85.4 ± 6.7 (10)
0.1 ppm	81.1 ± 8.5 (5)	61.6 ± 11.1 (7)	68.6 ± 10.3 (8)	74.6 ± 8.5 (9)

[a] Number of samples in parentheses.

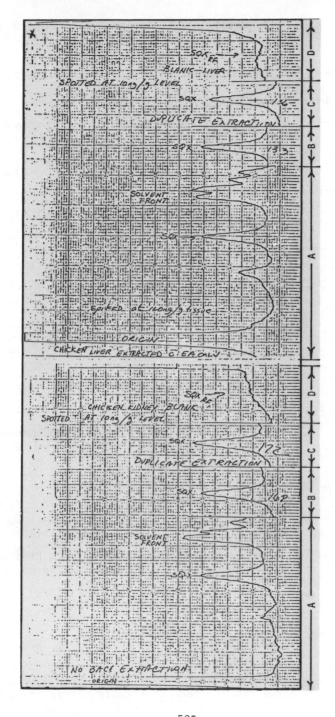

Figure 3. Panel 1—Typical scan of a TLC plate from a chicken kidney extract; Panel 2—Typical scan of a TLC plate from chicken liver extract. Regions: A—scan of entire channel, B—scan of just SQX peak and integration, C—scan of SQX peak from duplicate sample, D—scan of tissue blank concentrated 10 times more than sample when spotted. The arrow indicates SQX R_f.

TABLE III. MEAN CONCENTRATIONS OF SULFADIAZINE (ppm) IN PLASMA
AND TISSUES FROM THREE CALVES KILLED ON EACH OF DAYS 1, 3, AND 7
     FOLLOWING FIVE DAILY DOSES OF TMP/SDZ (30 mg/kg)

Time of Kill Following Last Dose (Day)	Calf	Kidney	Liver	Psoas Muscle	Quadriceps Muscle	Plasma
	A1	7.62	1.39	2.61	2.19	11.60
	A2	5.61	1.64	1.57	1.61	7.45
	A3	3.28	0.61	2.11	1.23	4.31
1						
	Mean	5.50	1.21	2.10	1.68	7.79
	+ SD	+2.2	+0.54	+0.52	+0.48	+3.7
	B1	0.30	0.03	0.11	0.07	0.21
	B2	0.39	0.10	0.15	0.20	0.52
	B3	0.97	0.29	0.26	0.35	1.53
3						
	Mean	0.55	0.14	0.17	0.21	0.75
	+ SD	+0.36	+0.13	+0.08	+0.14	+0.69
	C1	0.03	<0.01	<0.01	0.01	<0.01
	C2	0.06	<0.01	0.02	0.04	0.08
	C3	0.13	0.07	0.14	0.08	0.20
7						
	Mean	0.07	0.02	0.05	0.04	0.09
	+ SD	+0.05	+0.04	+0.08	+0.04	+0.10

without SDZ and the 12 other sulfonamides are shown in Figures 8
and 9, respectively. Similar results were achieved for muscle,
liver, urine, and plasma.

     DISCUSSION

For the routine assay of sulfonamide concentrations in the clinical laboratory, adaptations of the B-M procedure provide a reliable means for measuring the "bacteriostatically active fraction"

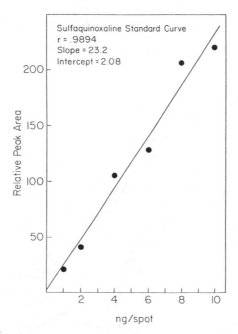

Figure 4.   Standard curve

(6). However, the B-M procedure should not be used for generation of data for complex pharmacokinetic investigations or studies
of residue depletion unless some steps are taken to improve the
sensitivity and validate the specificity. For most sulfonamides
detailed knowledge about metabolism is not available and, therefore, it has not been established how the relative amounts of
parent drug and B-M positive metabolites change with time during
the course of a pharmacokinetic study.

     To partially overcome the limitations of the B-M method, a
chromatographic separation step is often included in the sample
preparation scheme. TLC was used by Phillips and Trafton (8) for
screening sulfonamide residues in poultry tissues and by Bevill
et al. (9) for quantitating sulfamethazine excreted in lamb urine.
The current method is an advancement over these procedures,
because the sulfonamide is determined in situ on the separation
medium. Following a relatively simple sample preparation scheme,
quantities in the low nanogram/gram or milliliter (0.01 ppm)
range can be specifically measured. Application of this

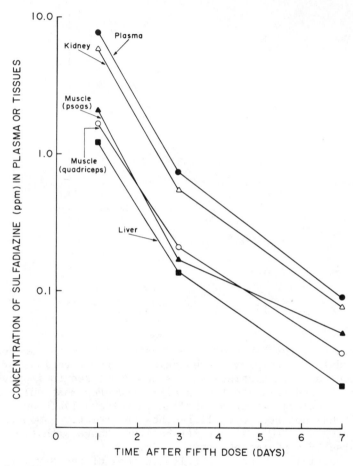

Figure 5. Mean concentrations of SDZ in plasma and tissue from three calves killed on each of days 1, 3, and 7 following five daily doses of TMP/SDZ.

quantitative TLC procedure has established that based on the depletion of intact SDZ from the various calf tissues, kidney is the marker tissue.

As concentrations of intact SDZ in plasma show a consistent relationship to the tissue concentrations (Figure 5), monitoring

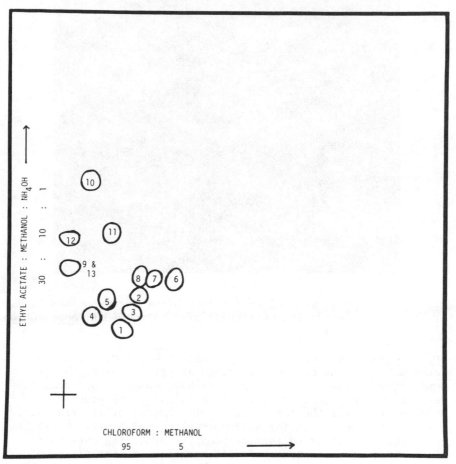

Figure 6. Schematic drawing of two-dimensional chromatogram (silica gel 60 plate) showing the identity of the 13 sulfonamides that might be present in calf tissue: SDZ (1), SMT (2), SMR (3), STZ (4), SQX (5), SBR (6), SEP (7), SDX (8), SCP (9), SNL (10), SPD (11), SGD (12), and SIX (13). Reprinted with permission of J. Assoc. Off. Anal. Chem.

Figure 7.   Chromatogram of calf kidney blank (arrow points to R_f of SDZ).   Reprinted with permission of J. Assoc. Off. Anal. Chem.

SDZ in plasma should provide an excellent preslaughter surveillance method.   The obvious advantage of using a preslaughter monitoring system is that if animals have unacceptable drug concentrations in plasma, they can be held until the drug is depleted to below the tolerance level.   Bourne et al. (10) have advocate this approach for sulfamethazine, and Bevill et al. (11) have recently supported this argument by providing data that showed that concentrations of sulfamethazine in the plasma of 60 swine were closely correlated (see Figure 10) with concentrations in the kidney.   The sulfamethazine concentrations were measured by quantitative TLC utilizing fluorescamine derivatization.

        Currently, the most widely used approach for surveillance of sulfonamide residues in edible tissues has been to use the B-M procedure to monitor postmortem samples.   Even though there has been dissatisfaction with this approach, because of the poor sensitivity and lack of specificity, it has been difficult to establish alternative procedures.   In theory, appropriate systems

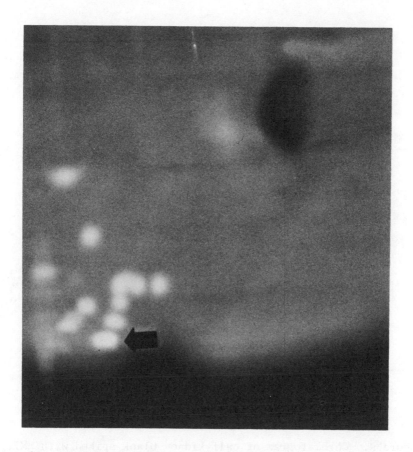

Figure 8.   Chromatogram of calf kidney blank spiked with SDZ, SMT, SMR, STZ, SQX, SBR, SEP, SDX, SCP, SNL, SPD, SGD, and SIX, each at 0.1 ppm (arrow points to R_f of SDZ).   Reprinted with permission of J. Assoc. Off. Anal. Chem.

using HPLC, GLC, or TLC can be developed; however, to validate each chromatographic approach for all sulfonamides and major metabolites would be a major undertaking. In our laboratory, studies have been concentrated on developing screening methods for one sulfonamide, SDZ, and qualitative TLC has been used for the following three reasons: (1) Quantities of SDZ and other sulfonamides down to 0.1 ng can be detected by the eye after uv irradiation of a chromatogram treated with fluorescamine. (2) The supplies and equipment are minimal--mainly a consistent supply of

Figure 9.   Chromatogram of calf kidney blank spiked with SMT, SMR, STZ, SQX, SBR, SEP, SDX, SCP, SNL, SPD, SGD, SIX, each at 0.1 ppm (arrow points to R_f of SDZ).   Reprinted with permission of <u>J</u>. <u>Assoc</u>. <u>Off</u>. <u>Anal</u>. <u>Chem</u>.

uniform TLC plates and a means of illuminating the plates with uv light.   (3) The screening technique for qualitative identification can be extended for use as a quantitative tissue residue assay.

Although the extraction procedure was optimized for SDZ and a two-dimensional TLC system was devised to distinguish SDZ from the 12 other sulfonamides that might be present in calf tissue, the general approach should be applicable to other sulfonamides. Application chemists at Whatman Inc. have shown that all thirteen of the compounds used above can be separated on a Whatman plate

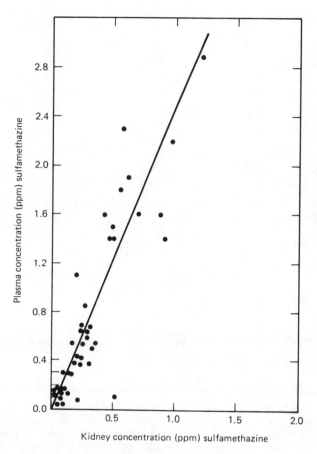

Figure 10. The concentration of sulfamethazine in the plasma and kidney of swine being slaughtered for human consumption (11). Reprinted with permission of J. Ag. Food Chem. and Dr. R. F. Bevill. Copyright by the American Chemical Society.

that has a 3 × 20 cm strip of K C_{18}, with the remainder of the plate being the K5 silica formulation (see Figure 11). Further work is now warranted towards developing a simple sample preparation scheme for tissue and plasma samples that could lead to a practical and efficient way to monitor sulfonamide residues.

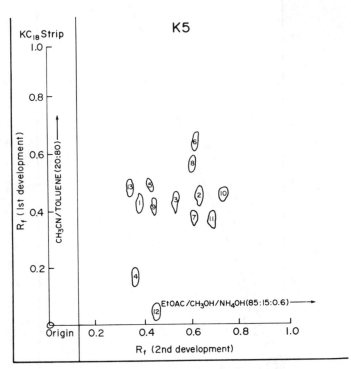

Figure 11. Schematic drawing of two-dimensional chromatogram. (Whatman combination KC_{18} and K5 plate) showing separation of 13 sulfonamides that might be present in calf tissue: SDZ (1), SMT (2), SMR (3), STZ (4), SQX (5), SBR (6), SEP (7), SDX (8), SCP (9), SNL (10), SPD (11), SGD (12), and SIX (13).

The advantages of a specific and sensitive method for assay of sulfonamides are clear. In answering pharmacokinetic questions, the specificity of the TLC technique provides additional detailed data, both for the parent drug and any other B-M positive metabolite. The specificity and sensitivity are again necessary in answering regulatory questions that require both differentiation among several sulfonamides and combinations, as well as sensitivity in the low ppm range. Qualitative and quantitative TLC, which offer a multitude of combinations of adsorbants and developing systems, provide a flexible, efficient, inexpensive, and reliable method for sulfonamide analysis.

ACKNOWLEDGMENTS

The authors thank Peter Flanagan and Dr. Colin Wells of the Wellcome Research Laboratories, Berkhamsted, England for their contributions to the Tribrissen residue depletion study in calves and Helen Hughes of those laboratories for technical assistance in the assay of samples. The authors thank Dr. Thomas E. Beesley of Whatman, Inc., Clifton, New Jersey for helpful discussions concerning the development of TLC procedures, and Michael Mazzei and Philip Griffiths of the Applications Department of Whatman, Inc., for use of the data shown in Figure 11.

REFERENCES

1. M. K. Perez, J. Assoc. Off. Anal. Chem. 61, 1183 (1978).
2. A. C. Bratton and E. K. Marshall, J. Biol. Chem. 128, 537 (1939).
3. D. P. Goodspeed, R. M. Simpson, R. B. Ashworth, J. W. Shafer, and H. R. Cook, J. Assoc. Off. Anal. Chem. 61, 1050 (1978).
4. C. W. Sigel, M. E. Grace, C. A. Nichol, and G. H. Hitchings, J. Pharm. Sci. 63, 1202 (1974).
5. C. W. Sigel, J. L. Woolley, and C. A. Nichol, J. Pharm. Sci. 64, 973 (1975).
6. C. W. Sigel and J. L. Woolley, Jr., in Densitometry in Thin Layer Chromatography, edited by J. C. Touchstone and J. Sherma (Wiley, New York, 1979), chap. 30.
7. J. L. Woolley, O. Murch, and C. W. Sigel, J. Assoc. Off. Anal. Chem. 61, 545 (1978).
8. W. F. Phillips and J. E. Trafton, J. Assoc. Off. Anal. Chem. 58, 44 (1975).
9. R. F. Bevill, R. M. Sharma, S. H. Meachum, S. C. Wozniak, D. W. Bourne, and L. W. Dittert, Am. J. Vet. Res. 38, 973 (1977).
10. D. W. A. Bourne, R. F. Bevill, R. M. Sharma, R. P. Gural, and L. W. Dittert, Am. J. Vet. Res. 38, 967 (1977).
11. R. F. Bevill, K. M. Schemske, H. G. Luther, E. A. Dzierak, M. Limpoka, and D. R. Felt, J. Agric. Food Chem. 26, 1201 (1978).

CHAPTER 36

Flame Ionization Detection System for Thin layer Chromatography of Lipids

T. Itoh, M. Tanaka and H. Kaneko

Extensive advances in chromatographic methods for lipid analysis have been made in recent years. The advent of these modern chromatographic techniques has made the study of lipids one of the most interesting and fruitful fields in biochemistry. Thin layer chromatography (TLC), because of its simplicity and excellent resolving power, has rapidly evolved as one of the most powerful and widely used techniques for the qualitative analyses of a variety of substances including lipids. There have been a number of attempts to extend the usefulness of TLC as a quantitative tool. These include densitometric quantification of either charred or stained samples on a chromatogram, or quantification of the recovered components from the adsorbent scraped from the plate. However, a sensitive and universal detecting reagent for all classes of lipids is still lacking. Because the hydrogen flame ionization detector (FID) has relatively high sensitivity and a wide range of linearity for the detection of organic compounds, it has been quite extensively used in the field of gas-liquid chromatography. Several groups have examined the possibility of using the flame ionization detector for the analysis of organic compounds resolved by thin layer chromatography (1-4). Most of these studies, however, require a delicate and often tedious technique for the chromatographic and detection procedure.

Recently, Okumura and coworkers developed adsorbent sintered chromatoplates (5). Their usefulness in lipid chromatographic analyses was established (6). Similar methods were used for the development of adsorbent sintered thin layer rods (7,8). These rods are now commercially available under the trade name of "Chromarod" (Iatron Labs, Inc., Tokyo). An autoanalyzer (Iatroscan) based on the flame ionization detector principle was also developed in conjunction with the adsorbent thin layer rod. The quality of this thin layer rod is the key to the success of this TLC-FID system. Initially, these rods were coated using the same procedure as that used for the preparation of regular TLC plates. The coated rods were fragile and the reproducibility of the chromatogram was often very poor. Now, very stable and uniform adsorbent sintered rods can be purchased. Furthermore, this company recently made some high-performance silica gel sintered rods that were prepared with fine and highly uniform silica gel powder. The combination of these new thin layered sintered rods and the autoanalyzer has facilitated the quantitative thin layer chromatography of chemical materials. This TLC-FID system has already been applied to lipid research by several groups (9-14).

Adsorbent sintered rods have many advantages compared to the adsorbent coated ones. Generally, sintered rods are reusable without reconditioning. Additionally, because the sintered rods are sturdy, the properties of the adsorbent can be modified by immersing in a reagent solution in order to achieve resolution of specific compounds. In this manner, reagent-impregnated rods can be prepared routinely. In this report, we discuss the application of the TLC-FID system for the quantitative analysis of a variety of lipids.

EXPERIMENTAL

Rods

Adsorbent-sintered quartz rods are 0.9 mm in diameter and 152 mm in length. They support a 75-micron thin layer of silica gel or other adsorbents fused with fine glass powder as a binding agent. The following rods are commercially available: Chromarod S, which is a regular silica gel powder sintered rod; Chromarod S II, which is a uniformly sized silica gel powder (5 microns in diameter) sintered rod; and Chromarod A, which is an alumina powder sintered rod. Silica gel powder-alumina powder (1:1, w/w) wintered rods, Florisil powder-silica gel powder (1:9, w/w) sintered rods, and Florisil powder sintered rods, which are custom made, are also used in this experiment.

Lipid Samples

Tripalmitin and 1-monostearin were purchased from Nippon Glyceride Kogyo Co., Ltd. (Tokyo); cholesterol from Kanto Kagaku Co., Ltd. (Tokyo); stearic acid from Nippon Oil & Fat Co., Ltd. (Tokyo); triolein, trilinolein, 1,2-distearin, and 1,3-distearin from Gaschro Kogyo Co., Ltd. (Tokyo); and 2-monopalmitin from Serdary

Research Laboratory (Ontario, Canada). Cholesteryl palmitate was chemically synthesized. Five kinds of fats (palm, coconut, cocoa butter, beef tallow, and lard), five kinds of oils (sunflower oil, tung oil, soybean oil, grape seed oil, and linseed oil), and commercial monoglyceride emulsifier were generous gifts from Dr. N. Hatsui (Kao Soap Co., Ltd.) and Dr. K. Yoshitomi (Nisshin Oil Co., Ltd,). Phosphatidylcholine, phosphatidylethanolamine, lyso-phosphatidylcholine and sphingomyeline were prepared from egg yolk (15). Cardiolipin (from bovine heart) was obtained from Dr. T. Takahashi (Institute of Kitasato, Tokyo). Phosphatidylserine and phosphatidylinositol were purified from beef brain and yeast lipids, respectively (16). Total lipid extracts of yeast and extremely thermophilic acidophilic bacteria were prepared by the procedure described previously (17).

Chemicals

Organic solvents used for the analysis were further purified by distillation in all-glass apparatus. Other organic and inorganic reagents were of analytical grade or of the highest quality commercially available and were used without further purification. Industrial grade potassium dichromate and concentrated sulfuric acid were used to clean the used silica gel sintered rods.

Sample Spotting and Developing

Just prior to use, all rods, except the argentation rods, were activated by passing through the automatic FID scanner. About 1 μl (20-30 μg lipids/chloroform) of sample was then applied at the origin of the rod by a microsyringe. The spotted rods were developed by a suitable solvent system until the solvent front traveled 10 cm from the origin. The developed rods were placed over phosphorus pentoxide (P_2O_5) in a vacuum desiccator for 5 minutes, and then scanned by an autoanalyzer.

Apparatus and Operating Conditions

The instrument used was an Iatroscan TFG-10. Differential and integral curves were recorded by a Hitachi 056 two pen recorder. Automatic scanning of sample rods by the autoanalyzer was performed under the following conditions: flow rate of hydrogen--125 ml/minute; flow rate of air--2000 ml/minute; chart speed--200 mm/minute; voltage of detector--50 mV; and voltage of recorder--100 mV.

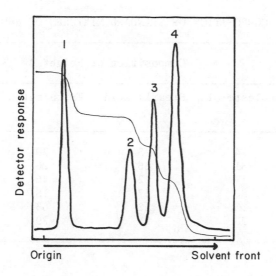

Figure 1. A typical chromatogram of neutral lipid mixture.
Solvent system: petroleum ether-ethyl ether-acetic acid (90:10:1,
v/v). 1--cholesterol; 2--stearic acid; 3--tripalmitin; 4--choles-
teryl palmitate.

RESULTS AND DISCUSSION

Neutral Lipids

Six kinds of standard neutral lipid mixtures containing different
amounts of free sterol, free fatty acid, triglyceride, and sterol
ester were used for evaluation purpose (Table I). A typical
chromatogram of neutral lipid mixture (A) is shown in Figure 1.
The components were completely separated by the solvent system
(a), petroleum ether-diethyl ether-acetic acid (90:10:1, v/v).
Recently, Ackman et al. (14) reported another solvent system for
the separation of neutral lipid mixtures on the Chromarod TLC
system--petroleum ether-diethyl ether-98% formic acid (97:3:1,
v/v). The mobility of free fatty acid in this latter solvent
system was higher than that of triglyceride. Each standard mixture was analyzed about 30 times. The relationships between the
peak area ratios and that of weight ratios for cholesterol,
stearic acid, and cholesteryl palmitate to tripalmitin are summarized in Figure 2. The detector response of each component was
found to be linear over a range of 2-10 μg with a maximum standard

TABLE I. COMPOSITION OF STANDARD MIXTURES OF NEUTRAL LIPIDS

Sample	Composition of Weight %			
	Cholesterol	Stearic Acid	Tripalmitin	Cholesteryl Palmitate
A	24	23	30	24
B	31	26	20	22
C	17	22	46	15
D	20	27	15	28
E	13	12	63	12
F	25	21	29	26

deviation of about 5% of the mean. As little as 0.1 μg of
material could be detected. In the case of stearic acid, the
weight ratios and peak area ratios were directly correlated, but
in the case of cholesterol and cholesteryl palmitate, the peak
area ratios were slightly higher than the weight ratios. The
response of the flame ionization detector for organic compounds
depends on the portion of carbon atoms that contain little or no
oxygen (18). The weight percentage of noncarbonyl carbon of
tripalmitin and palmitic acid were slightly lower than that of
cholesterol and cholesteryl palmitate. In our experience, the
TLC-FID system also gave the same tendency of detector response
for a neutral lipid mixture. If a correction factor is needed,
it can be obtained from Figure 2. In general, we use this system
without a correction factor. A reduction in chromatographic resolution was sometimes found after the sintered rods were used
several times. This may be due to the small amounts of salts or
inorganic compounds remained on the thin layer rods. After five
runs, the rods were reconditioned by immersing them overnight
in chromic-sulfuric acid cleaning solution (saturated potassium
dichromate solution-concentrated sulfuric acid--1:1, v/v), washing them in tap water for 1 hour, and then washing them with
distilled water.

Phospholipids

Nine kinds of standard phospholipid mixtures, which contained
different amounts of lyso-phosphatidylcholine, phosphatidyl-

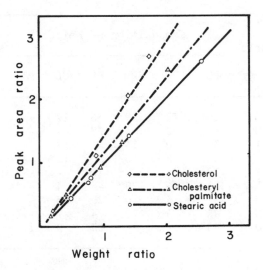

Figure 2. The relationships between the ratio of peak area and that of weight for cholesterol, stearic acid, and cholesteryl palmitate to tripalmitin.

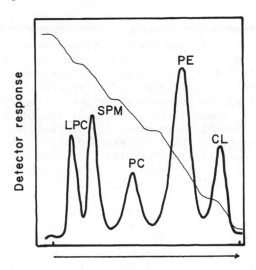

Figure 3. A chromatogram of phospholipid mixture. Solvent system: system: chloroform-methanol-water (65:25:4, v/v). Abbreviations are as in Table II.

TABLE II.   COMPOSITION OF STANDARD MIXTURES OF PHOSPHOLIPIDS[a]

Sample	Composition of Weight %						
	PC	LPC	SPM	PE	CL	PI	PS
G	16	12	19	36	17	--	--
H	66	2	4	28	--	--	--
I	36	10	14	23	17	--	--
J	46	8	13	20	13	--	--
K	77	1	2	20	--	--	--
L	51	16	29	--	4	--	--
M	37	5	6	29	23	--	--
N	47	15	27	4	7	--	--
O	43	--	--	23	9	17	7

[a] Abbreviations: PC-phosphatidylcholine; LCP--lyso-phosphatidyl-
choline; SPM--sphingomyelin; PE--phosphatidylethanolamine; L
CL--cardiolipin; PI--phosphatidylinositol; PS--phosphatidyl-
serine.

choline, sphingomyelin, phosphatidylethanolamine, phosphatidylinositol, phosphatidylserine, and cardiolipin, were analyzed to
standardize this method for phospholipid analyses (Table II). The
following solvent systems were used to resolve phospholipid mixtures: (b) chloroform-methanol-water (65:25:4),v/v); (c) chloro-
form-methanol-28% ammonia (65:35:5, v/v); and (d) chloroform-
acetone-methanol-acetic acid-water (50:20:10:10:5, v/v). A typical chromatogram of phospholipid mixture (G) is shown in Figure 3.
Each component was clearly separated by the solvent system (b).
The separation of phosphatidylserine and phosphatidylinositol from
lyso-phosphatidylcholine and sphingomyelin was, however, very difficult on the Chromarod S by any of the above solvent systems. It
was previously reported that Florisil was effective for the separation of phosphatidylserine and phosphatidylinositol (19). Therefore, sintered rods were made from Florisil powder and Florisil
silica gel mixed powder. However, these rods were still unable to
resolve phosphatidylserine and phosphatidylinositol in a phospholipid mixture. Samples (G) - (N) were analyzed about 30 times.
The relationships between the peak area ratios and that of weight

ratios for lyso-phosphatidylcholine, sphingomyelin, phosphatidyl-ethanolamine, and cardiolipin to phosphatidylcholine are summarized in Figure 4. The peak area and weight of each component were almost directly proportional.

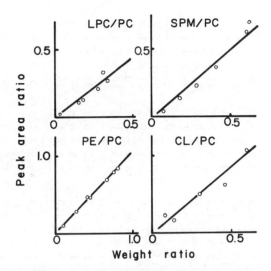

Figure 4. The relationships between the ratio of peak area and that of weight for LPC, SPM, PE, and CL to PC. Abbreviations are as in Table II.

Analysis of Lipid Extracts from Natural Sources

The TLC-FID system was applied for the estimation of neutral lipids of 30 strains of yeast (20). A typical chromatogram of total lipid of Debaryomyces hansenii MY-45 is shown in Figure 5. About 100 μg of lipid was applied to the rod. Total polar lipids were retained at the origin and each neutral lipid was clearly separated. The relative amounts of free sterol, free fatty acid, triglyceride and sterol ester were calculated from the chromatogram (Table III), and the relative amounts of total polar lipid and total neutral lipid were also estimated. Each yeast strain contained at least four kinds of neutral lipids--triglyceride, sterol ester, free sterol, and free fatty acid. In many strains, triglyceride constituted the major component (50-90%) of the total neutral lipid fraction, with sterol ester, free sterol, and free

TABLE III. NEUTRAL LIPID COMPOSITION OF 30 YEAST SPECIES[a]

Species	S (%)	FA (%)	TG (%)	SE (%)
Ascosporogenous yeast				
Saccharomyces cerevisiae OC-2	3.5 ± 0.6	2.5 ± 1.6	29.9 ± 6.3	52.9 ± 5.2
Saccharomyces rosei W-70	6.6 ± 0.9	T[b]	14.4 ± 1.4	76.4 ± 2.0
Saccharomyces carlsbergensis BHl-3	11.1 ± 1.6	10.7 ± 0.4	19.2 ± 0.7	57.1 ± 1.7
Saccharomyees rouxii MS 1-3	8.9 ± 1.5	3.0 ± 1.2	47.0 ± 2.5	38.1 ± 3.6
Schizosaccharomyces pombe IAM-4863	4.2 ± 0.6	4.0 ± 0.9	54.4 ± 5.0	28.9 ± 1.7
Kluyveromyces polysporus EC12-4	6.7 ± 0.3	2.8 ± 1.0	26.0 ± 1.7	54.8 ± 5.2
Schwanniomyces occidentalis IAM-4332	3.0 ± 1.5	7.2 ± 0.1	76.0 ± 5.5	4.9 ± 2.7
Debaryomyces hansenii MY-45	13.8 ± 1.7	10.0 ± 1.8	65.8 ± 5.2	1.8 ± 0.8
Debaryomyces nilssonii Z-9-6	21.2 ± 2.0	2.1 ± 0.5	41.2 ± 1.0	34.4 ± 3.0
Pichia membranaefaciens IV-5-1	3.8 ± 0.9	4.4 ± 1.3	70.4 ± 6.7	6.5 ± 0.6
Pichia farinosa WH 3-1	25.9 ± 2.5	3.5 ± 1.3	59.6 ± 4.3	4.5 ± 1.0
Hansenula anomala WH 16-2	11.9 ± 4.0	T	74.0 ± 1.1	7.6 ± 4.0
Lipomyces starkeyi IAM-4763	2.4 ± 0.7	3.2 ± 0.8	93.4 ± 1.8	T
Lipomyces lipoferus 0673	1.8 ± 0.2	7.6 ± 0.6	85.4 ± 0.6	T
Saccharomycodes ludwigii Shin 3-5	3.8 ± 1.9	3.8 ± 1.0	36.1 ± 5.7	46.1 ± 4.7
Ballistosporogenous yeast				
Sporobolomyces salmonicolor WF-174	7.2 ± 1.9	15.9 ± 1.3	66.0 ± 6.2	5.0 ± 1.0
Asporogenous yeast				
Cryptococcus neoformans CBS-132	3.6 ± 0.1	1.2 ± 0.2	90.7 ± 0.3	3.1 ± 0.1
Cryptococcus laurentii 2-6-5	6.3 ± 1.5	67.2 ± 1.3	21.3 ± 1.5	2.5 ± 0.8
Torulopsis colliculosa J-5	6.7 ± 1.5	1.7 ± 0.9	39.0 ± 1.6	46.6 ± 2.8
Torulopsis candida MYA-3	3.1 ± 0.9	1.1 ± 0.4	92.4 ± 2.3	1.0 ± 0.2
Candida krusei WF-16	19.4 ± 2.0	4.5 ± 1.0	71.1 ± 1.1	4.9 ± 2.3
Candida tropicalis Shin 1-3	18.4 ± 0.8	4.8 ± 0.8	69.7 ± 1.7	7.0 ± 0.8
Candida mycoderma WF-8	6.5 ± 1.1	1.6 ± 0.2	70.9 ± 2.5	1.3 ± 0.2

Species	S (%)	FA (%)	TG (%)	SE (%)
Asporogenous yeast				
Candida pulcherrima 33C	30.3 ± 2.5	T	62.5 ± 2.9	2.0 ± 0.7
Candida utilis 7005	3.1 ± 1.4	2.2 ± 0.6	87.6 ± 3.4	2.3 ± 3.0
Trigonopsis variabilis S-3-9	27.7 ± 1.6	14.3 ± 1.2	32.0 ± 2.8	6.9 ± 0.6
Trichosporon cutaneum KC4-3	1.8 ± 0.6	T	93.4 ± 2.1	2.5 ± 1.4
Kloeckera apiculata KK-3	7.1 ± 1.7	29.3 ± 3.4	51.2 ± 3.8	4.8 ± 1.4
Rhodotorula glutinis H3-9-1	6.7 ± 1.2	7.0 ± 1.4	79.7 ± 3.7	4.5 ± 1.0
Rhodotorula rubra AY-2	3.6 ± 0.3	8.0 ± 0.8	84.4 ± 1.7	1.7 ± 0.3
Rhodotorula rubra Np. 2-17-4B	1.3 ± 0.9	6.6 ± 0.3	90.9 ± 0.3	0.8 ± 0.1

a Abbreviations as in Figure 5.

b Less than 1%.

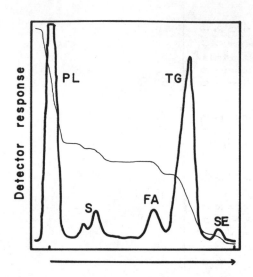

Figure 5. Chromatogram of total lipid extract from Debaryomyces hansenii MY-45. Solvent system: petroleum ether-ethyl ether-acetic acid (90:10:1, v/v). Abbreviations: PL, polar lipid; S, free sterol; FA, free fatty acid; TG, triglyceride; and SE, sterol ester.

fatty acid constituting 2-5%, 3-7%, and 5%, respectively. However, Saccharomyces, K. polysporous, and T. colliculosa did not fit the above generalization. They contained sterol derivatives as the major neutral lipid (50-80%). Cr. laurentii and Kl. apiculata contained unusually high levels of free fatty acids (about 70%, 30% respectively).

The relationship of yeast growth stages and yeast cellular lipids was also monitored by the TLC-FID system (21). As the growth of Cr. neoformans progressed, the amount of total cellular lipid increased remarkably, due mostly to the accumulation of triglyceride.

It has been reported that extremely thermophilic acidophilic bacteria (Sulfolobus) contain unusual neutral lipids, phospholipids, and glycolipids in which alkyl groups are bound through ether linkages to polyols (22). These lipids may play an important role in the membrane structure. The distribution pattern of these lipid classes is shown in Figure 6. The relationship between the growth temperature (50° - 85° C) and the distribution of these lipid groups was also examined by the TLC-FID system (23). Thus,

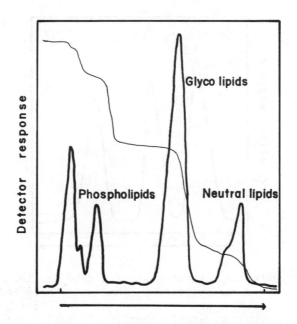

Figure 6. Chromatogram of total lipid extract from thermophilic acidophilic bacteria (Sulfolobus). Solvent system: chloroform-methanol-water (80:20:2, v/v).

this TLC-FID system could be applied to quantitative analyses of a variety of lipid samples. It is suggested that the system in conjunction with an internal standard method could estimate not only relative amounts but also absolute amounts of lipids.

Modifications

Argentation Chromatography (24). Silver nitrate impregnated silica gel sintered rods were prepared by immersing the clean Chroma-rod S in 12.5% silver nitrate solution, followed by activating the dried rods at 120° C for 2 hours. The activated rods were stored in a dark desiccator over anhydrous calcium chloride until use. After using, the argentation rods were regenerated by immersing them overnight in warm concentrated nitric acid solution followed by washing them in water. The washed rods could be treated with silver nitrate solution as described above before reuse. The following solvent systems were used for argentation chromatography: for triglyceride--(e) benzene-chloroform-acetic

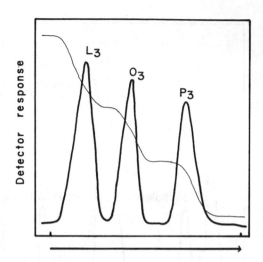

Figure 7. Argentation chromatogram of a triglyceride mixture.
The argentation rod was prepared by immersing Chromarod S in 12.5%
silver nitrate solution and activating at 120° C for 2 hours.
Solvent system: benzene-ethyl ether (98:2, v/v). Sample: a mixture of tripalmitin (P$_3$)-triolein (O$_3$)-trilinolein (L$_3$) (30:30:40,
w/w).

acid (90:10:1, v/v), and (f) benzene-diethyl ether (98:2, v/v);
and for phosphatidylcholine and phosphatidylethanolamine--(g)
chloroform-methanol-water (65:25:3, v/v). A standard triglyceride
mixture that contained tripalmitin, triolein, and trilinolein
(30:30:40, w/w) was used for the purpose of evaluation. A typical
argentation chromatogram of standard mixtures is shown in Figure
7. Solvent system (e) was effective for the resolution of a triglyceride mixture having a low degree of unsaturation, and solvent system (f) was better for the triglyceride mixture having a
high degree of unsaturation. A direct correlation was found
between the peak area ratios and weight ratios of each triglyceride. The chromatograms for the argentation chromatography of
five kinds of neutral fats and six kinds of neutral oils are
shown in Figures 8 and 9. The fat samples were composed of mixtures of triglycerides that contained a low degree of unsaturation, and the oil samples were composed of a mixture of highly
unsaturated triglycerides. The distribution patterns of molecular species of triglycerides in oils and fats were also clearly

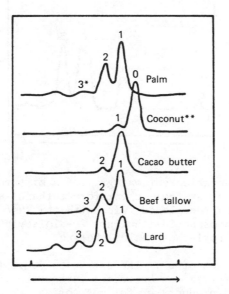

Figure 8. Argentation chromatograms of five kinds of fats.
Solvent system: benzene-ethyl ether (98:2, v/v). The argentation rod was activated at 180° C for 3 hours. Asterisks indicate
the number of double bonds in the molecular species.

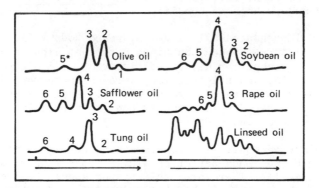

Figure 9. Argentation chromatograms of six kinds of oils.
Solvent system: benzene-chloroform-acetic acid (90:10:1, v/v).
Asterisks indicate number of double bonds in the molecular species.

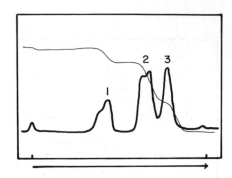

Figure 10.   Argentation chromatogram of a mixture of phosphatidyl-
cholines.   Solvent system: chloroform-methanol-water (50:35:4,
v/v).   The rod was developed twice with the same solvent system.
1--dilinoleoyl-phosphatidylcholine; 2--dioleoylphosphatidylcho-
line; 3--dipalmitoyl-phosphatidylcholine.

different from each other.   These results suggest that the argentation TLC-FID system can be useful for monitoring the quality of
oil and fat food products or for monitoring the source of oil and
fat products.
     The mixture of dipalmitoylphosphatidylcholine, dioleoylphos-
phatidylcholine, and dilinoleoylphosphatidylcholine was separated
to each molecular species on the silver nitrate impregnated
Chromarod S-II by solvent system (g) (Figure 10).   Molecular
species of phosphatidylcholine and phosphatidylethanolamine
obtained from yeast cellular lipids (Cr. neoformans) were also
separated under the same conditions.

Boric Acid Impregnated Silica Gel Sintered Rods (25).   Partial
glyceride mixtures, which contained different amounts of 1-mono-
glyceride, 2-monoglyceride, 1,2-diglyceride, 1,3-diglyceride,
free fatty acid, and triglyceride, were chromatographed using a
3% boric acid impregnated Chromarod S-II.   Each component was
completely separated by solvent system (i) petroleum ether-
diethyl ether-acetic acid (82:12:1, v/v).

Alumina Sintered Rods.   The separation of choline-containing
phospholipids was much better on alumina rods than on silica gel
rods (Figure 11).   Therefore, this system should be useful for
estimating the ratios of phosphatidylcholine and sphingomyelin in
blood.

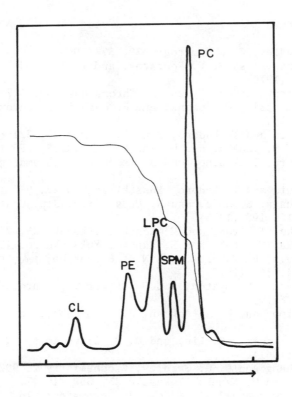

Figure 11.   Chromatogram of phospholipid mixture on alumina rod.
Solvent system: chloroform-methanol-water-15 N ammonium
hydroxide-pyridine (65:27.5:4:2:2, v/v).   Rod: Chromarod A.
Abbreviations as in Table II.

ACKNOWLEDGMENTS

The authors would like to thank Dr. Robert K. Yu and Dr. Thomas N.
Seyfried (Department of Neurology, Medical School of Yale University, New Haven, Connecticut) for their valuable suggestions for
preparing the manuscript.

REFERENCES

1. F. B. Padley, J. Chromatogr. 39, 37 (1969).
2. J. J. Szakasits, P. V. Peuritoy, and L. A. Woods. Anal. Chem. 42, 351 (1970).
3. T. Cotgreave and A. Lynes, J. Chromatogr. 30, 117 (1967).
4. K. D. Mukherje, H. Spaans, and E. Haahti, J. Chromatogr. 61, 317 (1971).
5. T. Okumura and T. Kadono, Bunseki Kagaku 21, 321 (1972).
6. T. Itoh, M. Tanaka, and H. Kaneko, Lipids 8, 259 (1973).
7. T. Okumura, T. Kadono, and A. Iso'o. J. Chromatogr. 108, 329 (1975).
8. T. Okumura and T. Kadono, Bunseki Kagaku 22, 980 (1973).
9. M. Tokunaga, S. Ando, and N. Ueda, Proc. Jap. Conf. Bioch. Lipids 15, 195 (1973).
10. M. Tanaka, T. Itoh, and H. Kaneko, Yukagaku 25, 263 (1976).
11. M. Tanaka, T. Itoh, and H. Kaneko, Yukagaku 26, 454 (1977).
12. E. Gantois, F. Mordret, and N. LeBarbancho, Rev. France Corps Gras. 24, 167 (1977).
13. D. Vandamme, V. Blaton, and H. Peeters, J. Chromatogr. 145, 151 (1978).
14. J. C. Sipos and R. G. Ackman, J. Chromatogr. Sci. 16, 443 (1978).
15. H. A. I. Newman, C. Liu, and D. B. Zilversmit, J. Lipids Res. 2, 403 (1961).
16. T. L. Chang and C. C. Sweeley, Biochemistry 2, 592 (1963).
17. T. Itoh and H. Kaneko, Yukagaku 23, 350 (1974).
18. R. G. Ackman and J. C. Sipos, J. Chromatogr. 16, 298 (1964).
19. A. N. Siakotos and G. Rouser, J. Am. Oil Chem. Soc. 42, 913 (1965).
20. H. Kaneko, M. Hosohara, M. Tanaka, and T. Itoh, Lipids 11, 837 (1976).
21. T. Itoh, H. Waki, and H. Kaneko, Agric. Biol. Chem. 39, 2365 (1975).
22. T. Furuya, T. Nagumo, T. Itoh, and H. Kaneko, Agric. Biol. Chem. 41, 1607 (1977).
23. T. Furuya, T. Nagumo, T. Itoh, and H. Kaneko, Proc. Jap. Conf. Bioch. Lipids 20, 5 (1978).
24. M. Tanaka, T. Itoh, and H. Kaneko, Yukagaku, in press.
25. M. Tanaka, T. Itoh, and H. Kaneko, 16th Annual Meeting of Japan Oil Chemists Society, Osaka (1977).

Index

A